Lecture Notes in Mechanical Engineering

Jarosław Stryczek · Urszula Warzyńska
Editors

Advances in Hydraulic and Pneumatic Drives and Control 2023

Springer

Editors
Jarosław Stryczek
Wrocław University of Science
and Technology, Faculty of Mechanical
Engineering
Wrocław, Poland

Urszula Warzyńska
Wrocław University of Science
and Technology, Faculty of Mechanical
Engineering
Wrocław, Poland

ISSN 2195-4356 ISSN 2195-4364 (electronic)
Lecture Notes in Mechanical Engineering
ISBN 978-3-031-43001-5 ISBN 978-3-031-43002-2 (eBook)
https://doi.org/10.1007/978-3-031-43002-2

This Springer imprint is published by the registered company Springer Nature Switzerland AG
The registered company address is: Gewerbestrasse 11, 6330 Cham, Switzerland

Paper in this product is recyclable.

Preface

The first conference of the Polish fluid power society was held on May 25–26, 1962, in Wrocław with an attendance of about 160 participants. Subsequent conferences in this series were held every 2–3 years until modern times. The organizers have always been the community of fluid power professionals from the Wroclaw University of Science and Technology, metal works in Wroclaw, and the Association of Polish Mechanical Engineers.

The aim of these conferences was and still is to integrate the environment of fluid power society and to present its achievements in the field of hydraulics and pneumatics. This work was presented usually in the form of conference books published in Polish. Recent conferences are international. During the current conference entitled "Hydraulic and Pneumatic Drives and Controls 2023", articles from Austria, Czech Republic, Kazakhstan, Germany, Romania, Ukraine, and Poland are presented. The articles are prepared in English and present the up-to-date and original achievements of the authors. The papers are grouped into three chapters: Hydraulic Components, Hydraulic Systems, and New Materials and Special Solutions in Fluid Power Technology.

An international team of reviewers, recognized authorities in the field of fluid power, took care of a solid substantive level of this monograph. At the same time, Editor Publishing House Springer ensured the high editorial level of the publication. We would like to thank the authors, reviewers, and the publisher. We present this monograph to the readers in the hope that they will find articles of interest to them.

Jarosław Stryczek
Urszula Warzyńska

Organization

Program Committee Chairs

Stryczek, Jarosław	Wroclaw University of Science and Technology, Poland
Warzyńska, Urszula	Wroclaw University of Science and Technology, Poland

Program Committee Members

Bogdevicius, Marijonas	Vilnius Gediminas Technical University, Lithuania
Bonanno, Antonino	CNR-ITAE, Italy
Castilla, Roberto	Universitat Politècnica de Catalunya (UPC), Spain
Deptuła, Adam	Opole University of Technology, Poland
Dindorf, Ryszard	Kielce University of Technology, Poland
Domagała, Zygmunt	Wroclaw University of Science and Technology, Poland
Doudkin, Mikhail	D. Serikbayev East Kazakhstan Technical University
Dumitrescu, Catalin	National Institute of Research & Development for Optoelectronics/INOE 2000, Romania
Fiebig, Wiesław	Wroclaw University of Science and Technology, Poland
Filo, Grzegorz	Cracow University of Technology, Poland
Gamez-Montero, Pedro Javier	Universitat Politecnica de Catalunya, Spain
Handroos, Heikki	Lappeenranta University of Technology, Finland
Ivanović, Lozica	University of Kragujevac, Serbia
Jaroń, Ryszard	Collins Aerospace, Poland
Jasiński, Ryszard	Gdansk University of Technology, Poland
Karpenko, Mykola	Vilnius Gediminas Technical University, Faculty of Transport Engineering, Vilnius, Lithuania
Kollek, Wacław	Wroclaw University of Science and Technology, Poland
Kosucki, Andrzej	Lodz University of Technology, Poland
Kovalev, Vasyl	National Technical University of Ukraine
Kudźma, Sylwester	Rotary Power, South Shields, UK

Contents

Hydraulic Components

New Combined Centrifuge for Cleaning the Working Fluids of Hydraulic Drives from Mechanical Contaminations

Georgy Guryanov, Mikhail Doudkin, Alina Kim(✉), and Olga Vasilieva

D. Serikbayev East Kazakhstan Technical University, Ust'-Kamenogorsk, Kazakhstan
alinakim3107@mail.ru

Abstract. The contamination of the working fluid in the hydraulic drives of machines can exceed the permissible values by several times, while the actual time between failures of hydraulic units is significantly lower than the standard. This indicates the need to improve the processes of cleaning the working fluids of hydraulic drives of machines and equipment from mechanical contaminants, for which it is necessary to improve the means of cleaning. Based on this, a new design centrifuge with a hydrojet rotor rotation drive is developed, in which a synergistic combination of several different ways to increase work efficiency was used to increase the efficiency and quality of cleaning the working fluid. A theoretical substantiation of the main parameters of the new centrifuge is made, a prototype is made, its experimental studies and production tests are carried out, some results of which are given in this paper. Because of the experiments, graphical functional dependences of the parameters of the centrifugation process on the main influencing factors are obtained, indicating the high efficiency of the new centrifuge and its ability to reduce the wear of hydraulic units. In addition, the experiments showed a satisfactory accuracy of the obtained theoretical dependencies. Production tests of a prototype of the new centrifuge confirmed the results of laboratory studies and the high efficiency of its operation. The use of a new centrifuge in a hydraulic drive can have a positive economic effect. The new centrifuge can also be used to remove mechanical impurities from a wide variety of technical liquids in various machine and equipment systems.

Keywords: Hydraulic drive · Cleaning · Working fluid · Centrifuge · Contamination

1 Introduction

In modern mobile machines and industrial equipment, hydraulic drives are widely used, the complexity of the designs of which leads to increased requirements for the performance of the working fluids used in them, in particular, for their purity, on which the operational reliability of hydraulic drives and machine performance largely depend. [1–4].

The study of the working fluid of hydraulic drives of machines shows that its contamination can exceed the permissible one by several times, which is one of the main reasons

J. Stryczek and U. Warzyńska (Eds.): NSHP 2023, LNME, pp. 3–13, 2024.
https://doi.org/10.1007/978-3-031-43002-2_1

for premature failures of hydraulic units and intensive wear of their parts [1–4]. At the same time, in order to increase the working efforts of the actuators of the hydraulic drive, they tend to increase the working pressures, which is achieved by reducing the gaps in the friction pairs of the hydraulic units, and this significantly increases the requirements for the purity of the working fluids.

The complex problem of increasing the operational reliability of hydraulic units with the tightening of requirements for the purity of working fluids can be solved largely by improving the quality of cleaning of the latter.

This requires appropriate cleaners, the requirements for the efficiency of which are constantly increasing. The currently used cleaning devices, due to standard approaches to their creation, cannot fully ensure the required purity of the working fluids of hydraulic drives, reduce the wear of hydraulic units, and require improvement.

Experience shows that it can be effective to use centrifuges [5–9], which, more than other units for cleaning working fluids, meet the requirements for cleaners in terms of cleaning efficiency [5–7, 10, 14], have the best combination of advantages and disadvantages and have ample opportunities for their improvement [11, 12].

The idea of the work and its scientific novelty are to create a combined centrifuge of a new design that provides an increased quality of purification of working fluids compared to existing ones, theoretical justification and experimental determination of its main parameters and performance indicators, based on previously developed general principles for improving centrifuges and new non-standard approaches [13, 14].

2 New Combined Centrifuge

A new centrifuge has been developed at the East Kazakhstan Technical University, in which, to increase the efficiency of its operation, three-stage cleaning, a complex force field, a thin-layer circulating movement of liquid in the rotor and a hydrojet drive with the effect of "floating" are synergistically combined. The new centrifuge is shown in Fig. 1.

The centrifuge works as follows. The centrifuge is installed in a hydraulic drive and connected through axis 2 to the drain line of the working fluid. The liquid from the line through axis 2 and column 8 enters the rotor 1 of the centrifuge, where, moving in a thin layer flow, it passes through three purification stages 3, 4, 5, being exposed to complex combined force fields - centrifugal, inertial, electrostatic, and making circulation movements. Particles of mechanical impurities under the action of increased forces are released from the fluid flow and settle on the rotor stages. Multi-stage purification and flow circulation provide a high purification factor, and a complex force field and thin-layer movement organization ensure a high degree of purification. Leaving the nozzles 6, the liquid creates a reactive force that rotates the rotor 1 and provides the effect of "floating". The rotor is installed in the housing and closed with a lid on top (not shown in the figure).

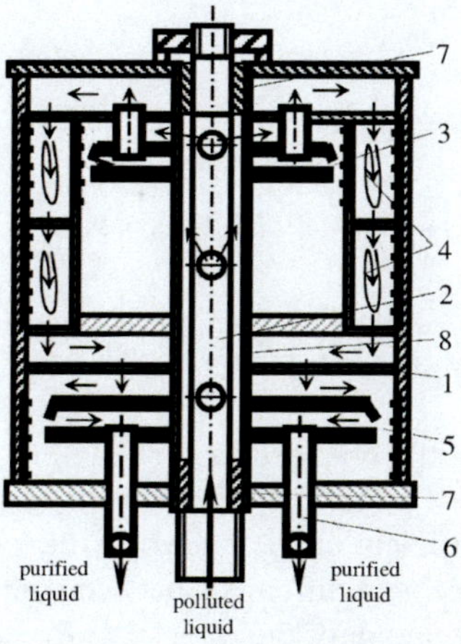

1 - rotor; 2 - axis; 3 - the first stage of cleaning (combination of three forces); 4 - the second stage of cleaning (combination of two forces and circulation movement); 5 - the third stage of purification (combination of three forces); 6 - nozzles; 7 - bearings; 8 - column.

Fig. 1. - New combined centrifuge

3 Theoretical Determination of the New Centrifuge Parameters

To substantiate the effective operating modes of the centrifuge, theoretical studies were carried out and dependences were obtained to determine its main parameters [15–18].

Liquid flow through the centrifuge rotor, which excludes the entrainment of particles exceeding the calculated size:

$$Q_{kr} = \frac{\pi \cdot f \cdot d_{ch}^2 \cdot \Delta_p \cdot w^2 \cdot R_p \cdot (R_p^2 - r_o^2)}{144 \cdot 10^{-3} \cdot \mu \cdot (1 + k \cdot f)}, m^3/s \tag{1}$$

Calculation according to formula (1) gives the value of the critical flow rate of the working fluid through the cleaner $Q_{cr} = 0.0045\ m^3/s = 272$ l/min.

The critical diameter of the deposited particles for a centrifuge with a hydrojet rotor drive:

$$d_{ch} = \sqrt{\frac{18 \cdot \mu \cdot Q}{\Delta_p \cdot \omega^2 \cdot 2 \cdot \pi \cdot R_p^2 \cdot H}}, \text{ m.} \tag{2}$$

The minimum length (height) H of the rotor required for effective particle settling:

$$\mathrm{H} = \frac{18 \cdot \mathrm{Q} \cdot \mu}{2 \cdot \pi \cdot \mathrm{R}^2 \cdot d_{ch}^2 \cdot \Delta_p \cdot \omega^2}, \text{ M.} \tag{3}$$

With the particle settling time $t = 0.02$ s, the settling length equal to the minimum length of the rotor will be $H = 3 \times L = 3 \times 0{,}048 \approx 0{,}15$ m.

The rotational speed n of the centrifuge rotor with a hydrojet drive, depending on the pressure p or the flow rate Q of the liquid in the hydraulic drive drain line and the design parameters of the centrifuge [18]:

$$K_1 \cdot \frac{\pi^2}{900} \cdot n^2 + ((1,8 \cdot c \cdot r_{cp}^2 + 4 \cdot r_c^2) \cdot p_{zh} \cdot f_a \sqrt{\frac{p \cdot (1-\psi)}{p_{zh}}} + K_2) \cdot \frac{\pi}{30} \cdot n - 8 \cdot f_a \cdot r_c \cdot p \cdot (1-\psi) = 0 \quad (4)$$

$$K_1 \cdot \frac{\pi^2}{900} \cdot n^2 + ((0,45 \cdot k \cdot r_{cp}^2 + r_c^2) \cdot p_{zh} \cdot Q + K_2) \cdot \frac{\pi}{30} \cdot n - \frac{Q^2 \cdot p_{zh} \cdot r_c}{2 \cdot f_a} = 0 \quad (5)$$

Determining the rotor speed using the obtained Eqs. (4) and (5) is best done on a computer.

The formulas indicate: f - the coefficient of friction; d_{ch} - the nominal diameter of a solid particle of pollution, m; Δ_ρ - density difference between the working fluid and particle, kg/m^3; ω - angular speed of rotation of the rotor, s^{-1}; R_p - radius of the rotor, m; r_o - the radius of the location of a liquid particle, m; μ - dynamic coefficient of fluid viscosity, Pa·s; k - the coefficient of proportionality; Q - liquid flow rate, m^3/s; H - the rotor height, m; K_1 and K_2 - generalized coefficients that take into account the geometrical parameters of the centrifuge and the acting forces and are determined from the ratio of driving forces and resistance forces; c - the coefficient of fluid removal from the rotor to the nozzles; r_{cp} - the average radius of the liquid layer, m; r_c - the arm of the reactive force, m; ρ_{zh} - the density of the working fluid, kg/m^3; f_a - the area of nozzle openings, m^2; p - the liquid pressure in front of the centrifuge, Pa; ψ - pressure loss coefficient.

4 Experimental Studies of the New Centrifuge

A prototype centrifuge, a test bench were made for research, and a technique was developed [19, 20].

The purpose of experimental studies is to determine the performance indicators and the main parameters of the centrifuge, including those that cannot be theoretically determined.

The following are only some of the most important and revealing results of the experiments.

Experiments were carried out on hydraulic oil using artificial pollutants. In the general case, by passing the working fluid through the purifier, one of the factors was changed and, by taking and analyzing samples of the fluid after the centrifuge using special equipment, the parameter under study was determined, obtaining graphical dependencies. The scheme and general view of the stand for the study of centrifuges are shown in Fig. 2.

With the use of planning, an experimental assessment of the efficiency of the centrifugation process was carried out, regression equations were obtained and, on their basis, flow rates Q (100 ... 260) l/min, pressure p (0.1 ... 0.46) MPa and rotor radius R_p (100...180) mm, which agree quite well with the results of theoretical studies.

After that, the main parameters characterizing the efficiency of the centrifuge were experimentally determined.

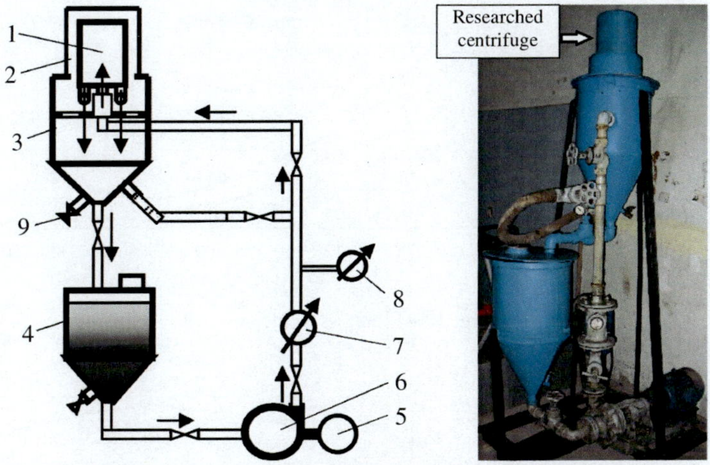

1 - centrifuge; 2 - cap; 3 - top tank; 4 - lower tank; 5 - electric motor; 6 - pump; 7 - flow meter; 8 - manometer; 9 - tap for fluid sampling

Fig. 2. - Stand for the study of the centrifuge

The experimental and theoretical dependences of the fineness of cleaning, that is, the maximum diameter d_{ch}, microns, of particles of contaminants remaining in the liquid after it passes through the centrifuge, on the flow rate Q of the liquid through the rotor are shown in Fig. 3. It follows from the graphs that the centrifuge has a high fineness of cleaning (up to 5…10 μm) at liquid flow from 150 to 300 l/min. The experimental values of the fineness of cleaning are worse than the theoretical ones, since the ideal model for the deposition of pollution particles was adopted in theoretical studies.

An important indicator of the operation of the centrifuge is the purification factor K_{och}, that is, the ratio of the difference between the volume concentrations of contaminants in the original and purified working fluid to the volume concentration of contaminants in the source fluid. The results of the experiments are shown in Fig. 4. It can be seen from the graphs that in the range of flow rates Q from 150 to 300 l/min, the purifier shows a very high cleaning coefficient (0.9 ... 0.96).

The histogram of sludge distribution (u_i, %) over the stages of the cleaner is shown in Fig. 5. It can be seen from the histogram that the use of two additional stages doubles the efficiency of cleaning the working fluid ($u_2 + u_3 = 52\%$) compared to one stage of purification (u_1=48%).

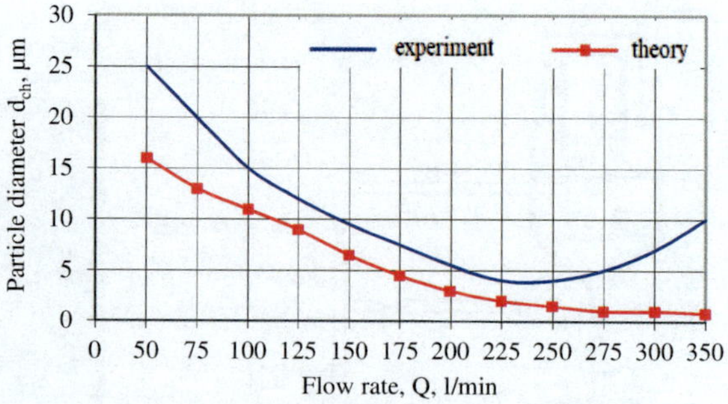

Fig. 3. - Fineness of cleaning (d_h) centrifuge

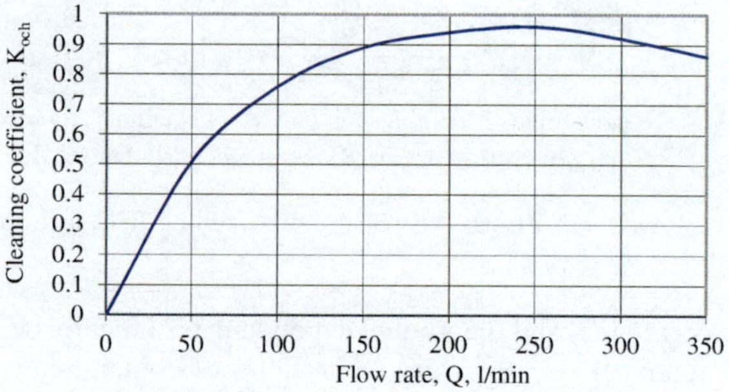

Fig. 4. - Cleaning factor (K_{och}) of the centrifuge

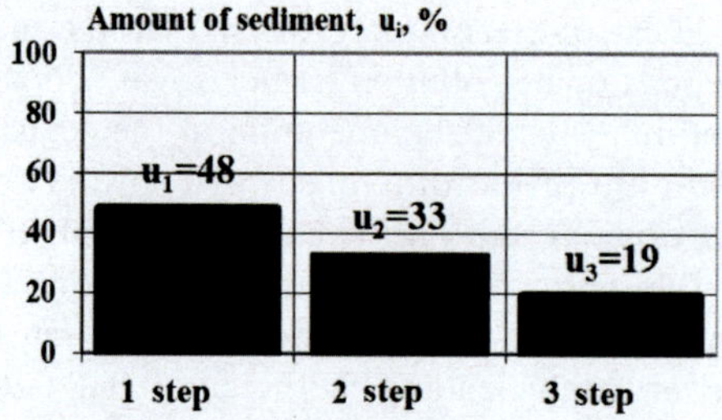

Fig. 5. - Sediment distribution over the centrifuge rotor stages

The results of experiments to determine the intensity of the contaminant release from the working fluid, that is, changes in the weight concentration (q, %) of contaminants in the working fluid from the number of its passes through the purifier are shown in Fig. 6. The graph shows that the weight concentration of contaminants in the working fluid reaches a normal value (0.002… 0.003% by weight - 12… 13 purity class) after 3… 4 passes of the liquid through the purifier, and then maintained at an acceptable level with a gradual decrease in concentration to class 11 purity.

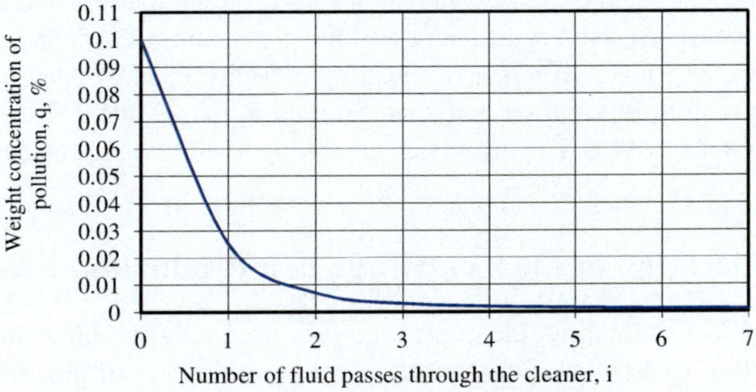

Fig. 6. - **The intensity of the contaminant release from the working fluid**

Figure 7 shows graphs of the experimental and theoretical dependence of the rotational speed (*n*, rpm) of the centrifuge rotor on the flow rate *Q* and the pressure *p* of the liquid. The discrepancy between the theoretical and experimental values of the rotor speed ranges from 10 to 18%, that is, the theoretical conclusions and the assumptions made are sufficiently correct, and the resulting equations accurately describe the speed mode of the centrifuge.

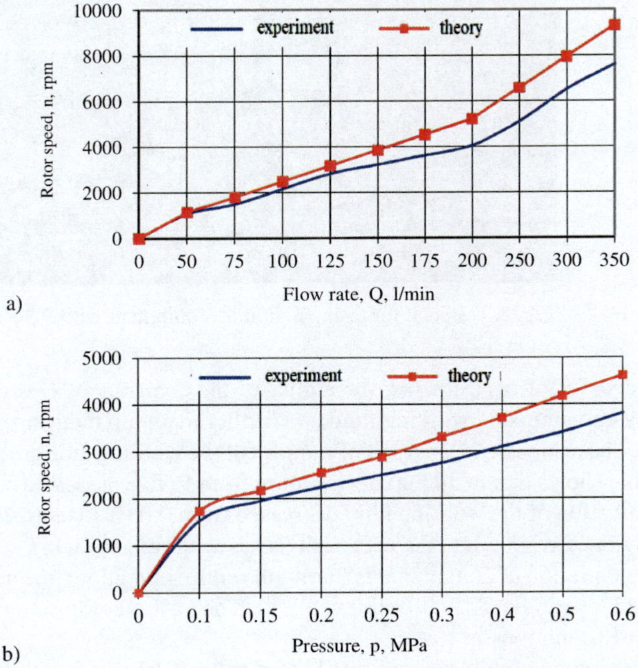

Fig. 7. - Frequency *n* of the rotor rotation on the flow rate (a) and pressure (b) of the working fluid

The conducted experimental studies have shown that the operation of the new combined centrifuge is very efficient and allows obtaining a fineness of cleaning up to 5 microns, and the coefficient of cleaning capacity reaches 0.95. At the same time, the concentration of pollution particles is not more than 0.001… 0.002% by weight, which corresponds to 11… 12 purity class according to GOST 17216–2001 with an acceptable purity class 13.

5 Operation of the Centrifuge in a Production Environment

For production testing, the centrifuge was used as a working fluid cleaner of a stand for testing hydraulic equipment of construction and road machines at the LLP Irtysh Construction Mechanization Department (Ust-Kamenogorsk).

Studies of the contamination of the working fluid of the stand showed that it is 6 times higher than the standard. At the same time, the duration of operation of pumps and regular filters of the stand is 1.5… 2.5 times lower than the standard one.

The centrifugal cleaner of the working fluid was installed on top of the tank with the working fluid and was connected to the drain line. The stand with the centrifuge installed on it is shown in Fig. 8.

Fig. 8. - Bench for tesing hydraulic equipment with a centrifuge

In a series of experiments, the ability of the centrifuge to restore the purity of the initially contaminated working fluid and further maintain it during operation was determined. The results of the analysis of samples of the working fluid are presented in Table 1. The table shows that in 30 min of operation (four to five passes through the cleaner), the contamination of the working fluid decreased from 0.03% to 0.0022% by weight, which practically corresponds to the 12th purity class according to GOST 17216-2001, and then the concentration of contaminants during the entire operation time was maintained at the achieved level (about 12 cleanliness class). The vast majority of contaminant particles are up to 10 microns in size.

Thus, production tests confirmed the results of laboratory studies and showed that the centrifuge is able to effectively restore the purity of the working fluid to the norm and then maintain this level during the operation of the stand.

Table 1. Results of the analysis of the working fluid of the test bench

Working fluid sampling time	Amount of contaminants in % by weight	Number of particles, in % of their total number				
		5–10 μm	10–20 μm	20–30 μm	30–50 μm	>50 μm
Before centrifuge installation	0,03	32	27	26	10	5
After installing the centrifuge:						
after 15 min of work	0,0048	71	15	8	4	2
after 30 min of work	0,0022	89	9	2	–	–
after 50 h of operation	0,0018	84	14	2	–	–
after 100 h of operation	0,0023	81	15	3	1	–
after 150 h of operation	0,002	78	16	6	–	–
after 200 h of operation	0,0026	83	12	5	–	–
after 250 h of operation	0,0025	86	10	2	2	–

Checking the selectivity of the centrifuge, i.e. the ability to isolate from the working fluid, first of all, the most dangerous particles of inorganic pollution from the point of view of wear, was carried out by analyzing the components of the pollution isolated by the centrifuge from the working fluid of the bench, which showed that most of the pollution caught by the centrifuge contains particles of atmospheric dust (34%) and metal (35.5%), as well as relatively dense particles of abrasive paste (9%), rubber (10%), paint, plastics, etc. (6%) with a relatively small amount of organic impurities (5.5%).

This confirms the high selectivity of the centrifuge, and hence its ability to effectively deal with the wear of hydraulic units.

6 Conclusion

To combat the contamination of working fluids of hydraulic drives, a combined centrifuge of a new design with increased cleaning efficiency was developed; its theoretical and experimental studies were carried out. The obtained theoretical dependences make it possible to calculate the main parameters of the centrifuge, and the experimental results indicate the high efficiency and quality of the centrifuge and the achievement of the expected values of the fineness of cleaning (about 5 microns) and the cleaning coefficient (up to 0.95), as well as the ability to deal with the wear of hydraulic units. This is also confirmed by the results of production tests of the centrifuge. The use of the new centrifuge will improve the reliability and durability of hydraulic units, the reliability of the operation of machines and equipment. The new centrifuge can also be used for cleaning other technical liquids.

Based on the positive research results obtained and the prospects of the proposed approaches to improving centrifuges, in the future it is supposed to develop and investigate new effective designs of centrifuges based on various combinations of cleaning methods and methods for organizing the complex movement of liquid in their rotors, to theoretically substantiate and develop a generalized approach to improving centrifuges, to develop a new approach to the theoretical description of the centrifugation process, to obtain a generalized engineering method for calculating centrifuges, to develop ways to reduce the resistance to rotation of the rotors, and to justify the use of new materials for the manufacture of centrifuge rotors from the standpoint of their strength and deformability.

References

1. Vernigora, V.A., Grigoruk, I.E.: Rezervy povysheniya effektivnosti tekhnicheskoj ekspluatacii mashin s gidroprivodom. Mekhanizaciya stroitel,stva, vol. 2, pp. 27–28 (2007)
2. Zorin, V.A., Krohin, S.A.: Analiz izmeneniya rabotosposobnosti rabochih zhidko-stej gidrosistem stroitel'no-dorozhnyh mashin v ekspluatacii. Sovershenstvovanie ekspluatacii stroitel'no-dorozhnyh mashin na sovremennom etape. Sbornik nauchnyh trudov. M., pp. 53–57 (1999)
3. Kandyba, S.V.: Vliyanie svojstv rabochej zhidkosti na iznos agregatov gidravli-cheskoj sistemy ekskavatorov. Mekhanizaciya stroitel'stva, vol. 7. pp. 7–9 (2004)
4. Saveliev, A., Zhileykin, M., Kustarev, G., Grib, V.: Icreasing the reliability of the autograder metal construction by modeling and re-assembling of the working equipment. News Nat. Acad. Sci. Repub. Kaz. Ser. Geol. Tech. Sci. **6**(438), 276–286 (2019)
5. Konovalov, V.M., Skrickij, V.Y.A., Rokshevskij, V.A.: Ochistka rabochih zhidkostej v gidroprivodah stankov. M., Mashinostroenie, p. 288 s (1996)
6. Kovalenko, V.P., Il'inskij, A.A.: Osnovy tekhniki ochistki zhidkostej ot mekha-nicheskih zagryaznenij. M., Himiya, 272 p. (2002)
7. Grigorev, M.A.: Ochistka masla v dvigatelyah vnutrennego sgoraniya. M., Mashino-stroenie, p.148 s (2003)
8. Pichugin, S., et al.: The analysis of road machine working elements parameters. World Appl. Sci. J. **23**(2), 151–158 (2013)
9. Kim, A., et al.: Mathematical and experimental study of deformations of a steel roll of a road roller with a variable geometry of a contact surface. Prod. Eng. Arch. **25**, 1–7 (2020). https://doi.org/10.30657/pea.2019.25.01
10. Vencel, S.V.: Smazka i dolgovechnost' dvigatelej vnutrennego sgoraniya. Kiev, p. 238 s (1997)
11. Sakimov, M.A., Ozhikenova, A.K., Ozhiken, A.K., Azamatkyzy, S.: Finding allowable deformation of the road roller shell with variable curvature. News Nat. Acad. Sci. Repub. Kaz. Ser. Geol. Tech. Sci. **3**(429), 197–207 (2018)
12. Belyanin, P.N.: Centrobezhnaya ochistka rabochih zhidkostej aviacionnyh gidrosi-stem. M., Mashinostroenie, p. 192 s (1996)
13. Apshikur, B., Asangaliyev, E.A., Mlynczak, M.: Development of an installation for shear ground testing in the railway track construction. News Nat. Acad. Sci. Repub. Kaz. Ser. Geol. Tech. Sci. **6**(438), 22–35 (2019)

14. **Guryanov, G., Bugaev, A., Rogovsky, V., et al.: Process modeling and experimental verification of the conditions of ice coverage destruction of automobile roads. J. Mech. Eng. Res. Dev.** **42**(4), 1–8 (2019)
15. Guryanov, G.A., Sidorenko, V.N.: Teoreticheskoe obosnovanie parametrov centrifugirovaniya rabochej zhidkosti v centrobezhnom ochistitele. Vestnik Vostochno-Kazahstanskogo gosudarstvennogo tekhnicheskogo universiteta im. D. Serikbaeva. Nauchnyj zhurnal **2**(20), зз 23–32 (2003)
16. Fadeyev, S.N. et al.: Studying the machines for road maintenance. Life Sci. J. **10**(SPL.ISSUE 12), 134–138, 24 (2013)
17. Doudkin, M., Kim, A., Moldakhanov, B.: Determination of forces in the working unit of the milling-rotary snow cleaner. Int. Rev. Mech. Eng. **15**(2), 96–105 (2021). https://doi.org/10.15866/ireme.v15i2.20859
18. **Vavilov, A., Kim, A., Guryanov, G., Likunov, A.: New technology of the steel fiber manufacturing from technogenic waste.** Int. J. Mech. Prod. Eng. Res. Dev. **10**(3), 611–622 (2020). https://doi.org/10.24247/ijmperdjun202056
19. Vedenyapin, G.V.: Obshchaya metodika eksperimental'nogo issledovaniya i obrabotki opytnyh dannyh. M., vol. 221, p. з (2007)
20. Savelyev, A., Gribb, V., et al.: Modernization of the metal structure of the grader working equipment. Int. Rev. Mech. Eng. **14**(1), 1–8 (2020)

Spool Valve Stability and Sealing Land Deformation

Rudolf Scheidl(✉) and Bernhard Manhartsgruber

Johannes Kepler University Linz, Altenbergerstraße 69, 40404 Linz, Austria
Rudolf.scheidl@jku.at

Abstract. In the development of spool valves with narrow sealing lands problems with spool sticking were encountered. In previous investigations of this phenomenon bending deformation has been identified as cause. Bending creates a conical sealing gap destabilizing the centered spool position. This leads to substantial lateral forces pressing the spool to the bore and creating high friction forces. In a patent a sealing land modification was proposed which is able to reverse this effect and to stabilize the centric spool position.

In the current paper thorough analyses of the sealing land elastic deformations by the front side and by the sealing gap pressure, their influences on the gap pressure field and on resulting lateral forces are presented. Non-dimensionalization is applied to lower the number of system parameters and to obtain results as generic as possible. Finite Element models for deformation and gap pressure are developed to solve the associated field problems.

Results show that shear deformation of sealing lands with a relatively small radial extension can create a stabilizing conical gap in its center section. General results are presented showing the magnitude of this effect in dependence on design and material parameters, also in comparison to the effect of the gap pressure field and to manufacturing tolerances.

Keywords: Spool valve spool stability · Sticking valve · Elasto-hydrostatics · Sealing land deformation

1 Introduction

The first experience of the authors with valve problems due to an unfavorable sealing gap pressure field was related to injectors of common rail diesel injectors. The theoretical investigations in [1] showed that pressure induced elastic deformations of the two components forming the gap lead to a conical gap converging towards the higher pressure. Such a gap makes the centered spool position unstable and creates a substantial lateral force in the eccentric position pressing the spool onto the bore. Similar effects were found for a special switching valve with slender sealing gaps [2]. Their bending due to different pressure on their both front sides causes a destabilizing conical gap. The related lateral forces caused valve sticking at higher pressures. In both cases design modifications of one of the involved components could be found which inverted the effect leading to a gap which stabilizes the centered spool position also against some

J. Stryczek and U. Warzyńska (Eds.): NSHP 2023, LNME, pp. 14–23, 2024.
https://doi.org/10.1007/978-3-031-43002-2_2

other lateral forces. These achievements motivated the utilization of such effects by a sleeve for the lubricated guidance of a piston rod [3] to get rid of dry friction. In [4] rotating, textured radial shaft seals using elasto-hydrodynamic effects are shown to create an oil film which can even pump back oil from air to oil side against pressure to make leakage almost zero. A self-adjusting, slender cylindrical seal made of plastics to realize a non-contacting sealing for spool valves by exploiting an elasto-hydrostatic principle is studied in [5].

The problem of gap pressure induced valve sticking due to manufacturing tolerances is considered in many fluid power text books. An analytical treatment of the related valve stability problem for the case of conical type spools is given in [6].

In this paper the pressure induced elastic deformation of spool valve sealing lands and its effect on lateral spool valve stability is revisited. It is a generalization of the results of [2] which were limited to a specific spool design. The main effects and the influence of the major system parameters are described exploiting a rigorous nondimensionalization. First, a simple axisymmetric model based on an assumed pressure distribution in the gap is developed to analyze the main effects and present them in a comprehensible form. Second, a Finite Element model is built which can solve the full elasto-hydrostatic problem and exemplary cases are analyzed.

2 The Gap Seal Model

The situation which is studied in this paper is depicted in Fig. 1. The sealing land is an annular ring being part of the spool or being firmly attached to it, such hat it has no relative motion at the mating surface. The other part of the spool is considered fully stiff. In practice this part has some deformation too, but its consideration involves many more design parameters which complicates a general investigation.

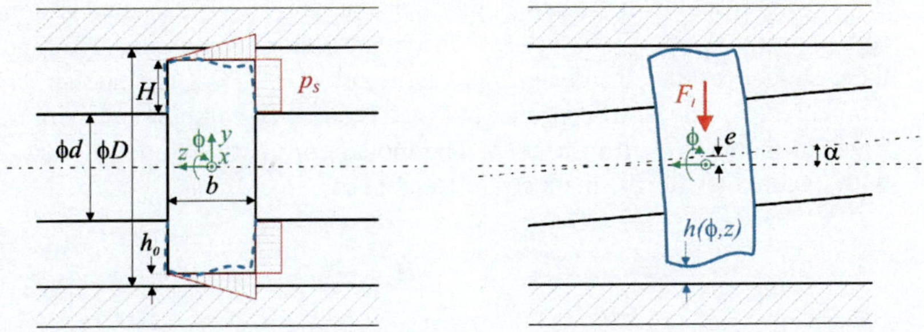

Fig. 1. Left: Sketch of the sealing land model with pressure loading on a front side and in the gap for the undeformed situation and a typical deformed shape of the land. Right: Gap geometry due to spool displacement and land elastic deformation.

On the left side the sealing land of a spool valve is shown in its centered position. The sealing gap is uniformly h_0, the spool diameter is denoted by D, the spool with width b is part of or mounted on the connecting rod with diameter d. The problem is formulated in a Cartesian and, for the sealing gap related issues, cylindrical coordinates

$z-\phi$; Fig. 1 gives the coordinates' names. The pressure p_S acts on the right side of the land and a linearly declining gap pressure on its circumference. A feasible deformation pattern is also sketched in the figure. It changes the gap field. The right hand side of this figure depicts the gap situation with an additional spool eccentricity e and inclination α. Gap $h(\phi, z)$ and pressure field $p(\phi, z)$ couple by elasto-hydrostatic interaction. The pressure field may create a resulting lateral force F_l. If it acts against the eccentricity, as shown in in Fig. 1, it stabilizes the centered spool position; otherwise the centered position is unstable.

It is clear that the role of elastic gap deformation depends highly on its size relative to the initial gap h_0.. Therefore, very small gaps, high pressure, and soft land material (such as plastics) strengthen this role. From a practical viewpoint, the deformation must also be related to the manufacturing tolerances, which obviously are related to the nominal gap size. This statistical feature is only indirectly treated in this paper by the declaration that the effect of elastic land deformation is only dominating if it exceeds the manufacturing tolerances. The tolerance problem includes also surface roughness which is neglected here too, yet with the warning that the considered gap situations must exceed asperity heights.

Throughout this paper only equilibrium conditions are considered. This is justified by the assumption that the frequency bandwidth of all changes of deformation and pressure are below the lowest natural frequency of the sealing land.

2.1 Sealing Land Deformation – Simplified Case

In the perfect situation (uniform gap) according to Fig. 1, i.e. a uniform gap, the pressure is known. The pressure on the left is assumed being zero. If a constant pressure is added to obtain the general situation it causes a very small volumetric shrinkage of the full spool, the effect of which is neglected here. Thus, p_S means the pressure difference of both sides.

The deformation of the land depends on the ratio *H/d*. This actual system parameter's role is partly skipped here in order to obtain a more comprehensive representation of the essential results. To this end, the assumption $H/d \ll 1$ is made and a planar plane strain model of the land according to Fig. 2 is used. To compute the deformation a Finite Element model is built in Matlab. The model core is the Matlab library 'Matfem' [7] with its modules for 2D plain strain elements.

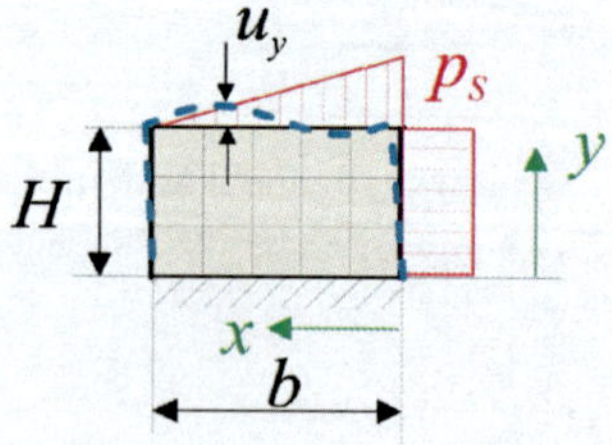

Fig. 2. Simplified sealing land model with pressure loads and exemplary deformation.

Nondimensionalization of this case is done by relating p_S to the modulus of elasticity and all spatial coordinates and displacements to land width b. Then in view of the mathematical nature of the three basic equations of linear elasticity theory in the static case, i.e. equilibrium conditions, stress strain relations, and strain deformation relations, the quantities of the dimensioned variables can be computed by the following simple rules,

$$u = \overline{u}\, b \underbrace{\frac{p_S}{E}}_{\psi_S}\ , \quad \varphi = \overline{\varphi}\frac{p_S}{E}\ , \quad \sigma = \overline{\sigma} p_S \tag{1}$$

u and p mean a displacement or stress component, φ a rotation and a superimposed bar the dimensionless version of a quantity.

This dimensionless model constitutes a two parameter problem family with the dimensionless parameters *H/b* and the Poisson ratio ν, which has a noticeable effect on the results as will be shown below.

Value ranges of the two quantities ψ_s (the nondimensional pressure according to Eq. (1)) and ν for typical hydraulic pressures and for steel and plastic as land material are given in Table 1.

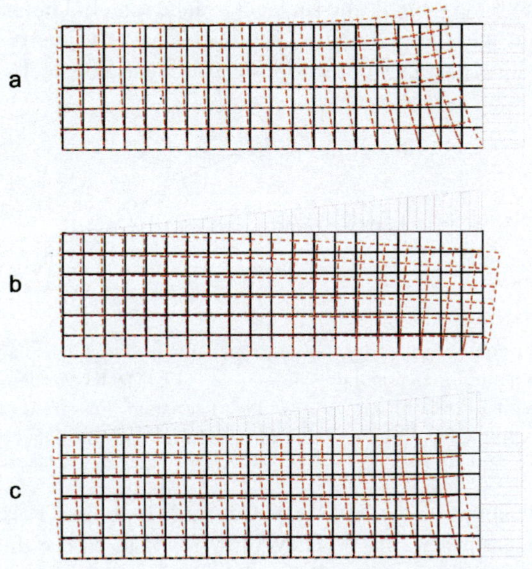

Fig. 3. FE-Mesh of undeformed and deformed state for $H/b = 0.5$, a Poisson ration $\nu = 0.38$, and a nondimensional pressure $\psi_S = 1$; a) constant pressure on left front, b) linear declining pressure on upper edge, c) combination of pressure loads of a) and b); displacement magnification factor: 0.2.

The land deformations due to front side or gap pressures and for their combination are depicted by Fig. 3 for a land aspect ratio of $H/b = 0.5$, and a Poisson number

Table 1. Typical values of ψ_s and ν for a system pressure $p_S = 210\,bar$ and three different land materials; values for PEEK taken from [8], for POM from [9].

Land material	*E/MPa*	$\psi_S = p_S/E$	ν
Steel	2.1 105	10–4	0.3
PEEK	3675	5.7 10^{-3}	0.38
POM	3000	7 10^{-3}	0.37

$\nu = 0.38$. . A pressure load at the front (case a) causes a deformation with a gap increasing towards the falling pressure, while the declining radial gap pressure causes a gap narrowing towards lower pressure side. The combination of both pressure loads creates an S-shaped gap profile, converging in an inner part but diverging at the margin sections.

The radial displacements which affect the gap can be assessed from the graphs in Fig. 4. They show the values for $\psi_s = 1$. Taking the typical values from Table 1 and assuming a land-width value of $b = 10$ mm we obtain radial displacements in the range of 0.1 μm for steel and 7 μm for POM. For an initial radial gap of $h_0 = 10$ μm, a typical value for low performance proportional or on-off valves, these elastic gap changes play a very small role for steel as land material but are significant for the much softer plastic materials. Of course, for much smaller values of h_0 in the range of 1–2 μm the deformation plays a significant role also in case of steel.

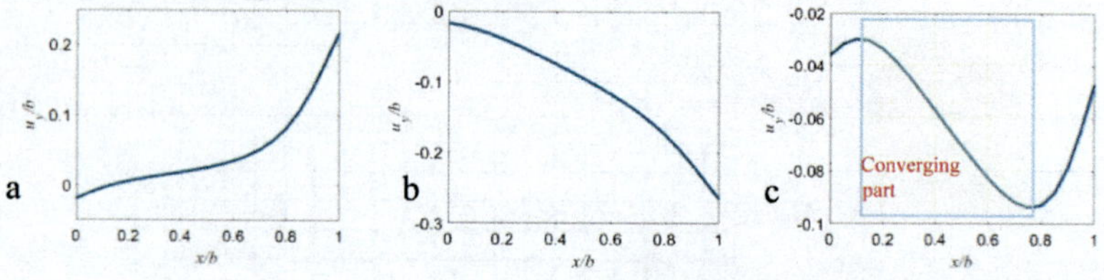

Fig. 4. Outer contour given by the nondimensional vertical displacements (u_y/b) of the upper seal edge for a nondimensional pressure $\psi_s = 1$. FE-Mesh of undeformed and deformed state for $H/b = 0.5$ and a Poisson ration $\nu = 0.38$; a) constant pressure on left front, b) linear declining pressure on upper edge, b) combination of pressure loads of a) and b).

The influences of the sealing aspect ratio H/b land Poisson number ν are visualized by the graphs in Fig. 5. They show for load case c the outer contour (left plot) and radial displacement extent of the inner, converging part (right plot), which ranges approximately from $\frac{1}{4}$ to $\frac{3}{4}$ of the full width b for the two numbers of ν given in Table 1. The graph exists only for small values of H/b, since for higher values no converging part exists. This converging part is a kind of conicity value and a measure for the self-centering capability of the land, if the margin zones, where the resulting gap would be diverging, are chamfered.

The mentioned conicity value is relatively small compared to the diverging contour inclination for higher aspect ratios. Those stem from a plate bending effect due to the front

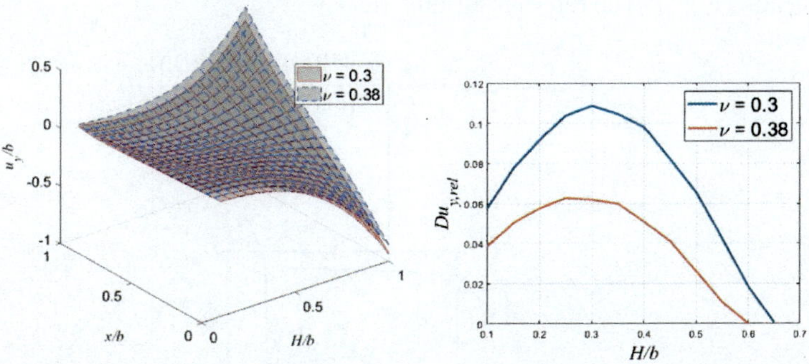

Fig. 5. Left plot: Outer contour given by the nondimensional vertical displacements (Du_y, $rel = u_y/b$) of the upper seal edge for $\psi_s = 1$ over aspect ratio H/b and for two Poisson ratios. Right plot: Extent of radial displacement leading to a converging gap towards low pressure side.

side pressure and increase progressively with *H/d*. The Poisson ratio ν has a significant influence on this value and influences also the optimum aspect ratio, where conicity is maximum.

A significant gap change makes the presented approach invalid, since it changes also the gap pressure field. This will be studied in the next section.

2.2 Stabilizing Effect by a Chamfered Land

Figure 4c shows the converging gap in the center of the gap (see Fig. 6). If both margin sections are hydrostatically deactivated by chamfering the sealing land there, the centered position of the spool is stabilized. This idea is protected as a subclaim of the patent [10]. The range of the margin sections depends on H/b and ν. They can be estimated from the plots in Fig. 6. The optimum aspect ratio which exhibits the highest converging radial displacement depends slightly on ν but is close to 0.3.

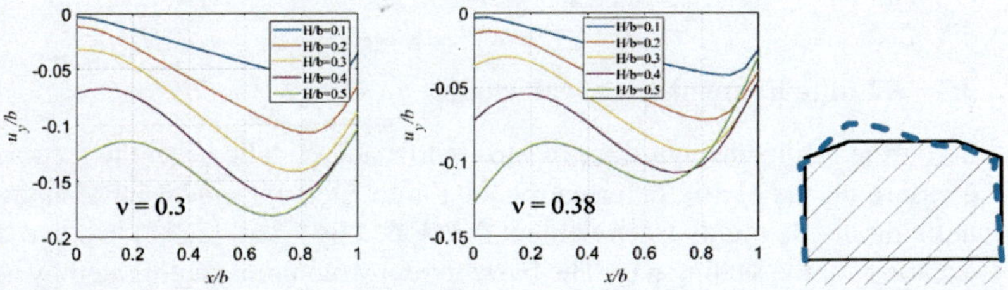

Fig. 6. Outer contours for different aspect ratios H/b and for two values of ν and the chamfering of the land.

For the stiffness of this stabilizing effect an approximate formula can be derived, combining the results of the analysis presented so far and the theory of the lateral hydrostatic force F_{lat} on an eccentrically placed, stiff conical pistons presented in [11];

see also Fig. 7. The relevant formula reads:

$$F_{lat} = \frac{5\pi\, L^2\,(4\mu\alpha + 2k\mu\zeta_0 + k\alpha\lambda)}{16\mu^2(10+\lambda^2)}(p_1 - p_2) \tag{2}$$

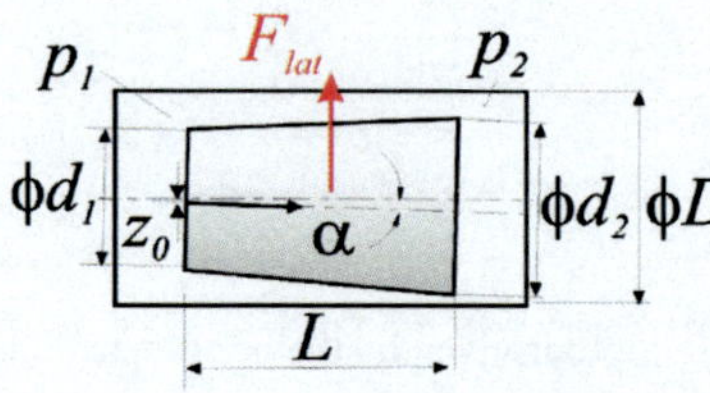

Fig. 7. Geometry of a conical body in a cylindrical bore exhibited to a pressure difference $p_1 - p_2$ and lateral force F_{lat} created by the sealing gap pressure field.

The meanings of the nondimensional values of (2) as far as not obvious from Fig. 7 are defined in (3).

$$k = \frac{d_2 - d_1}{L}, \quad \mu = h_0/D = \frac{(1 - d_1/D)}{2}, \quad \zeta_0 = z_0/h_0\,, \quad \lambda = \frac{L}{R} \tag{3}$$

The conicity k can be directly taken from the right plot in Fig. 5. A pure lateral shift without a rotation α of the spool axis ($\alpha = 0$), a renaming of the variables from those of [11] to those in this paper ($L = b$, p_2-$p_1 = p_S$, $k = Du_{y,rel}$), and the consideration of the scaling rules (1) yields the following formula:

$$F_{lat} = \frac{10\pi\, b^2\, Du_{y,rel}\; \psi_S \zeta_0}{16\mu(10+\lambda^2)} p_S \tag{4}$$

Pressure p_s has actually a quadratic influence, due to its involvement in ψ_s. This formula is limited to those cases, where the relative gap change due to the elastic deformation is small. $Du_{y,rel}$ must be taken from Fig. 5.

2.3 A Finite Element Gap Seal Model

So far the results disregard the elasto-hydrostatic coupling. i.e. the feedback of the gap pressure on the elastic deformation. A Finite Element model in Abaqus is developed utilizing the Reynolds-User element developed by Gradl [3, 12] to solve the Reynolds-equation for the sealing gap. The basic model structure is highlighted by a cross section in Fig. 8. The land is modeled by 8 node brick elements representing only of 180° (see also Fig. 9) of the full land, with appropriate symmetry conditions at both circumferential ends. The inner nodes are rigidly connected to a master node to which the radial shift z_0 is applied to model eccentricity.

Abaqus provides the reaction force resulting from the gap pressure field at this node. The deformed geometry of a steel land with $d = 6\ mm$, $D = 12\ mm$, $p_S = 250\ bar$, $h_0 = 20\ \mu m$ (uniform) and $z_0 = 20\ \mu m$ is visualized by Fig. 9.

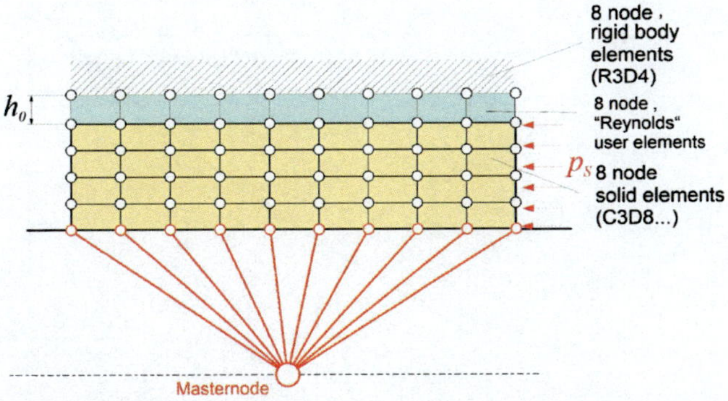

Fig. 8. Main features of the Abaqus Finite Element Model

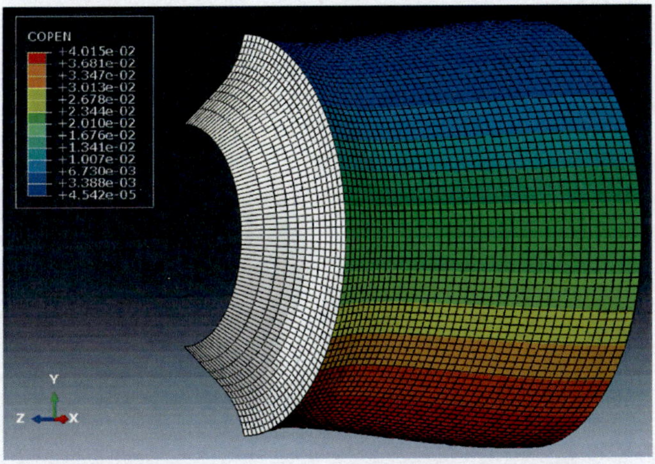

Fig. 9. Deformed shape of a steel sealing land computed by an Abaqus Finite Element model; displacement magnification factor: 5000; colors show the remaining radial gap.

The resulting stabilizing force of this semi-model as function of eccentricity z_0 is given in Fig. 10. The full land creates twice this force. This result is for the case without the chamfering measure reported in Sect. 2.2 which would intensify this effect. The elastic radial deformations are of order O(10^{-4} mm), thus dominating the situation only for a very precise manufacturing.

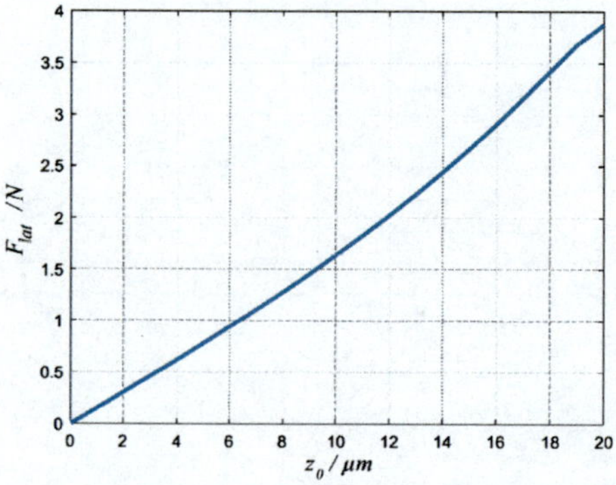

Fig. 10. Lateral force F_{lat} resulting from the gap pressure as function of eccentricity z_0 for the semi model.

3 Summary

The effect of pressure induced spool valve sealing land deformation is analyzed by simple analytical and a Finite Element model. Low lands have a dominating shear deformation which can create a radial gap stabilizing the centric spool position. The effect can be intensified by chamfering the land. Stiff materials like steel require very precise manufacturing to utilize this effect practically, while plastics exhibit this effect with moderate manufacturing tolerances.

Acknowledgement. This work was sponsored by the COMET K2 Symbiotic Mechatronics and is funded by the Austrian Federal Government, the Federal State Upper Austria, and its Scientific Partners.

References

1. Mitter, R.: Rechnerische Untersuchung eines Verschleißproblems im Injektor eines Common Rail Dieseleinspritzsystems auf Basis eines elasto-hydrodynamischen Modells – Ventilkolbenführung und Dichtfunktion gegen Hochdruck. Diploma thesis, Johannes Kepler University Linz, Austria (2000)
2. Winkler, B., Mikota, G., Scheidl, R., Manhartsgruber, B.: Modelling and simulation of the elastohydrodynamic behavior of sealing gaps. Aust. J. Mech. Eng. **2**(1), 72–86 (2005)
3. Gradl, C., Scheidl, R.: A combined hydrostatic hydrodynamic bearing based on elastic deformation. In: Proceedings of the Proceedings of the 9th FPNI Ph. D. Symposium on Fluid Power: 26–28 October 2016, Florianópolis, SC, Brazil (2016)
4. Schulz, M., Baumann, M., Bauer F., Haas, W.: Influence of different shaft surface finishes on the tribological and functional behaviour of radial shaft seals. In: Proceedings of 11th IFK, 19–21 March 2018 Aachen
5. Matthias, S., Rudolf, S., Lukachev, E.: A new elastic non contacting sealing concept for valves. Int. J. Fluid Power **23**(3), 433–452 (2022)
6. Rituraj, R., Scheidl, R.: Stability analysis of spools with imperfect sealing gap geometries. Int. J. Fluid Power **21**(3), 383–404 (2020)

7. GID Simulation homepage. https://www.gidsimulation.com/downloads/educational-finite-element-codes-matfem/. Accessed 15 Feb 2023
8. Drozdov, A.D., deClaville Christiansen, J.: Thermo-mechanical behavior of poly (ether ether ketone): experiments and modeling. Polymers (Basel) **13**(11), 1779 (2021)
9. MatWeb homepage. https://www.matweb.com/search/datasheet.aspx?matguid=0c568e14e05c4dba9dc2944331a042cd. Accessed 15 Feb 2023
10. US Patent #7537026, Filed: Apr 5, 2005, Date of Patent: May 26, 2009. Inventors: Dantlgraber, J., Försterling, H., Scheidl, R., Winkler, B.
11. Scheidl, R., Resch, M., Scherrer, M., Zagar, P.: An approximate, closed form solution of sealing gap induced lateral forces for imperfect sealing land geometries. In: Advances in Hydraulic and Pneumatic Drives and Control 2020, pp. 102–111. Springer International Publishing. (2021). https://doi.org/10.1007/978-3-030-59509-8_9
12. Gradl, C.: Hydraulic stepper drive: Conceptual study, design and experiments. PhD thesis, Johannes Kepler University Linz, 2017

Modelling of the Hydrostatic Relief in the Cylinder Block of an Axial Piston Pump

Tadeusz Złoto(✉) and Grzegorz Klimkowski

Institute of Technology, State University of Jan Grodek in Sanok, Sanok, Poland
{tzloto,gklimkowski}@up-sanok.edu.pl

Abstract. Axial piston pumps are positive displacement units with the highest possible pressures that can be attained. One of the fundamental pairs affecting the energy efficiency of the hydraulic machine is the cylinder block –valve plate. The work presents the construction of an axial piston pump. Three possible designs of the valve plate were analyzed, namely the design with positive overlap, zero overlap and the one, that is most often used, i.e. with relief grooves. Relationships for the distribution of hydrostatic pressure on the valve plate in the zone high pressure on the inner, outer and middle rings in dependence on the angle of rotation of the cylinder block were developed. Subsequently the location of the resultant hydrostatic forces occurring on the valve plate in the zone high pressure and the upper and lower transition zones relief the cylinder block during its rotation were depicted. Using the MathCad 2001 [5] program, computational simulations of the movement trajectories of the resultant pressing forces acting on the bottoms of the cylinder block located in the zone high pressure and the resultant hydrostatic forces relief the cylinder block during its rotation were performed. For example, the trajectory of the movement of hydrostatic forces pressing and relief the cylinder block for the zero overlap of the valve plate was presented. The work is completed with substantive conclusions.

Keywords: axial piston pump · hydrostatic relief · trajectories of movement of hydrostatic forces

1 Introduction

In many branches of industry axial piston pumps are used. These devices, due to the possibility to work at high pressures and at high power, feature high values of energy efficiency coefficients, determined by the ratio of power to weight or volume [4]. This type of displacement machines is most often used in drives of devices with complex functions and high requirements in terms of efficiency and performance. This, in turn, stimulates a continuous progress towards an increase of the operational indicators of these machines by improvement of the structure.

The range of application of piston units is wide and it is constantly expanding. The most important fields of application (e.g. by Parker, Bosch-Rexroth and others) of hydraulic piston machines include:

J. Stryczek and U. Warzyńska (Eds.): NSHP 2023, LNME, pp. 24–32, 2024.
https://doi.org/10.1007/978-3-031-43002-2_3

- aerospace (aircraft), automotive (presses), heavy (rolling mills), construction (excavators),
- forestry and agricultural machinery (forest cranes, harvesters) and military traction systems.

The widely used types of construction of piston machines with axial pistons include piston pumps with an swash plate (Fig. 1). The cylinder block 4 rotates with the pistons 6, while the valve plate 5 and the swash plate 8 are stationary. By changing the inclination angle of the swash plate, a variable value of the distance of the pistons in the reciprocating motion is obtained, what leads to a variable flow rate of the working liquid.

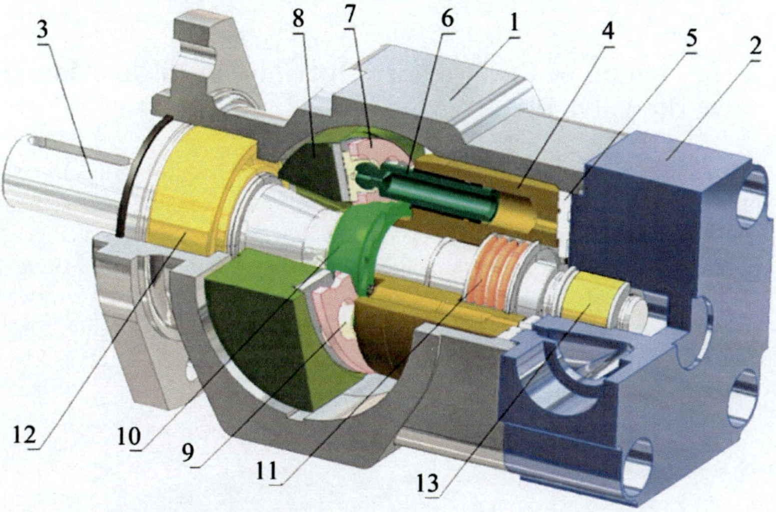

Fig. 1. The main structural elements of an axial piston pump: 1 - body, 2 - cover, 3 - shaft, 4 - cylinder block, 5 - valve plate, 6 - piston, 7 - separator, 8 - swash plate, 9 - slipper, 10 - spherical support, 11 - central spring, 12, 13 - rolling bearings

The fundamental groups of assemblies that cause energy losses in axial piston pumps are the sliding pairs: piston-cylinder, slipper-swash plate and cylinder block-valve plate. Three possible designs of the valve plate were considered in the work, i.e. the design with a positive overlap, the one with a zero overlap with the maximum length of the suction and pressure kidneys, and the solution which is most often used i.e. with relief grooves [6], as shown in Fig. 2. The characteristic radii of the valve plate are: $r_1 = 0.0284$ m, $r_2 = 0.0304$ m, $r_3 = 0.0356$ m, $r_4 = 0.0376$ m, the angles. $\alpha_m = 44.674°$and $\alpha_c = 26.109°$, the number of pistons z = 9.

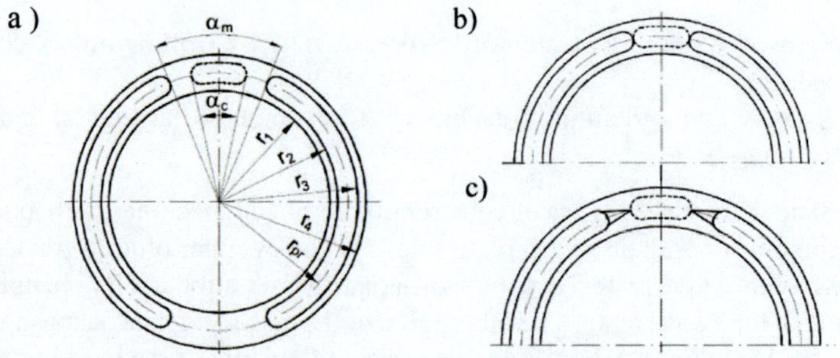

Fig. 2. Design solutions of the valve plate: a) with positive overlap, b) with zero overlap, c) with relief grooves

2 Modelling of the Pressure Distribution in the Main Zone of the Valve Plate

In the gap between the valve plate and the cylinder block there is a complex distribution of pressure which reliefs the cylinder block. For further considerations, the model pressure distribution on the positively overlapped valve plate was adapted (Fig. 3). The analysis of pressure distribution concerns the main pressure zoneψ (φ) for three design variants of the valve plate with overlays: positive, zero and with relief grooves. In the inner zone between the radii r_1 and r_2 in the range of the changing angle ψ (φ) of the high pressure p_t *[Pa]* during the rotation of the cylinder blok, there exists a logarithmic pressure distribution shown in Fig. 4 [3].

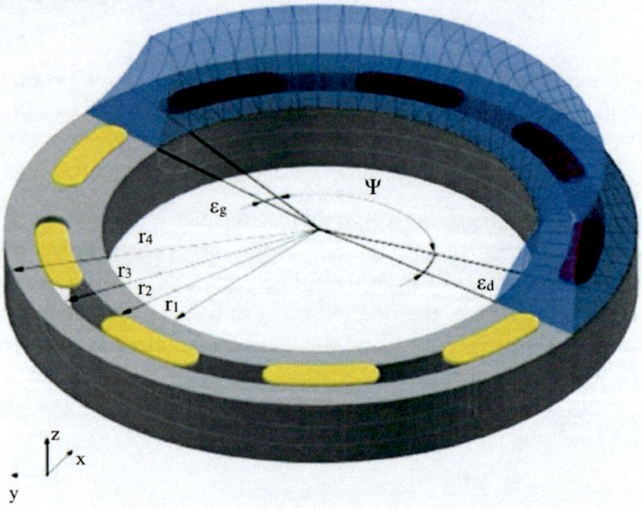

Fig. 3. Model pressure distribution on the valve plate

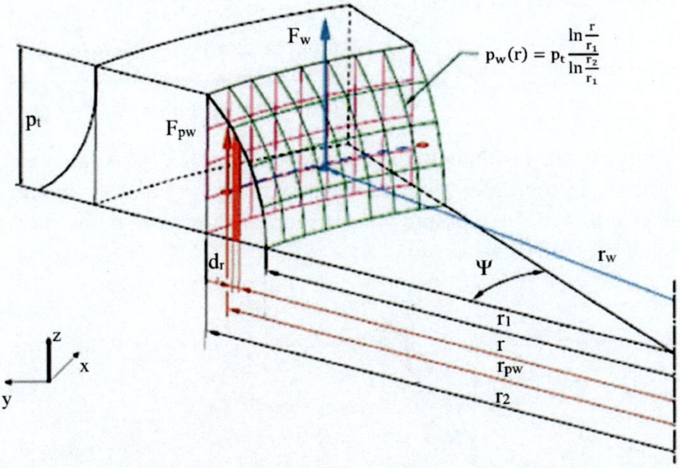

Fig. 4. Model distribution of high pressure in the inner zone of the valve plate

By integrating the pressure distribution function $p_w(r)[Pa]$ (Fig. 4), the unit hydrostatic force was obtained according to the relationship (1) in the radial cross-section of the inner zone:

$$F_{pw} = \int_{r_1}^{r_2} p_t \frac{\ln\frac{r}{r_1}}{\ln\frac{r_2}{r_1}} \cdot dr = p_t\left(r_2 + \frac{r_1 - r_2}{\ln\frac{r_2}{r_1}}\right) \tag{1}$$

The hook radius r_{pw} *[m]* for the unit hydrostatic force F_{pw} *[N/m]* of the pressure distribution in the radial section (Fig. 4) can be calculated from the following relation:

$$r_{pw} = \frac{\int_{r_1}^{r_2} p_t \frac{\ln\frac{r}{r_1}}{\ln\frac{r_2}{r_1}} \cdot r \cdot dr}{\int_{r_1}^{r_2} p_t \frac{\ln\frac{r}{r_1}}{\ln\frac{r_2}{r_1}} \cdot dr} = \frac{1}{4} \cdot \frac{r_2^2\left(2\ln\frac{r_2}{r_1} - 1\right) + r_1^2}{r_2\left(\ln\frac{r_2}{r_1} - 1\right) + r_1} \tag{2}$$

The resultant hydrostatic force F_w $(\varphi)[N]$ of the pressure distribution in the inner zone within the variable angle ψ $(\varphi)[°]$ of the pressure zone is:

$$F_w(\varphi) = F_{pw} \cdot l_{\psi w}(\varphi) = F_{pw} \cdot r_{pw} \cdot \psi(\varphi) \cdot \frac{\pi}{180} \tag{3}$$

Using dependencies (1) and (2), the following relationship was obtained:

$$F_w(\varphi) = p_t\left(r_2 + \frac{r_1 - r_2}{\ln\frac{r_2}{r_1}}\right) \cdot \frac{r_2^2\left(2\ln\frac{r_2}{r_1} - 1\right) + r_1^2}{r_2\left(\ln\frac{r_2}{r_1} - 1\right) + r_1} \cdot \frac{\pi}{720} \cdot \psi(\varphi) \tag{4}$$

The center of gravity of the force F_w (φ) located on the radius r_w $(\varphi)[m]$ was calculated from the following expression:

$$r_w(\varphi) = \frac{r_{pw} \cdot l_{cw}(\varphi)}{l_{\psi w}(\varphi)}, \text{ where } l_{cw}(\varphi) = 2r_{pw} \cdot \sin\left(\frac{\psi(\varphi)}{2}\right) \tag{5}$$

Finally, the radius r_w is:

$$r_w(\varphi) = \frac{90}{\pi} \cdot \frac{r_2^2\left(2\ln\frac{r_2}{r_1} - 1\right) + r_1^2}{r_2\left(\ln\frac{r_2}{r_1} - 1\right) + r_1} \cdot \frac{\sin\left(\frac{\psi(\varphi)}{2}\right)}{\psi(\varphi)} \tag{6}$$

Similarly, the resultant hydrostatic force F_z (φ)*[N]* was calculated from the pressure distribution in the outer zone between radii r_3 *[m]* and r_4 *[m]* in the variable angle rangeψ $(\varphi)[^\circ]$ of the pressure zone. Figure 5 shows the model pressure distribution in the outer pressure zone ψ (φ).

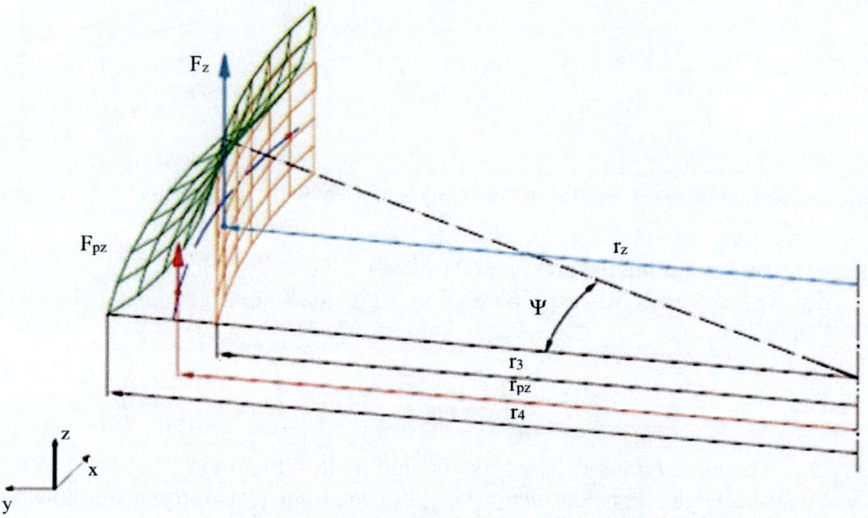

Fig. 5. Model pressure distribution in the external pressure zone

In the radial section with a logarithmic pressure distribution, there is a unitary hydrostatic force F_{pz} *[N/m]* determined from the integral relationship:

$$F_{pz} = \int_{r_3}^{r_4} p_t \frac{\ln\frac{r_4}{r}}{\ln\frac{r_4}{r_3}} \cdot dr = p_t\left(\frac{r_4 - r_3}{\ln\frac{r_4}{r_3}} - r_3\right) \tag{7}$$

The unit hydrostatic force F_{pz} occurs at the radius r_{pz} *[m]* determined from the relation:

$$r_{pz} = \frac{\int_{r_3}^{r_4} p_t \frac{\ln\frac{r_4}{r}}{\ln\frac{r_4}{r_3}} \cdot r \cdot dr}{\int_{r_3}^{r_4} p_t \frac{\ln\frac{r_4}{r}}{\ln\frac{r_4}{r_3}} \cdot dr} = \frac{1}{4} \cdot \frac{r_3^2\left(2\ln\frac{r_3}{r_4} - 1\right) + r_4^2}{r_3\left(\ln\frac{r_3}{r_4} - 1\right) + r_4} \tag{8}$$

The resultant hydrostatic force F_z (φ) for the pressure distribution in the outer zone within the variable angle ψ (φ) is given as:

$$F_z = F_{pz} \cdot l_{\psi z} = F_{pz} \cdot r_{pz} \cdot \frac{\pi}{180} \cdot \psi(\varphi) \tag{9}$$

Using dependencies (7) and (8), the following expression was obtained:

$$F_z = p_t \left(\frac{r_4 - r_3}{\ln \frac{r_4}{r_3}} - r_3 \right) \frac{r_3^2 \left(2 \ln \frac{r_3}{r_4} - 1 \right) + r_4^2}{r_3 \left(\ln \frac{r_3}{r_4} - 1 \right) + r_4} \cdot \frac{\pi}{720} \cdot \psi(\varphi) \tag{10}$$

The radius of the location of the center of gravity $r_z\ (\varphi)[m]$ of the force $F_z\ (\varphi)$ was calculated from the following relationship:

$$r_z(\varphi) = \frac{r_{pz} \cdot l_{cz}(\varphi)}{l_{\psi z}(\varphi)}, \text{ where } l_{cz}(\varphi) = 2r_{pz} \cdot \sin(\frac{\psi(\varphi)}{2}) \tag{11}$$

Finally, radius r_z is given as:

$$r_z(\varphi) = \frac{90}{\pi} \cdot \frac{r_3^2 \left(2\ln \frac{r_3}{r_4} - 1 \right) + r_4^2}{r_3 \left(\ln \frac{r_3}{r_4} - 1 \right) + r_4} \cdot \frac{\sin\left(\frac{\psi(\varphi)}{2} \right)}{\psi(\varphi)} \tag{12}$$

In the middle area between the radii r_2 and r_3 of the valve plate within the variable angle $\psi\ (\varphi)$ of the pressing zone, there is a constant high pressure, as shown in Fig. 6.

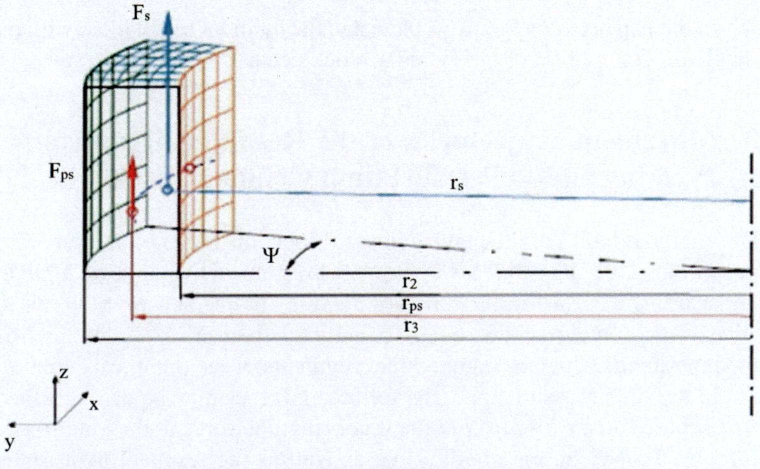

Fig. 6. Modelled pressure distribution in the central pressure zone

In the radial section of the pressure distribution, acts the unit hydrostatic force F_{ps} *[N/m]*, which is defined by the following formula:

$$F_{ps} = p_t(r_3 - r_2) \tag{13}$$

The center of gravity of the unit hydrostatic force F_{ps} is located in the middle of the central zone on the radius r_{ps} *[m]*:

$$r_{ps} = \frac{r_2 + r_3}{2} \tag{14}$$

The resultant hydrostatic force F_s (φ) *[N]* of the high pressure distribution in the middle zone within the variable angle ψ (φ) *[°]* of the pressure zone was calculated according to the following relationship:

$$F_s(\varphi) = F_{ps} \cdot l_{\psi s}(\varphi) = F_{ps} \cdot r_{ps} \cdot \frac{\pi}{180} \cdot \psi(\varphi) \tag{15}$$

Using dependencies (13) and (14), the following expression was obtained:

$$F_s(\varphi) = p_t\left(r_3^2 - r_2^2\right) \cdot \frac{\pi}{360} \cdot \psi(\varphi) \tag{16}$$

The center of gravity of the resultant hydrostatic force F_s (φ) located on the radius r_s (φ) *[m]* was calculated from the following relationship:

$$r_s(\varphi) = \frac{r_{ps} \cdot l_{cs}(\varphi)}{l_{\psi s}(\varphi)}, \text{ where } l_{cs}(\varphi) = 2r_{ps} \cdot \sin(\frac{\psi(\varphi)}{2}) \tag{17}$$

Finally, the radius r_s is given as:

$$r_s(\varphi) = \frac{180}{\pi}(r_2 + r_3) \cdot \frac{\sin\left(\frac{\psi(\varphi)}{2}\right)}{\psi(\varphi)} \tag{18}$$

In the gap between the valve plate and the cylinder block filled with oil, it is assumed that hydrostatic relief of the cylinder block occurs.

3 Movement Trajectories of the Resultant Hydrostatic Force Pressing and Relief the Pump Cylinder Block

During the rotation of the pump cylinder block, the range of pressure distribution in the gap of the valve plate changes in the pressure zone in the range of angle ψ (φ) and in the upper transition zone in the range of angle ε_g (φ)*[°]* as well as in the lower transition zone in the range of angle ε_d (φ)*[°]*. Figure 7 shows the resultant hydrostatic forces of the pressure distribution in individual segments of the pump valve plate.

In the high pressure zone, the following forces may be distinguished: the resultant hydrostatic force F_w (φ)*[N]* at the inner ring, the force at the outer ring F_z (φ)*[N]*, the force F_s (φ)*[N]* on the middle ring, as well as the resultant hydrostatic force F_n (φ) *[N]* from pressure occurring in the active kidneys (the last one is not considered in the present work).

In the upper transition zone, the resultant hydrostatic force F_{wg} *[N]*on the inner ring, the force F_{zg} *[N]* on the outer ring and the force F_{sg} *[N]* on the middle ring are distinguished.

Respectively, in the lower transition zone, the resultant hydrostatic force F_{wd} *[N]* on the inner ring, the force F_{zd} *[N]* on the outer ring and force F_{sd} *[N]* on the middle ring are distinguished.

Taking into account the dependencies from 1 to 18 and other dependencies for the pressure distribution in the upper and lower transition zones which are not presented in this work, considering various solutions of the valve plate, several calculation models of the trajectory of the displacement of the resultant hydrostatic forces relief the cylinder block were formulated. Using the MathCad 2001 [5] program, computational simulations of the movement trajectories of the resultant pressing forces acting on the bottoms of the cylinder block located in the pressure zone and the resultant hydrostatic forces relief the cylinder block during its rotation were performed.

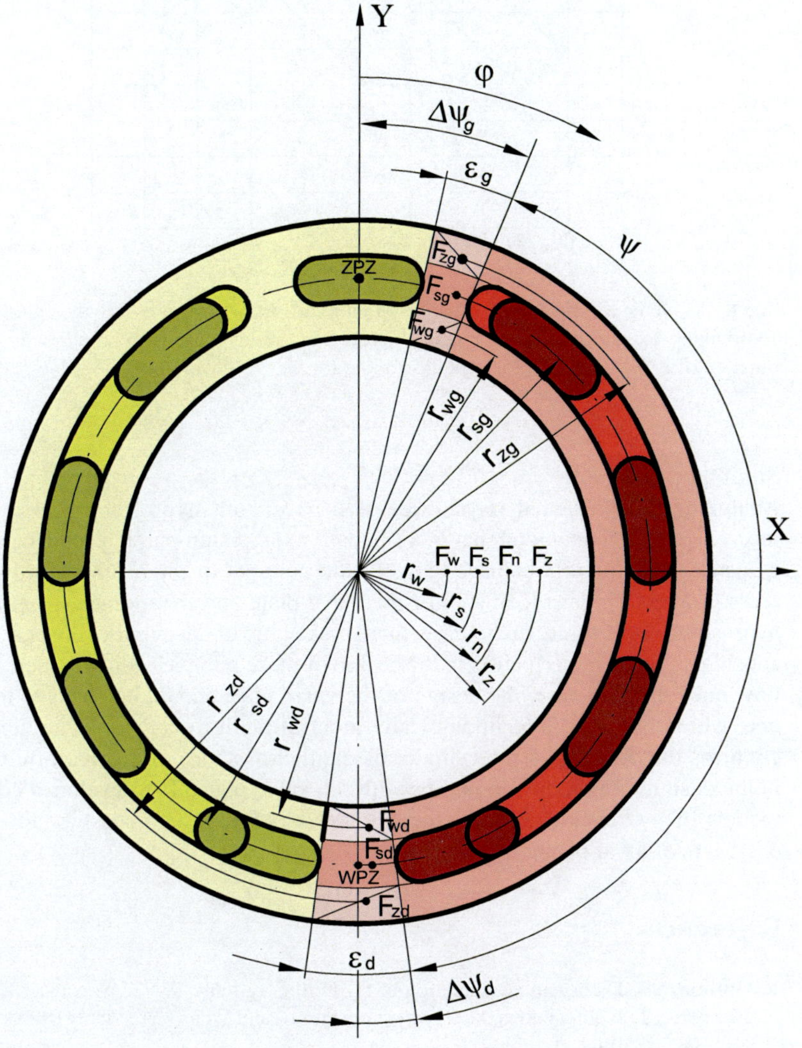

Fig. 7. Location of the resultant hydrostatic forces occurring in the pump valve plate

Figure 8 shows exemplary trajectories for the displacement of the resultant hydrostatic forces pressing the cylinder block against the valve plates as well as trajectories of the resultant hydrostatic forces relief the cylinder block as functions of the rotation angle for the cylinder block for the zero overlap of the valve plate.

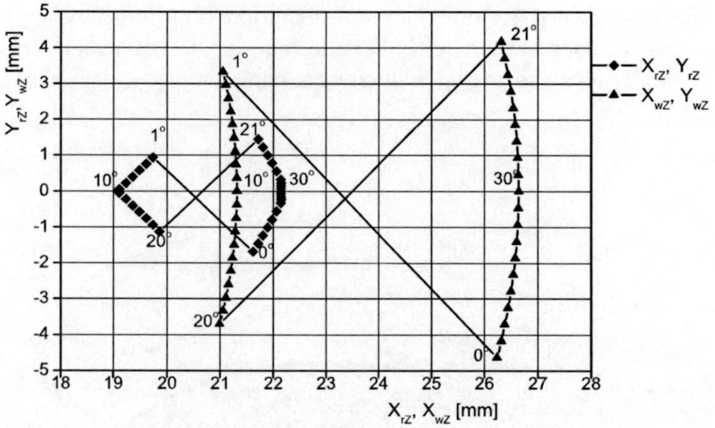

Fig. 8. Movement trajectories of the resultant hydrostatic force pressing the cylinder block with coordinates X_{wZ}, Y_{wZ} and the resultant hydrostatic relief force with coordinates X_{rZ}, Y_{rZ} as a function of φ the cylinder block rotation angle for the zero overlap of the valve plate.

Summary

Analyzing the presented trajectories of the resultant hydrostatic pressing and relief forces, it can be concluded that in a standard axial piston pump, produced in the Polish industry, there is an imbalance between the moment of the resultant hydrostatic force pressing the cylinder block against the valve plate and the moment from the resultant hydrostatic relief force. In this case, there is skewing of the cylinder block, which may be considered as an imperfection in the design of these pumps leading to increased leakage flow rates. In such case, the design of the valve plate should be changed to a spherical one, which however is technologically more difficult to perform. The literature [1, 2] confirms the skewing of the cylinder block, the existence of this harmful phenomenon in the design of axial piston pumps with flat valve plate. The developed computational models of hydrostatic relief of the cylinder block may be necessary for the work of designers of axial piston units.

References

1. Guillon, M.: Theory and Calculation of Hydraulic Systems. WNT, Warsaw (1966)
2. Ivantysyn, J., Ivantysynova, M.: Hydrostatic Pumps and Motors. Academy Books International, New Delhi (2001)
3. Jang, D.S.: Verlustanalise an Axialkolbeneinheiten. Dissertation RWTH, Aachen (1997)
4. Osiecki, A.: Hydrostatic Drive of Machines. WNT, Warsaw (1998)
5. Paleczek, W.: Mathcad Professional. Academic Publishing House EXIT, Warsaw (2003)
6. Stryczek, S.: Hydrostatic Drive. WNT, Warsaw (1995)

CFD Approach and Visualization of Fluid Flow in a Single Acting Vane Pump

Paulina Szwemin[1], Wiesław Fiebig[1](✉), Feng Wang[2], and Piotr Antoniak[1]

[1] Faculty of Mechanical Engineering, Wroclaw University of Science and Technology, Lukasiewicza Street 7/9, 51-370 Wroclaw, Poland
wieslaw.fiebig@pwr.edu.pl

[2] State Key Laboratory of Fluid Power and Mechatronic Systems, School of Mechanical Engineering, Zhejiang University, 38 Zheda Road, Hangzhou 310027, Zhejiang, P. R. China

Abstract. Vane pumps are widely used in many applications and often are the source of flow in hydrostatic power transmission systems. The paper deals with visualization methods of fluid flow inside a single-chamber vane pump. In this article, a stand for investigating flow phenomena in vane pumps was presented with the use of a high-speed camera. Due to the complex nature of the flow and limitations of the visual experimental method, a numerical model of the chosen pump was prepared. CFD flow simulations were performed for different values of suction pressure, discharge pressure and rotation speed. Comparison of the results of visual experimental investigation and computer simulation showed that CFD modeling can be successfully used for the prediction of fluid flow inside different vane pumps.

Keywords: CFD · vane pump · cavitation · flow visualization

1 Introduction

Vane pumps belong to a group of rotary positive displacement pumps. They operate at very low noise levels and with much lower pulsation compared to more popular gear pumps [1]. The construction of the vane pump greatly affects them by cavitation [2]. Cavitation negatively affects the operation of the pump, especially its efficiency. It can reduce the flow rate and erode the internal surfaces of components [8].

Because of different difficulties in the detection of cavitation with experimental methods, an alternative approach using CFD simulation can be effective. Computational fluid dynamics is a powerful tool for visualising fluid flow and identifying flow phenomena in hydraulic components. Therefore, it is commonly used by engineers to solve complex fluid flow problems.

Suzuki [4] conducted the CFD analysis taking into account the gases in the oil and stated that the pressure rise point is lagged behind as the pump rotational speed and gas content increase.

J. Stryczek and U. Warzyńska (Eds.): NSHP 2023, LNME, pp. 33–43, 2024.
https://doi.org/10.1007/978-3-031-43002-2_4

Rundo [5] investigated the filling capacity of the inter-vane chamber which depends on the ratio between the axial thickness and the diameter of the pump. At a low ratio between the axial thickness and the diameter of the pump, the filling ratio and volumetric efficiency are higher.

Frosina [6] presented a 3D CFD modelling approach which is a good tool for optimizing vane pumps for the vane pump for the mobile application. Lobsinger in [7] presented a two-dimensional multiphase study of a balanced vane pump concerning cavitation and feeding capacity of high-speed pumps. The analysis of vane loads and motion in a balanced vane pump has been presented in [9].

The simulation results presented in [10, 11] demonstrate the initiation, development, and collapse of cavitation during the rotation of the vane. The shape of the vane tip and the size of the gap are found to influence the size of cavitation and the flow rate of the vane pump. A numerical 3D CFD approach of a variable displacement vane pump has been conducted in [12] to predict vane detachment in the transition zones when unwanted pressure spikes are generated.

In this paper, the flow visualisation results and CFD simulation results in a single-acting vane pump have been compared. CFD simulations have been conducted for different values of suction pressure, discharge pressure and rotational speed.

2 CFD Modeling

The first step in the flow analysis simulation process was the extraction of the fluid domain. Based on the pump geometry created in CATIA V5, the fluid volume that fills the pump was created. In order to facilitate the finite element mesh generation, improve its quality and reduce the computation time, some simplifications in the model have been introduced. Small faces, edge filets, and chamfers have been removed, and the geometry of the inlet and outlet channels has been simplified.

The flow was assumed as incompressible and isothermal. In order to properly show turbulent character of fluid flow inside the vane pump the Shear-Stress Transport (SST) turbulence model was used. It combines precise and robust formulation of the κ–ω model in the near-wall region with the independence of the free flow of the κ-ε model in a more distant region [8]. The convergence of solution was judged by examining the residual levels and additional monitor points. The convergence criteria for the analysis has been reduced from the default value of 1e-3 for residual to 1e-5. The size of time step was set to 1e-6s.

2.1 2D Analysis

First, the 2D model was created to verify the correctness of the generated mesh, defined motion of the vanes, and used simulation settings.

The computational fluid dynamics software ANSYS CFX was used for calculations because it allows the use of the 'immersed solid' approach. The 'Immersed solid' solution was chosen as it allows simulation of the rigid motion of solid objects through fluid regions, avoiding any mesh deformation or remeshing.

The analysis was performed taking into account a number of assumptions.

- fluid is Newtonian,
- flow is incompressible,
- fluid properties are constant,
- temperature is constant.

Using the 2D model, numerical calculations were made for selected pump operating parameters (discharge pressure: 10 MPa, suction pressure: 0 MPa, rotational speed: 1500 rpm). Model validation was carried out on the basis of the generated pressure and velocity maps which are presented in Fig. 1.

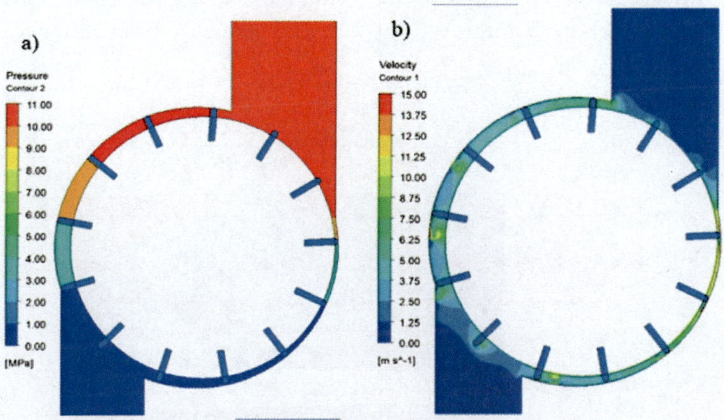

Fig. 1. 2D CFD model for a discharge pressure of 10 MPa and a rotational speed of 1500 rpm: a) pressure contour, b) velocity contour.

The analysis of distributions showed that they correctly reflect the theoretical phenomena that occur in the single-acting vane pump. Therefore, it was considered that the CFD model can be used to visualize the fluid flow phenomena in the single-acting vane pump.

Unfortunately, the 2D model cannot take into account the axial flow of the fluid, which is crucial in the suction area where cavitation may occur. Instead, it was used to investigate the flow and pressure distribution in the tip of the gap between the vane tip and the stator. A part of the geometry with one vane was used to perform the simulation.

It was assumed that the pressure difference between the front and back of the vane is 10 MPa as this is the maximum discharge pressure of the tested pump. A rotational speed of 1500 rpm for the vane movement and the stator wall was set. Two vane geometries were taken into account (Fig. 2).

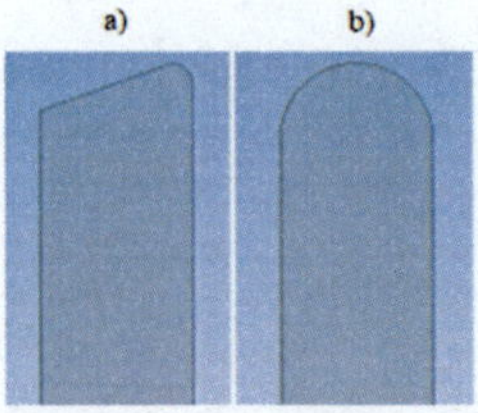

Fig. 2. Vane shapes: a) asymmetrical, b) rounded.

The simulation results obtained are shown in Figs. 3, 4, 5. Figure 3 shows that due to pressure distribution above the head of the vane, the total pressure force acting on the tip of the vane is smaller for the asymmetric shape. Moreover, the negative pressure covers a larger area for this type of vane and indicates the occurrence of cavitation.

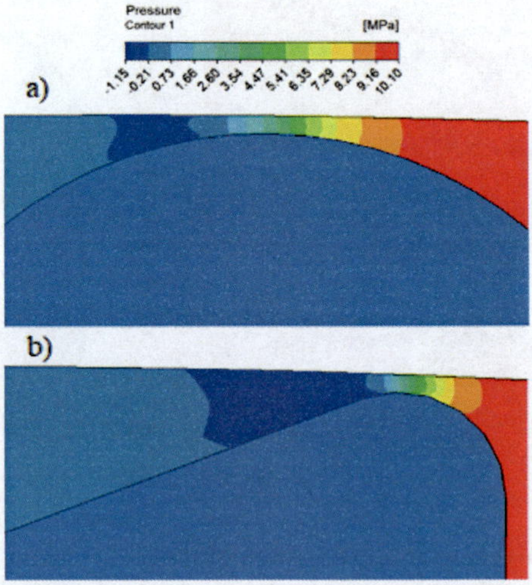

Fig. 3. Pressure distribution in the gap between the cam ring and the top of the vane for different vane shapes: a) rounded, b) asymmetrical.

The fluid velocity in the radial gap is higher when the asymmetric vane is used. This is visible on the velocity contour shown in Fig. 4. The moving vane leaves a fluid vortex behind. It is much smaller for the vane with rounded head (Fig. 5) due to the lower velocity values in the radial gap. A pressure drop and an increase in fluid velocity in the gap may indicate a developing cavitation phenomenon. Taking into account the results obtained, it can be concluded that the asymmetric vane is more susceptible to cavitation.

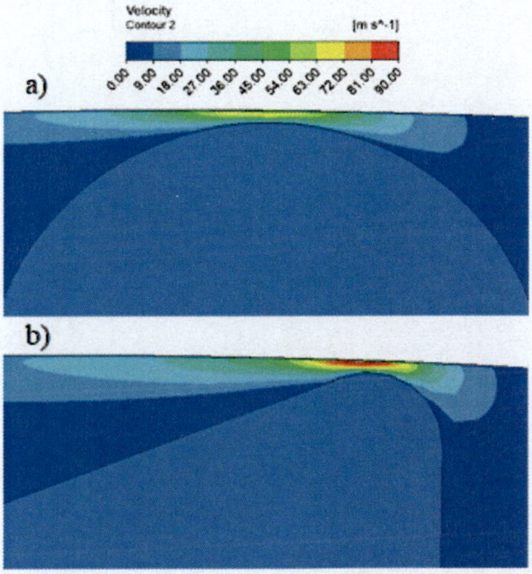

Fig. 4. Velocity distribution in the gap between the cam ring and the top of the vane for different vane shapes: a) rounded (max velocity 72 m/s), b) asymmetric (max. Velocity 90 m/s).

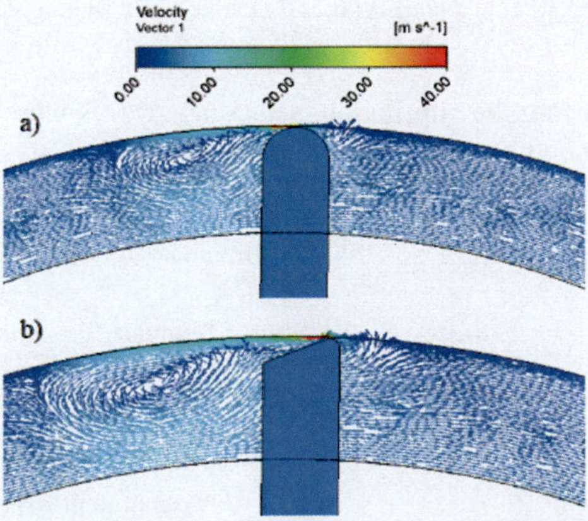

Fig. 5. Vector plot for different vane shapes: a) rounded, b) symmetric.

2.2 3D Analysis

To analyse the phenomena of fluid flow phenomena in all important areas and taking into account the axial flow, a 3D model was built (Fig. 6). It allowed us to analyse the flow through the kidneys and obtain more detailed results.

As before, the ANSYS CFX software and the "immersed solid" approach were used for the calculations. In this way, any problems related to mesh deformation were avoided. Simulations for different values of suction pressure, discharge pressure and rotational speed were performed to compare the results with the experimental research.

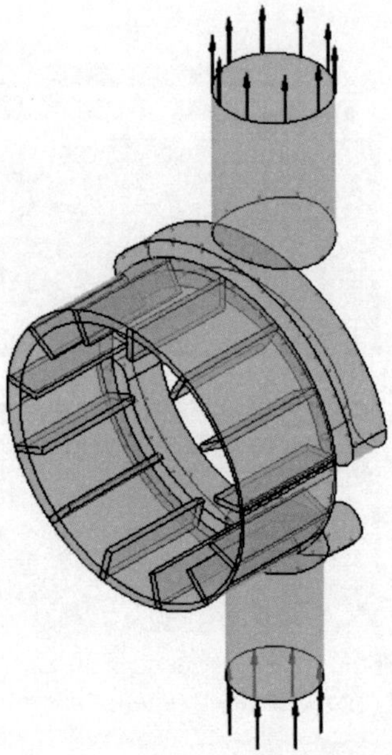

Fig. 6. CFD simulation model 3D.

Before starting the simulation, selected surfaces were named and boundary conditions were applied according to Table 1.

Table 1. Boundary conditions.

Surface	Boundary condition
Inlet	Pressure Inlet (0 MPa)
Outlet	Pressure outlet (discharge pressure)
Vanes	Immersed solid (rotating motion)
Housing/Rotor	No-slip moving wall
Rotor	Rotating wall (rotational speed)
Interfaces	General connection

3 Flow Visualization

3.1 Vane Pump

The pump for the tests is a single-acting vane pump (PV2V3 with 25 cm3/rev) produced by Ponar Wadowice. It belongs to a group of pumps with variable displacement and pressure control. The cross-sectional view of the tested pump is shown in Fig. 7.

The PV2V3 vane pump is composed of two groups of elements:

- Rotating parts: rotor and 13 vanes,
- stationary parts to which the rest of the components belong.

The principle of operation of the pump is the same as in the case of vane pumps from other manufacturers. Two vanes, the rotor, the stator, and the disc form the chambers where the fluid is transported. By moving the rotor to the right the volume of chambers 8 increases starting from the suction channel (S). When the maximum volume of these chambers is obtained, they are separated from the suction side by means of the control discs. As the rotor continues to rotate, the chambers begin to connect to the pressure side. They reduce their volume and this causes the fluid to push to the pressure channel (P).

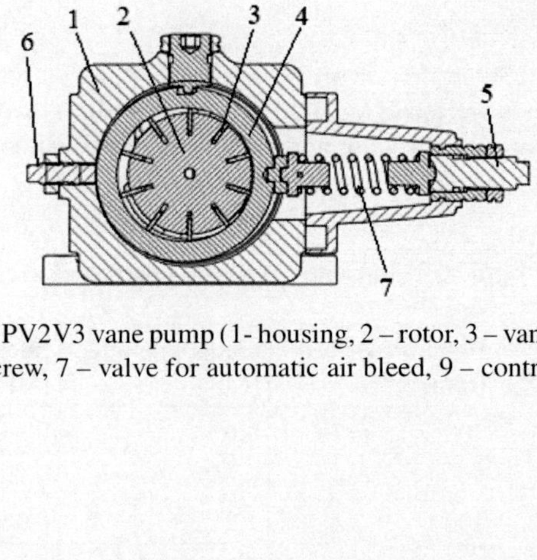

Fig. 7. The structure of PV2V3 vane pump (1- housing, 2 – rotor, 3 – vanes, 4 – stator, 5 – pressure regulator, 6 – volume screw, 7 – valve for automatic air bleed, 9 – control disc).

3.2 Test Bench

In order to visualize the real flow inside the vane pump, an experimental vane pump with transparent (plexi) bodies was designed and built. It was tested on a specially prepared test stand as shown in Fig. 8. The use of transparent materials makes the flow inside the pump visible [3]. Therefore, it can be recorded with a high-speed camera that delivers high-resolution images. The same research method for studying fluid flow phenomena was proposed by the authors in [3].

Experimental tests with the use of a high-speed camera allow for visualisation of the fluid flow and thus identification of the cavitation phenomenon occurring inside the pump. This method is effective, but has some limitations, which are:

- Due to the use of the plexiglass cover, tests can only be carried out in the range of low discharge pressures.

- It is possible to visualize the flow only in the spaces between the vanes.
- The gap over the vane head is too small to see the flow phenomena that occur there.

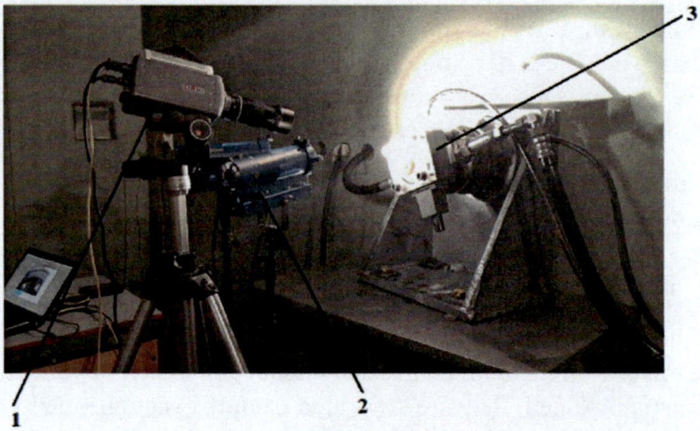

Fig. 8. Designed test bench (1–high-speed camera, 2-lighting, 3-vane pump)

3.3 Results

Visualization tests were carried out for various values of operating parameters, such as inlet pressure, rotational speed, and outlet pressure. All considered parameter sets used during experimental research are presented in Table 2.

Table 2. Parameters considered in the experimental research.

Rotational speed [rpm]	Outlet pressure [MPa]	Inlet pressure [MPa]
300	0,6	0/0,05
500	0,6/1,2/2	0
1000	1,2	0,05
1000	0,6/1,2/2/2,5	0
1000	1,2	−0,02
1500	1,5/2	0

In Fig. 9 the results of the experimental tests are presented. Different inlet pressure were considered in order to verify the influence of suction pressure on the number of air bubbles in the area between the vanes. During the analysis, it was noted that the number of bubbles increased as the suction pressure was lowered.

Moreover, on some of the registered movies, vane detachment has been identified. It has been found that when the pump works at lower pressures, the gap above the vane

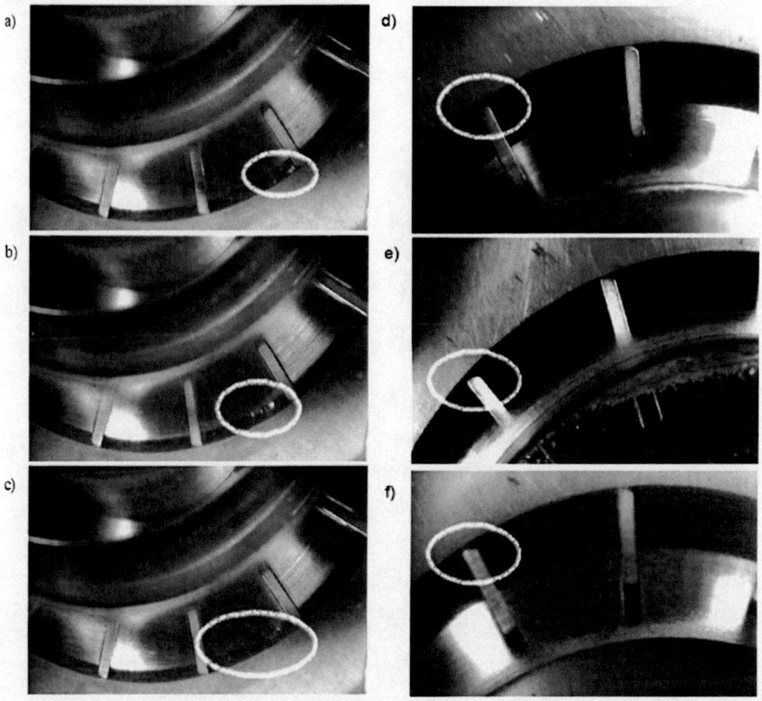

Fig. 9. Visual experimental method - visible air bubbles for different suction pressure values: a) 0,05 MPa, b) 0 MPa, c) -0,02 MPa, vane detachment at discharge pressure of 0,6 MPa and rotation speed: d) 300rpm, e) 500rpm, f) 1000 rpm.

head changes depending on the vane position. As is known, the size of this gap affects the leakage and leads to a reduction in pump performance. The observed vane detachment for different values of rotational speed is shown in Fig. 9d) e) f).

Taking into account the operating parameters of the pump at which the vanes detachment appears, it can be assumed that this phenomenon is caused by the lack of sufficient support. At higher delivery pressures, no vane detachment was observed, or the displacement was so small that it was not noticed.

3.4 Discussion of Results

The results obtained during the experimental research and simulation were compared. Images for the same pump operating parameters and vane position were selected and analysed. This enabled the verification of the 3D CFD model and the assessment of its use in future research.

The case for a discharge pressure of 0.6 Pa and a rotational speed of 1000 rpm was selected for comparison as it gave the most interesting results during experimental tests. In the area where air bubbles are visible, the velocity of the fluid increases, which is shown by the velocity vector distribution obtained from the simulation (Fig. 10). A vortex

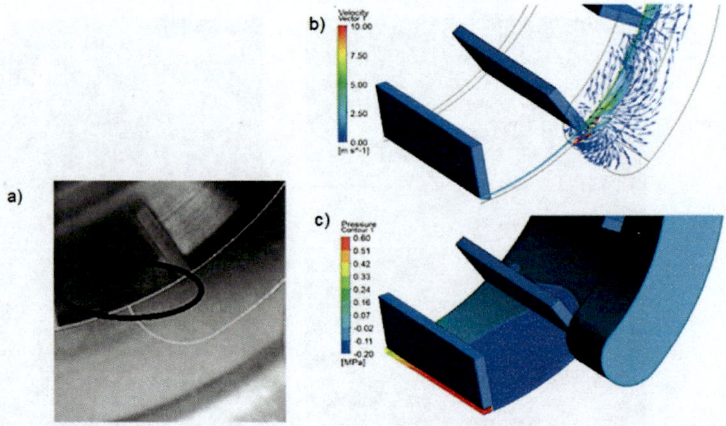

Fig. 10. Comparison of the results of experimental research and CFD simulation for a rotation speed of 1000 rpm, a suction pressure of 0 MPa and a discharge pressure of 0,6 MPa: a) experimental, b) velocity vectors, c) pressure distribution.

also appears in this zone. Moreover, a pressure drop is visible in the indicated area. An increase in fluid velocity and a decrease in pressure may suggest the development of cavitation.

4 Conclusions

Despite the quite simple structure of the single-vane pump, the flow phenomena appearing inside this pump are very complex. Experimental tests with the use of a high-speed camera are effective and give good results, but have many limitations that make it impossible to analyse the fluid flow in all relevant areas.

The use of numerical solution methods such as CFD allows for an in-depth analysis of the fluid flow taking into account different operating conditions of the pump. Moreover, computational fluid dynamics makes it possible to observe fluid flow phenomena in many areas of the pump. The results obtained depend on the accuracy of the geometric model and the complexity of the numerical solution. "Solid immersion" technique was used in CFD modelling. Among many advantages, this method also has some limitations, which may affect the results. For this reason, further simulation studies are planned with the use of the dynamic mesh approach. This will allow us to compare both methods and choose the one that gives better results for the considered issue.

Performed simulation and experimental tests showed that cavitation in a vane pump occurs in the area of connection between chamber volume and suction kidney. The intensity of this phenomenon depends on the operating parameters of the pump.

By experimental tests, it was noticed that during rotation, in a certain position, the vane detaches from the cam ring. Enlargement of the gap above the vane head can lead to leakage and thus reduce the volumetric efficiency of the pump.

References

1. Rundo, M., Altare, G., Casoli, P.: Simulation of the filling capability in vane pumps. Energies **12**, 283 (2019)
2. Zhang, Q., Xu, X.Y.: Numerical Simulation on Cavitation in a Vane Pump with Moving Mesh, ICCM2014 28–30th July, Cambridge, England
3. Stryczek, J., et al.: Visualisation research of the flow processes in the outlet chamber – outlet bridge – inlet chamber zone of the gear pumps, Arch. Civ. Mech. Eng. (2014)
4. Suzuki, K., Nakamura, Y., Yakabe, S., Watanabe, H., Nakamura, K.: Characteristics prediction of vane pump by CFD analysis. KYB Tech. Rev. Technol. **53**, 8–15 (2016)
5. Rundo, M., Altare, G.: CFD studies on radial feeding in high-speed vane pumps. In: 21st Australasian Fluid Mechanics Conference, Adelaide, Australia, 10–13 December 2018
6. Frosina, E.: A three dimensional CFD modelling methodology applied to improve hydraulic components performance. Energy Procedia **82**, 950–956 (2015)
7. Lobsinger, T., Hieronymus, T., Brenner, G.: A CFD investigation of a 2D balanced vane pump focusing on leakage flows and multiphase flow characteristics. Energies **13**, 3314 (2020)
8. Marinaro, G., Frosina, E., Senatore, A., Stelson, K. A., et al.: 3D CFD numerical analysis of vane dynamic effects on the pressure ripple in a variable displacement vane pump. In: E3S Web of Conferences, vol. 197, p. 07001 (2020). ATI Congres
9. Fiebig, W., Cependa, P., Kuczwara, H., Wang, F.: Analysis of vane loads and motion in a hydraulic double vane pump with integrated electrical drive. Arch. Civ. Mech. Eng. **21**, 112 (2021)
10. Mucchi, E., Agazzi, A., D'Elia, G., et al.: On the wear and lubrication regime in variable displacement vane pumps. Wear **306**(1–2), 36–46 (2013)
11. Alghamdi, A., Elashmawy, M.: Vane geometry effect on lubrication conditions between vane tip and cam-ring in hydraulic vane machines. Int. J. Mech. Eng. Appl. **3**(1), 1–10 (2014)
12. Frendo, F., Novi, N., Squarcini, R.: Numerical and experimentalanalysis of variable displacement vane pumps. In: International Conferenceon Tribology. 20–22 September 2006, Parma, Italy (2006)

The Methodology for Determining the Theoretical Based on the Characteristics of Effective Absorbency Versus Pressure Drop in the Motor

Paweł Śliwiński(✉)

Gdansk University of Technology, Narutowicza 11/12, 80-233 Gdansk, Poland
pawel.sliwinski@pg.edu.pl

Abstract. This article describes a method for determining the theoretical and actual working volume of a hydraulic motor. It has been shown that the actual working volume of the motor is the sum of the theoretical working volume and the additional volume that depends on the pressure drop in the working chambers of the motor. It was also shown that the volumetric losses in the hydraulic motor are not only a function of the pressure drop in the working chambers, but also a non-linear function of the motor speed. The results of the experimental tests of the satellite motor are also described. Analysis of the test results confirmed the correctness of the method for determining both the theoretical working volume and the actual working volume. It was also shown that the calculation of the partial efficiencies (i.e. mechanical efficiency and volumetric efficiency) of a hydraulic motor should be based on the actual working volume instead of the theoretical one.

Keywords: Theoretical Working Volume · Actual Working Volume · Hydraulic Motor · Volumetric Efficiency · Mechanical Efficiency

1 Introduction

Hydraulic positive displacement machines are the core of modern hydraulic system [1–5]. The pump is the most important element in this system [6–9]. But the behavior of the hydraulic motor also plays an important role in the overall performance of the hydraulic system [10, 11]. Knowledge of the basic characteristics of the hydraulic motor and pump, such as volumetric losses, mechanical losses and pressure losses, is important to assess volumetric efficiency and mechanical pressure efficiency [12, 16]. They are therefore important for both users and designers of a hydraulic system [17].

So far, the so-called theoretical working volume q_t has been used to evaluate the losses in the motor and its partial efficiency. The value of the theoretical working volume q_t is assumed to be constant over the entire range of the pressure drop Δp measured at the motor connections. Furthermore, the q_t is independent of the motor speed n [18, 19].

In practice, the theoretical working volume q_t is determined in a simplified way. The influence of the compressibility of the fluid is neglected. Similarly, the pressure drop

J. Stryczek and U. Warzyńska (Eds.): NSHP 2023, LNME, pp. 44–57, 2024.
https://doi.org/10.1007/978-3-031-43002-2_5

Δp_{ich} in the internal channels of the motor is treated as insignificant and is negligible. This means that the pressure drop Δp in a motor is equal to the pressure drop Δp_i in the working mechanism of that motor [18].

The pressure drop Δp in a hydraulic motor is mainly the result of its torque load M. This is because the working chambers are elastically deformed. Therefore, the working volume of a loaded motor is greater than that of an unloaded motor [21–23]. Osiecki observed similar phenomena in a prototype of an axial piston pump [24]. In this way, the working volume depends on the pressure drop Δp_i in the working chambers of the motor and was called the actual working volume q_r [21–23]. In works [21–23] was shown that if the influence of pressure in the working chambers of the engine on the increase in their volume is neglected, it is possible to obtain a value of volumetric or mechanical efficiency above one. For this reason, it is worth considering using actual working volume q_r instead of theoretical working volume q_t to calculate both volumetric and mechanical efficiency.

The first works on the methodology of determining the theoretical working volume appeared in the 1940s and 1960s. They were the works of Wilson (1949) 25, Schlosser and Hilbrands (1963) [26, 27]. Then the work of Toet and Balawender in the 1970s and again in 2019 [18, 19, 28, 29]. The Toeth method was also described by Post in [30]. The method proposed by Toet and Balawender is a method for determining the theoretical and actual working volume of a hydraulic motor based on the characteristics of the flow rate delivered into the hydraulic motor relative to the speed at a constant pressure drop in the motor. In the paper [21] a new view of the Balawender method is presented. Also, in [22] another new method for determining the working volume is proposed. This new method is based on the characteristics of the effective absorptivity of the motor.

Another original method for determining the theoretical working volume is proposed by Michael and Garcia-Bravo. This method was called Latin Hyperspace Sampling (LHS - a pseudo-randomized method for selecting experimental test points) [31].

Balawender has proved that it is possible to determine the theoretical working volume q_t of a hydraulic motor, based on the characteristics of the hydraulic motor's absorptive capacity q_e per revolution of the motor shaft as a function of the pressure drop Δp in the motor at constant speed n ($q_e = f(\Delta p)_{n=const.}$) [18, 19]. The q_e was defined as:

$$q_e = \frac{Q}{n} \tag{1}$$

where Q is the flow rate in the motor. The flow rate Q in the satellite motor was described by the following formula [22, 41, 42]:

$$Q = \underbrace{q_t \cdot n}_{Q_t} + \underbrace{\underbrace{\left(C_q \cdot \Delta p_i + \frac{C_{id}}{n^{0,5}}\right) \cdot m^2 \cdot H \cdot n}_{\Delta Q_L} + Q_{Lfg} + Q_{CU} + Q_{Le}}_{Q_L} \tag{2}$$

where:

- Q_t – the theoretical flow rate in the motor,
- Q_L – the volumetric losses in the motor,

- ΔQ_L – the component of volumetric losses depends on the compressibility of the fluid and speed of the motor,
- Q_{Lfg} – the flow rate in flat clearances of the working mechanism,
- Q_{CU} – the flow rate in the gaps of the commutation unit,
- Q_{Le} – the external leakage,
- Δp_i – the pressure drop in the working mechanism of the motor,
- m – the teeth module,
- H – the height of satellite mechanism,
- C_q, C_{id} – coefficients.

A precise description of the particular components of the above-mentioned formula can be found in [41, 42].

The methodology of determining Δp_i and Δp_{ich} is described in detail in the works [21, 22, 41–44]. The individual components of the above mathematical model were also described in detail in the publications [41, 42] and are not explained in detail here.

Based on the Balawender method, this article describes the proposed new method for determining the theoretical working volume q_t and the actual working volume q_r of a hydraulic motor. The correctness of this method has been verified experimentally.

Hydraulic motors with a satellite mechanism are the object of interest of many researchers and set a new trend in research in the field of hydraulic displacement machines. This is proved by the works [20, 32–40]. These motors are also the main object of scientific interest of the author. Therefore, a satellite motor was chosen for the experimental test (see Sect. 3.1).

2 Proposed Method for Determining Theoretical and Actual Working Volume

For a loaded motor the theoretical working volume q_t is not equal to the actual working volume q_r. The actual working volume q_r is described by the following formula [21, 22]:

$$q_r = q_t + \underbrace{\left(C_{q1} \cdot \Delta p_i + C_{q2} \cdot \Delta {p_i}^2 + C_{q3} \cdot \Delta {p_i}^3\right) \cdot m^2 \cdot H}_{\Delta q_p} \tag{3}$$

where:

- Δq_p is the change in working volume of the loaded motor;
- C_{q1}, C_{q1}, C_{q1} – coefficients.

The Δq_p cannot be the cause of the mechanical and volumetric losses in a motor. Therefore the ΔQ_L should take the form:

$$\Delta Q_L = \underbrace{\left(\frac{C_{id}}{n^{0.5}}\right) \cdot m^2 \cdot H}_{\Delta q_L} \cdot n \tag{4}$$

where Δq_L is the change in unit volumetric losses (that is volumetric losses per one revolution of the motor shaft).

Considering formula (3) the flow rate in a motor should be described by:

$$Q = \underbrace{q_r \cdot n}_{Q_r} + \underbrace{\Delta q_L \cdot n + Q_C + Q_{Lfg} + Q_{CU} + Q_{Le}}_{Q_L} \tag{5}$$

where Q_C is the component of the flow rate that depends on the compressibility of the fluid, defined as follows:

$$Q_C = Q \cdot \int_{p_2}^{p_H} \frac{1}{K_{Z(p)}} dp \tag{6}$$

and $K_{Z(p)}$ is the tangential isentropic bulk modulus [14, 21]. The influence of liquid compressibility on the working volume has been extensively described in publications [21, 22].

Directly from the measured data, the flow rate q_e per one revolution of the motor shaft should be calculated as follows:

$$q_e = q_r + \Delta q_L + \frac{Q_C + Q_{Lfg} + Q_{CU} + Q_{Le}}{n} \tag{7}$$

In known that the volumetric losses Q_L depend on the pressure drop Δp_i in the working mechanism. In this way, the flow rate Q should be related to pressure p_H in the high-pressure working chambers. Then [21, 22]:

$$Q_{(p_H)} = Q - Q_C \tag{8}$$

The flow rate $q_{e(pH)}$ in relation to the pressure p_H should be calculated as:

$$q_{e(pH)} = \frac{Q_{(pH)}}{n} \tag{9}$$

If the pressure drop Δp_i in the motor decreases then the volumetric losses Q_L decreases also. Thus:

$$\lim_{\Delta p_i \to 0} \left(q_{e(pH)}\right)_{(n)} = q_t + \Delta q_L = q_n \tag{10}$$

Leakage Q_L also decreases in general equilibrium of the flow rate in the motor when rotational speed n of this motor increases. Thus, the theoretical working volume can be formulated as follows:

$$\lim_{n \to \infty} (q_n) = q_t \tag{11}$$

The graphical representation of q_n and theoretical working volume q_t is shown in Fig. 1.

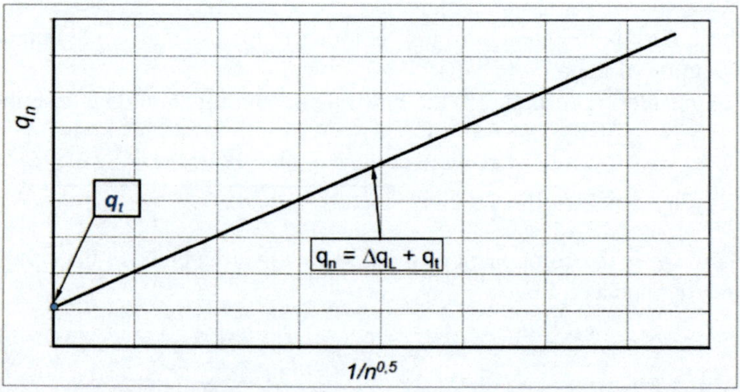

Fig. 1. The graphical representation of the new method.

Fig. 2. General view of hydraulic satellite motor (left) and working mechanism of this motor (right) [42, 45].

3 Results of the Experiment

3.1 Tested Motor and Test Rig

The prototype of a hydraulic satellite motor was chosen for the experimental tests. The operating mechanism of this motor is the satellite mechanism (Fig. 2.). This mechanism is well known in the literature and the working principle of this mechanism (and also of the motor) can be found in [41, 43–46].

The parameters of satellite working mechanism are following (from CAD documentation):

- the height of the working mechanism H = 25 mm;
- the minimum area of the working chamber $A_{min} = 26.11$ mm^2;
- the maximum area of the working chamber $A_{max} = 83.51$ mm^2;
- the geometrical working volume $q_g = 34.44$ cm^3/rev.

The satellite motor was tested using the Total Azolla 46 oil. The oil kinematic viscosity was $\nu = 40$ cSt and density $\rho = 873$ kg/m^3 (at the temperature in the inflow motor port $T_1 = 43$ °C).

The diagram of the measurement system of the test rig is shown in Fig. 3.

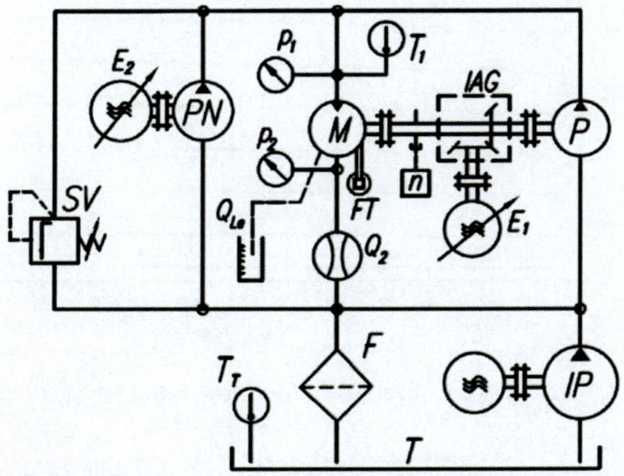

Fig. 3. Diagram of the test rig measurement system [21, 22, 46]: P – pump, M – tested motor, PN – pump for filling leaks in P and M, IP – impeller pump (pre-supply pomp), SV – safety valve, F – filter, T – reservoir, IAG – intersecting axis gear, E_1 and E_2 – electric motors with frequency converters, T_1, T_T – temperature sensors, Q – flowmeter, Q_{Le} – leakage measurement, FT – force transducer (torque measurement), n – inductive sensor (rotational speed measurement).

During the test of the motor the following parameters were measured:

- pressure p_1 in the inlet port (strain gauge pressure transducer, range $0 \div 10$ MPa and $0 \div 40$MPa, class 0.3);
- pressure p_2 in the outlet port (strain gauge pressure transducers, range $0 \div 2.5$ MPa, class 0.3);
- motor absorption Q (piston flowmeter, the volume of flowmeter measuring chamber is 0.63 dm^3, range $0 \div 200$ l/min, class 0.2);
- torque M (strain gauge force transducer FT mounted on arm 0.5m (arm is fixed to motor body), range $0 \div 100$ N, class 0.1);
- speed of rotation of the shaft n (inductive sensor, measurement accuracy ± 0.01 rpm);
- temperature T_1 of the liquid in the motor inlet port (RTD temperature sensor, class A, max. Measurement error 0.5 °C).

3.2 Characteristics of the Motor Output Flow Rate

The experimental test of the satellite motor was carried out in the range of pressure drop in the motor $\Delta p = 32$ MPa and in the range of speed $n = 50 \div 1500$ rpm. The characteristics $Q = f(\Delta p)$ and $Q = f(\Delta p_i)$ at n = const. Were shown in Fig. 4. And Fig. 5. The pressure drop Δp_i in the working mechanism was determined according to methodology precisely described in [21, 22, 42, 43].

The external leakage Q_{Le} in the tested motor was very low (less than 0.05 l/min in all ranges of Δp in the motor) and was neglected for further consideration.

The flow rate $Q_{(pH)}$, which is related to the pressure p_H in the high pressure chamber of the motor was calculated according to formula (8). The characteristics of $Q_{(pH)} = f(\Delta p_i)$ are shown in Fig. 6.

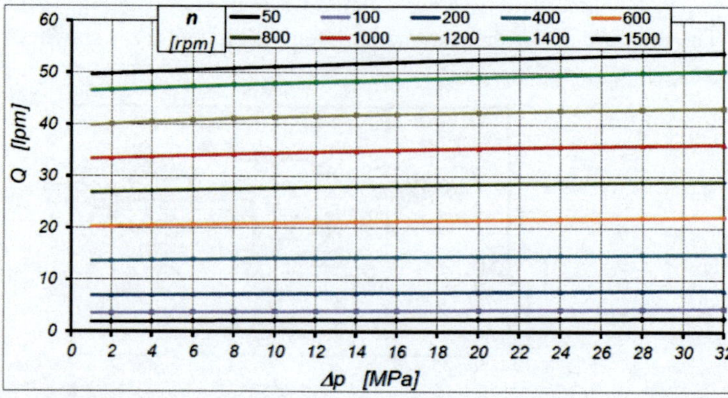

Fig. 4. Characteristics of motor output flow rate Q as a function of Δp at n = const.

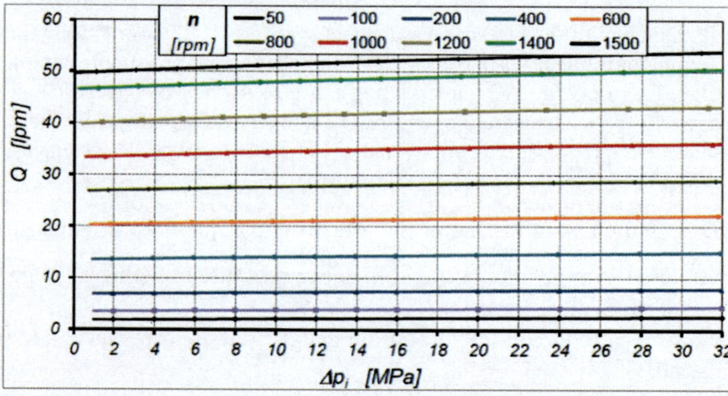

Fig. 5. Characteristics of motor output flow rate Q_2 as a function of Δp_i at n = const. The influence of liquid compressibility was omitted.

3.3 Theoretical Working Volume

According to the proposed new method, the characteristics of $q_{e(pH)} = f(\Delta p_i)$ (see Fig. 7.) were calculated considering the flow rate $Q_{(pH)}$ related to the pressure in the high-pressure working chamber (Fig. 6.). That is, the values of q_e were calculated according to formula (9). All characteristics of $q_{e(pH)} = f(\ \Delta p_i$ are described by equations and presented in the Table 1.

The characteristics of $q_n = f(n^{-0.5})$ are shown in Fig. 8.

The results of research shown that, according to the proposed method, the working volume q_n should be described by Eq. (10) and then (Fig. 8.):

$$q_n = 24.989 \cdot n^{-0.5} + 32.527 \tag{12}$$

That is, the theoretical working volume of satellite motor is $q_t = 32.527\ cm^3$/rev.

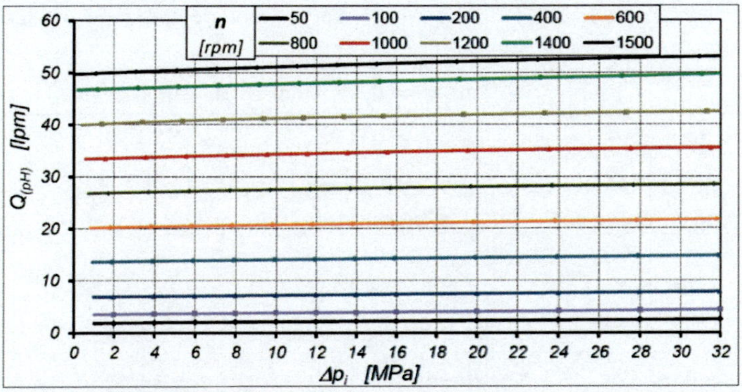

Fig. 6. Characteristics of motor output flow rate $Q_{(pH)}$ related to the pressure p_H in the high-pressure working chamber as a function of Δp_i at n = const.

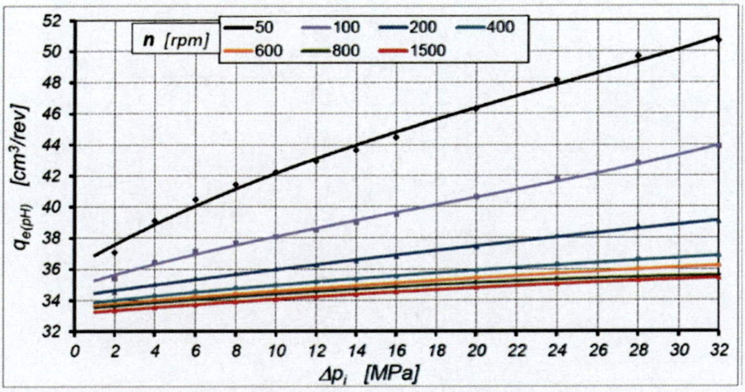

Fig. 7. haracteristics $q_{e(pH)} = f(\Delta p_i)$ at n = const. Equations in Table 1.

4 The Assessment of Actual Working Volume

Taking into consideration formulas (3) and the formulas from Table 1 is possible to assess the actual working volume q_r in tested hydraulic motor. The characteristic of $\Delta q_p = f(\Delta p_i)$ is shown in Fig. 9.

Therefore, according to formula (3), the actual working volume q_r is:

$$q_r = 32,527 + 0,11167414 \cdot \Delta p_i + 0,00174171 \cdot \Delta p_i^2 + 0,00001286 \cdot \Delta p_i^3 [\text{cm}^3/\text{rev.}] \quad (13)$$

Table 1. Equations $q_{e(pH)} = f(\Delta p_i^3)$ of motor output flow rate at n = const.

Lp	n [rpm]	$q_{e(pH)} = f(\Delta p_i^3)$ [cm³/rev]	q_n [cm³/rev.]	R^2
1	50	$q_{e(pH)} = 0.00021 \Delta p_i^3 - 0.0148 \Delta p_i^2 + 0.7251 \Delta p_i + 36.1738$	36.1738	0.9944
2	100	$q_{e(pH)} = 0.00015 \Delta p_i^3 - 0.0079 \Delta p_i^2 + 0.3839 \Delta p_i + 34.8875$	34.8875	0.9969
3	200	$q_{e(pH)} = 0.00001 \Delta p_i^3 - 0.0011 \Delta p_i^2 + 0.1797 \Delta p_i + 34.2685$	34.2685	0.9962
4	400	$q_{e(pH)} = 0.00002 \Delta p_i^3 - 0.0021 \Delta p_i^2 + 0.1445 \Delta p_i + 33.7095$	33.7095	0.9979
5	600	$q_{e(pH)} = 0.00001 \Delta p_i^3 - 0.0011 \Delta p_i^2 + 0.114 \Delta p_i^2 + 33.5305$	33.5305	0.9985
6	800	$q_{e(pH)} = -0.00001 \Delta p_i^3 - 0.0009 \Delta p_i^2 + 0.107 \Delta p_i + 33.4235$	33.4235	0.9979
7	1000	$q_{e(pH)} = - 0.0012 \Delta p_i^3 + 0.1108 \Delta p_i^2 + 33.2959$	33.2959	0.9983
8	1100	$q_{e(pH)} = 0.00003 \Delta p_i^3 - 0.0027 \Delta p_i^2 + 0.1224 \Delta p_i + 33.2802$	33.2802	0.9983
9	1200	$q_{e(pH)} = 0.00004 \Delta p_i^3 - 0.0037 \Delta p_i^2 + 0.1428 \Delta p_i + 33.2614$	33.2614	0.9991
10	1300	$q_{e(pH)} = 0.00002 \Delta p_i^3 - 0.0017 \Delta p_i^2 + 0.1078 \Delta p_i + 33.2487$	33.2487	0.9986

(*continued*)

Table 1. (*continued*)

Lp	n [rpm]	$q_{e(pH)} = f(\Delta p_i^3)$ [cm^3/rev]	q_n [cm^3/rev.]	R^2
11	1400	$q_{e(pH)} = 0.00001 \Delta p_i^3 - 0.0009 \Delta p_i^2 + 0.0902 \Delta p_i + 33.2403$	33.2403	0.9986
12	1500	$q_{e(pH)} = -0.00090 \Delta p_i^2 + 0.1006 \Delta p_i + 33.2245$	33.2245	0.9999

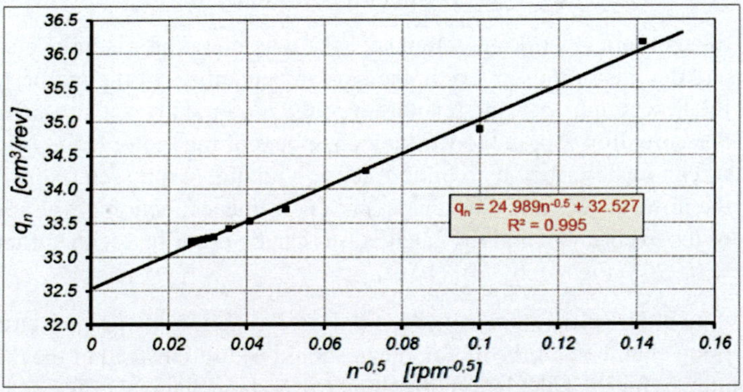

Fig. 8. Characteristics of working volume q_n as a function of speed n (according to the new method).

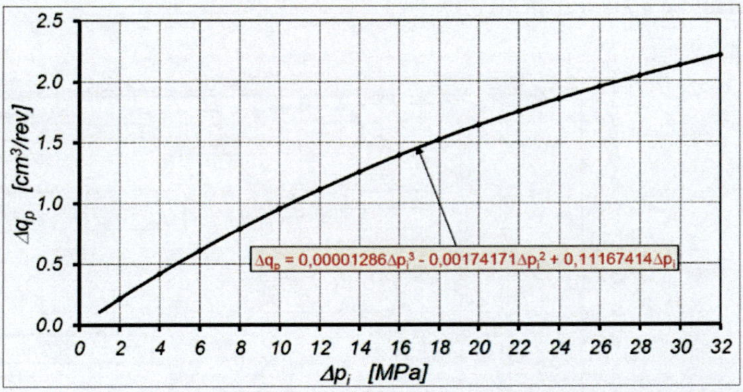

Fig. 9. Characteristic of $\Delta q_p = f(\Delta p_i)$.

5 Conclusions

The results of the experiment confirm the theoretical considerations. That is, the actual working volume q_r is a function of the pressure drop Δp_i in the working chambers of the motor. Therefore, the actual working volume q_r (instead of the theoretical working volume q_t) should be taken into accountfor for calculation of the volumetric and mechanical losses in the hydraulic motor (and the same volumetric and mechanical efficiency).

The results of the experimental investigation confirm that the characteristics $q_e = f(\Delta p)_{n=const}$ are non-linear. The value of the correlation coefficient R^2 of the non-linear function $q_e = f(\Delta p)_{n=const}$ is close to one ($R^2 > 0.99$, 0). Therefore, Eq. (7) describes very well the fluid flow per one revolution of the motor shaft.

The results of the experimental investigation confirm that:

a) the theoretical working volume q_t of a hydraulic motor is independent of its speed n and the pressure drop Δp_i in the working chambers of the motor;
b) the flow rate q_e per one revolution of the motor shaft is a non-linear function of the pressure drop Δp_i in the working chambers of the motor (Fig. 7.);
c) q_e can be described by a third order polynomial (expressed by formula (7));
d) the flow rate q_n (and the same q_e) is a non-linear function of the speed n. The results of the experiment and the calculations can be described with sufficient accuracy ($R^2 = 0.995$ – Fig. 8.) by Eq. (10).

Since the actual working volume q_r is a function of the pressure drop Δp_i in the working chambers of the motor, the q_r should be used instead of the theoretical working volume q_t for the calculation of volumetric and mechanical efficiency. Figure 10 shows the volumetric and mechanical efficiencies characteristics determined for q_t and q_r for comparison. Characteristics of these efficiencies are the test for the correctness of the determination of the theoretical working volume, as the values of these efficiencies cannot be greater than 1.

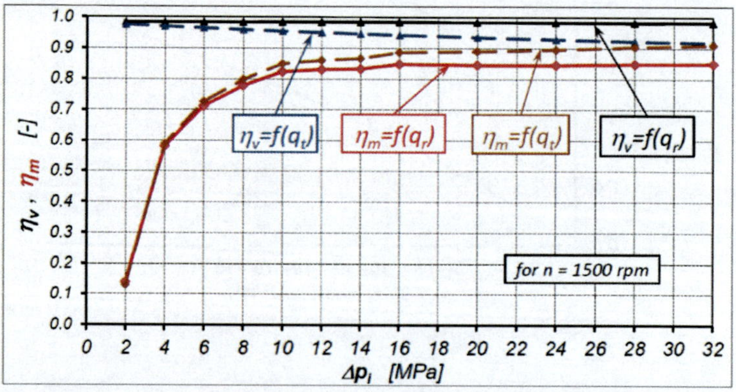

Fig. 10. Volumetric efficiency $\eta_v = f\ \Delta p_i)$ and mechanical efficiency $\eta_m = f\ \Delta p_i)$ of satellite motor at n = 1500 rpm for q_r calculated according to proposed method.

Volumetric efficiency, calculated for q_r, is nearly constant in all range of pressure drop Δp_i. It shows that axial clearances of satellites and rotor decreases under the influence of increasing the pressure drop Δp_i in the motor. Thus, the axial clearance compensation unit in motor operating correctly and the increasing in actual working volume q_r may be caused by small stiffness of curvature, rotor and teeth in these elements (also in satellites).

Therefore, the actual working volume q_r should be taken to calculate the volumetric and mechanical losses and also the volumetric and mechanical efficiency of the motor.

References

1. Guzowski, A., Sobczyk, A.: Reconstruction of hydrostatic drive and control system dedicated for small mobile platform. In: Proceedings of the 8th FPNI Ph.D Symposium on Fluid Power. 8th FPNI Ph.D Symposium on Fluid Power. Lappeenranta, Finland. June 11–13, 2014. V001T05A012. ASME (2014). https://doi.org/10.1115/FPNI2014-7862
2. Pobedza, J., Sobczyk, A.: Properties of high-pressure water hydraulic components with modern coatings. Adv. Mat. Res. Trans Tech. Publications Ltd **849** (2014). https://doi.org/10.4028/www.scientific.net/AMR.849.100
3. Stosiak, M., Zawislak, M., Nishta, B.: Studies of resistances of natural liquid flow in helical and curved pipes. Pol. Mar. Res. **3**(99) (2018). https://doi.org/10.2478/pomr-2018-0103
4. Lisowski, E., Filo, G., Rajda, J.: Analysis of the energy efficiency improvement in a load-sensing hydraulic system built on the ISO plate. Energies **14**, 6735 (2021). https://doi.org/10.3390/en14206735
5. Bak, M.: Torque capacity of multidisc wet clutch with reference to friction occurrence on its spline connections. Sci. Rep. **11**, 21305 (2021). https://doi.org/10.1038/s41598-021-00786-6
6. Osinski, P., Warzynska, U., Kollek, W.: The influence of gear micropump body asymmetry on stress distribution. Pol. Mar. Res. **24** (2017). https://doi.org/10.1515/pomr-2017-0007
7. Osinski, P., Deptula, A., Partyka, M.: Discrete optimization of a gear pump after tooth root undercutting by means of multi-valued logic trees. Arch. Civ. Mech. Eng. **4** (2013). https://doi.org/10.1016/j.acme.2013.05.001
8. Kollek, W., Osinski. P., Stosiak, M., Wilczynski, A., Cichon, P.: Problems relating to high-pressure gear micropump. Arch. of Civ. and Mech. Eng. 14(1). https://doi.org/10.1016/j.acme.2013.03.005 (2014)
9. Guo, S., Chen, J., Lu, Y., Wang, Y., Dong, H.: Hydraulic piston pump in civil aircraft: current status, future directions and critical technologies. Chin. J. Aeron. **33**, 16–30 (2020). https://doi.org/10.1016/j.cja.2019.01.013
10. Jasinski, R.: Analysis of the heating process of hydraulic motors during start-up in thermal shock conditions. Energies **15**(1), 55 (2022). https://doi.org/10.3390/en15010055
11. Jasiński, R.: Problems of the starting and operating of hydraulic components and systems in low Ambient Temperature (Part IV): modelling the heating process and determining the serviceability of hydraulic components during the starting-up in low ambient temperature. Pol. Mar. Res. **24**, 45–57 (2017). https://doi.org/10.1515/pomr-2017-0089
12. Patrosz, P.: Influence of properties of hydraulic fluid on pressure peaks in axial piston pumps' chambers. Energies **14**, 3764 (2021). https://doi.org/10.3390/en14133764
13. Patrosz, P.: Influence of gaps' geometry change on leakage flow in axial piston pumps. In: Stryczek J., Warzyńska U. (eds) Advances in Hydraulic and Pneumatic Drives and Control 2020. NSHP 2020. Lecture Notes in Mechanical Engineering. Springer, Cham (2021). https://doi.org/10.1007/978-3-030-59509-8_7

14. Zaluski, P.: Influence of fluid compressibility and movements of the swash plate axis of rotation on the volumetric efficiency of axial piston pumps. Energies **15**(1), 298 (2022). https://doi.org/10.3390/en15010298
15. Załuski, P.: Experimental research of an axial piston pump with displaced swash plate axis of rotation. In: Stryczek J., Warzyńska U. (eds) Advances in Hydraulic and Pneumatic Drives and Control 2020. NSHP 2020. Lecture Notes in Mechanical Engineering. Springer, Cham (2021). https://doi.org/10.1007/978-3-030-59509-8_12
16. Saheban Alahadi, M.J., Shirneshan, A., Kolahdoozan, M.: Experimental investigation of the effect of grooves cut over the piston surface on the volumetric efficiency of a radial hydraulic piston pump. Int. J. Fl. Pow. **18**(3) (2017). https://doi.org/10.1080/14399776.2017.1337440
17. Wu, X., Chen, C., Hong, C., He, Y.: Flow ripple analysis and structural parametric design of a piston pump. J. Mech. Sci. Tech. **31**, 4245–4254 (2017). https://doi.org/10.1007/s12206-017-0823-8
18. Balawender, A.: Analiza energetyczna i metodyka badań silników hydraulicznych wolnoobrotowych (Energy analysis and methodics of testing of low-speed hydraulic motors). Scientific book of the Gdansk University of Technology, Mechanika No. 54. Gdansk University of Technology Publishing House (1988)
19. Balawender, A.: Opracowanie metodyki wyznaczania teoretycznej objętości roboczej pomp i silników hydraulicznych wyporowych (The development of the methodology for the determination of the theoretical working volume of positive displacement pumps and hydraulic motors). PhD dissertation. Gdansk University of Technology, Gdansk (1974)
20. Zhang, B., Song, S., Jing, C., Xiang, D.: Displacement prediction and optimization of a noncircular planetary gear hydraulic motor. Adv. Mech. Eng. **13**(11), 1–13 (2021). https://doi.org/10.1177/16878140211062690
21. Sliwinski, P.: Determination of the theoretical and actual working volume of a hydraulic motor. Energies **13**, 5933 (2020). https://doi.org/10.3390/en13225933
22. Sliwinski, P.: Determination of the theoretical and actual working volume of a hydraulic motor – Part II (The method based on the characteristics of effective absorbency of the motor) Energies **14**, 1648 (2021). https://doi.org/10.3390/en14061648
23. Sliwinski, P.: The influence of pressure drop on the working volume of a hydraulic motor. Eksploatacja i Niezawodnosc – Maintenance and Reliability **24**(4) (2022). https://doi.org/10.17531/ein.2022.4.15
24. Osiecki, L.: Commutation units of hydraulics axial piston machines. Gdansk University of Technology Publishing House (2006)
25. Wilson, W. E.: Performance criteria for positive displacement pumps and fluid motors. Trans. ASME **71**(2) (1949)
26. Schlosser, W.M.J., Hilbrands, J.W.: Das theoretische Hubvolumen von Verdrangerpumpen. Olhydrauli und Pneumatik, no. 4 (1963)
27. Schlosser, W.M.J., Hilbrands, J.W.: Das volumetrische Wirkungsgrad von Verdrongerpumpen. Olhydrauli und Pneumatik, no 12 (1963)
28. Toet, G., Johnson, J., Montague, J., Torres, K., Garcia-Bravo, J.: The determination of the theoretical stroke volume of hydrostatic positive displacement pumps and motors from volumetric measurements. Energies **12**(3), 415 (2019). https://doi.org/10.3390/en12030415
29. Toet, G.: Die Bestimmung des theoretischen Hubvolumens von hydrostatischen Verdrangerpumen und Motoren aus volumetrischen Messungen. Olhydraulik Pnaumatik, no. 14 (1970)
30. Post, W.: Models for steady-state performance of hydraulic pumps: determination of displacement. In: Proceedings of the 9th Bath International Fluid Power Workshop, University of Bath, United Kingdom, vol. 9, 9–11 September 1996 (1996). https://research.tue.nl/en/publications/models-for-steady-state-performance-of-hydraulic-pumps-determinat

31. Michael, P., Garcia-Bravo, J.: The determination of hydraulic motor displacement. In: The 17th Scandinavian International Conference on Fluid Power SICFP 2021, 1–2 June 2021, Linkoping, Sweden (2021). https://ecp.ep.liu.se/index.php/sicfp/article/view/37/36
32. Wang, C., Luan, Z., Gao, W.: Design of pitch curve of internal-curved planet gear pump strain in type N-G-W based on three order ellipse. Adv. Mat. Res. **787** (2013). https://doi.org/10.4028/www.scientific.net/AMR.787.567
33. Li, D., Liu, Y., Gong, J., Wang, T.: Design of a noncircular planetary gear mechanism for hydraulic motor. Mat. Prob. Eng. **2021** (2021). https://doi.org/10.1155/2021/5510521
34. Luan, Z., Ding, M.: Research on non-circular planetary gear pump. Adv. Mat. Res. **339** (2021). https://doi.org/10.4028/www.scientific.net/AMR.339.140
35. Ding, H.: Application of non-circular planetary gear mechanism in the gear pump, Adv. Mat. Res. 591–593 (2012). https://doi.org/10.4028/www.scientific.net/AMR.591-593.2139
36. Oshima, S., Hirano, T., Miyakawa, S., Ohbayashi, Y.: Development of a rotary type water hydraulic pressure intensifier. JFPS Int. J. Fluid Pow. Sys. **2**(2) (2009). https://doi.org/10.5739/jfpsij.2.21
37. Oshima, S., Hirano, T., Miyakawa, S., Ohbayashi, Y.: Study on the output torque of a water hydraulic planetary gear motor. In: The Twelfth Scandinawian International Conference on Fluid Power, May 18–20, 2011. Tampere, Finland (2021)
38. Volkov, G., Fadyushin, D.: Improvement of the method of geometric design of gear segments of a planetary rotary hydraulic machine. IOP Conf. Series: J. Phys. **1889** (2021). https://doi.org/10.1088/1742-6596/1889/4/042052
39. Volkov, G., Smirnov, V.: Systematization and comparative scheme analysis of mechanisms of planetary rotary hydraulic machines. In: International Conference on Modern Trends in Manufacturing Technologies and Equipment (ICMTMTE 2018), vol. 02083 (2018). https://doi.org/10.1051/matecconf/201822402083
40. Volkov, G., Smirnov, V., Mirchuk, M.: Estimation and ways of mechanical efficiency upgrading of planetary rotary hydraulic machines. IOP Conf. Series: Mater. Sci. Eng. **709** (2020). https://doi.org/10.1088/1757-899X/709/2/022055
41. Sliwinski, P.: The influence of water and mineral oil on volumetric losses in hydraulic motor. Pol. Mar. Res. **93**(24), 213–223 (2017). https://doi.org/10.1515/pomr-2017-0041
42. Sliwinski, P.: Satellite displacement machines. Basic of design and analysis of power loss. Gdansk University of Technology Publishing House (2016)
43. Sliwinski, P.: Methods of determining pressure drop in internal channels of a hydraulic motor. Energies **14**, 5669 (2021). https://doi.org/10.3390/en14185669
44. Sliwinski, P., Patrosz, P.: The influence of water and mineral oil on pressure losses in hydraulic motor. In: Stryczek J., Warzyńska U. (eds) Advances in Hydraulic and Pneumatic Drives and Control 2020. NSHP 2020. Lecture Notes in Mechanical Engineering. Springer, Cham (2021). https://doi.org/10.1007/978-3-030-59509-8_10
45. Sliwinski, P.: Patent application P.437751: mechanizm satelitowy hydraulicznej maszyny wyporowej (Satellite operating mechanism of a hydraulic displacement machine) (2021). https://ewyszukiwarka.pue.uprp.gov.pl/search/pwp-details/P.437751
46. Sliwinski, P.: The influence of water and mineral oil on mechanical losses in a hydraulic motor for offshore and marine application. Pol. Mar. Res. **2**(106), 27 (2020). https://doi.org/10.2478/pomr-2020-0034

A New Approach in Hydrostatic Drives: "Digital" Pumps

Ahmed Zubair Jan[1], Krzysztof Kędzia[1(✉)], and Dariusz Prostański[2]

[1] Department of Mechanical Engineering, Wroclaw University of Science and Technology, Wroclaw, Poland
{ahmed.jan,krzysztof.kedzia}@pwr.edu.pl
[2] KOMAG- Institute of Mining Technology, Gliwice, Poland

Abstract. The use of digital pumps in place of traditional "analog" pumps is a fresh, promising strategy. A digital pump is made up of numerous sequential shifting valves and continuous pumps. During partial load function, digital pumps often perform more efficiently than traditional pumps. The paper will include a description of the valves structure and their connections suitable for three pumps on one shaft as well as discussion of digital hydraulic technology and digital valves as well as hydrostatic drive system.

Keywords: digital hydraulic technology · digital fluid power · digital hydraulic valves · digital pump · variable displacement pump · hydrostatic drive system

1 Introduction

William George Armstrong, known as the "grandfather of modern hydraulic power," started experimenting with hydraulics and creating a rotary engine 182 years ago. Humans have benefited significantly from hydraulic technologies across a wide range of industries [1]. To solve conventional hydraulic problems, high-band width servo or proportional valves are typically used. These valves are characterized by high efficiency, speed, good accuracy, and good controllability. However, just like every coin has two sides, conventional hydraulic problems have their own drawbacks, including cavitation, high power losses, sensitivity to contamination, high cost, etc. So, a new approach is required, and digital hydraulics is one innovative idea that has been proposed [2, 3]. The idea of digital hydraulics, in which systems use parallel-connected binary hydraulic components, was initially categorized by Linjama. During the investigation, the idea is reformulated as using an on/off valve rather than a proportional or servo valve to control flow or pressure [4]. The proportional valve loses less power than the on/off valve when compared to a technique controlled by a traditional valve. The digital hydraulics also has additional benefits, such as high efficiency, control flexibility, linearity, resilience, and it has been demonstrated that they are less susceptible to contamination than other servo control systems [5].

J. Stryczek and U. Warzyńska (Eds.): NSHP 2023, LNME, pp. 58–70, 2024.
https://doi.org/10.1007/978-3-031-43002-2_6

In digital hydraulic technology, a discrete digital signal that has been modulated controls a discrete fluid in order to get the active and intelligent authority of the output data. You can think of this technology as a control system. A word that can be used to characterize hydraulic components of high technological quality is "digital" hydraulic components. The technology is typically divided into three categories. The first category is traditional on/off technology, such as rotary pumps and motors; this type merely has two distinct valves and is frequently used in pneumatic systems. The second type is an on/off valve system constructed in parallel, and the third category is a switching technology-based system. The second type's main component is the parallel connection of on/off valves to create a Digital Flow Control Unit, which offers numerous control techniques, including bang-bang control, parallel connection, and serial connection, among others. The third category of switching technologies typically includes two switching converters and two switching configurations. Both configurations use high-speed on/off valves to link a fixed displacement pump in order to achieve varied fluid flow or pressure. The first design combines a fixed displacement pump with an on/off valve at the 6-pump output to direct the fluid back to the tank or achieve movements. Two pipelines, a switching valve, and a switching valve make up the first switching converter, called a wave converter. A switching valve, a spring-loaded cylinder, and an accumulator make up the second switching converter, called a resonance converter [6, 7].

2 Literature Review

The depletion of fossil fuel resources, rising fuel costs and increasingly stringent pollution regulations are driving designers of hydraulic drive systems to consider energy-saving solutions.

Williamson and Ivantysynova have researched displacement-controlled hydraulic pumps which drive hydraulic cylinders directly in mobile equipment, which eliminates many throttling losses present in load-sensing or other valve-controlled systems [8]. Another emerging option is digital hydraulics, which provides two distinct flow control paradigms: pulse code modulated (PCM) concepts, as provided by Linjama and Ploeckinger and pulse width modulated (PWM) systems- as presented by Ploeckinger and Long and Lumkes [9]. The second idea, which involves numerous pumps running continuously in parallel, serves as the backdrop for the research of digital pumps that is provided here. In recent years, digital valve approaches have been researched, replacing proportional valves with several simple on/off-valves operating in parallel to combine their unique flows in a stepwise, nearly continuous manner. Linjama and Holland have previously examined digital pump designs in which single cylinders of piston pumps are switched with valves to allow the distribution of flow in intervals between many outlets or idle. Schienbein and Kanter have patented an arrangement of constantly operating pumps coupled with multiple control valves and actuators via an array of directional valves. Linjama and Tammisto introduced a pump-motor-transformer that combines numerous continuously functioning pumps of various sizes with a variety of outlets [10, 11]. To reach the correct indicate pressure at the outlets of this system's pumps, which are repeatedly turned on and off, a capacitance volume is placed there to balance pressure and provide a steady flow toward the valves that regulate cylinder movement. To enable energy recovery and power stabilization, a second output was linked to an accumulator provide a thorough analysis of the many options for creating digital pump architectures, including a form in which output flow is directly connected to numerous actuators without the use of an intermediary control valve.

3 Digital Hydraulic Valves

The pressure and flow in digital hydraulic systems are controlled using solenoid-operated directional seat valves. On and off are the valves' two possible states. Digital valves are able to carry out all valve functions, including pressure regulation, direction, and flow control, utilizing only on/off valves, in contrast to traditional hydraulic systems that employ several types of valves to conduct a range of tasks. It takes at least 4 Digital Flow Control Units (DFCUs), which are made up of ON/OFF valves linked in parallel, to substitute a proportional valve. In a circuit for digital hydraulics, there could be a lot of control lines. As all those control lines is capable of being independently regulated, control is enhanced. This results in less power loss and more precise control (Fig. 1).

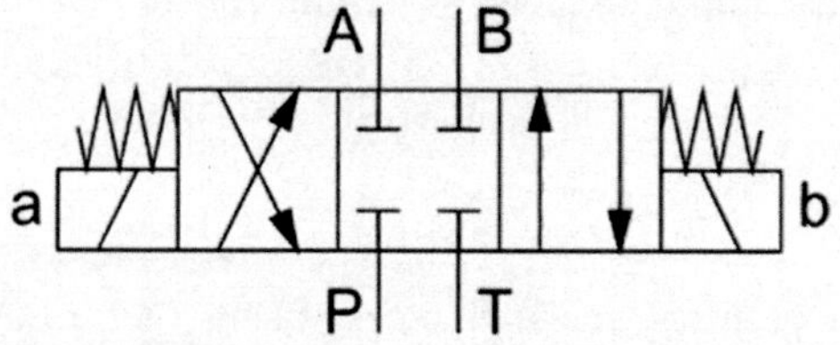

Fig. 1. 4 way- 3 position hydraulic valve: 'P'- pressure port, 'T' port which takes the spent fluid back to the hydraulic tank, 'A', 'B' ports, that delivers (spent fluids back) the fluid to the load (the tank); a,b- electric coils.

The disadvantage of typical servo valves is that they require high-frequency jitter for retracting the spool while it remains stationary in one position, making them less responsive and less trustworthy hydraulic components than ON/OFF valves [12]. Because of this, digital valve technology employs ON/OFF valves. These valves are frequently used in applications requiring quick switching periods and slow-moving loads. With the help of digital valves, we can accurately control the pressure used for moving or regulate machine elements. This is accomplished by using a set of ON/OFF valves.

3.1 Parallel Digital Hydraulic Valves

The digital flow control unit is another name for the parallel digital hydraulic valve. (DFCU). With coding the control of many switching valves that each are attached in parallel, it is possible to regulate the flow very precisely. The rate of flow of a DFCU is equal to the sum of all the switching valves operating in the "on" state. The DFCU's steady-state characteristics depend on the encoding mode of the switch valves' control signal and the total amount of switch valves (N). The three most popular coding methods being binary, Fibonacci, and PNM. (pulse number modulation). One way to characterize the state of DFCU is as a 2N pair of switching valves. Consequently, there are approximately 2N different kinds of flow output in various DFCU modes. A DFCU and a high-speed switching digital hydraulic valve are fundamentally different from one another since the former does not need frequently changing of a single valve between on and off to provide continuous system output [13, 14] (Fig. 2).

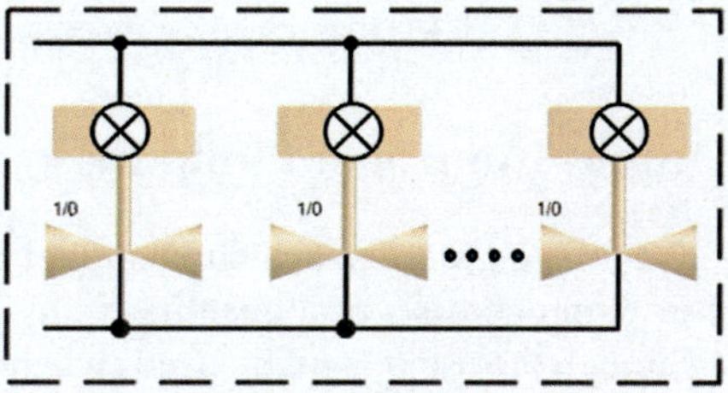

Fig. 2. Parallel digital two-way valve.

3.2 High-Speed Switching Digital Hydraulic Valve

The main goal of fast speeds switching hydraulic technology research is the creation of high-speed switching valves. Two categories make up this research:

- the study and development of novel high-speed switching valves.
- the high-speed switching valve method [16].

The fast speeds switching valve was first imagined and created around the end of the 1970s. High-speed solenoid switching valves were created by the British company Lucas by combining two types of electromagnets with specialized forms. The terms "Helenoid valve" and "Colenoid valve," respectively, were given to these electromagnets, which were of the spiral tube type and taper type. The challenge of the rate of acceleration of an armature being inversely related of the electro'magnetic force was resolved. However, because of the complex nature of their structure and the difficulties they present in processing and manufacturing, they are not used to a great amount. Various businesses and research institutions have released a variety of high-speed switching valves on the market [17] (Fig. 3). High-speed switching valves, also known as solenoid valves, are commonly used in various industrial applications to control the flow of fluids or gases. The parameters of high-speed switching valves can vary depending on the specific application and requirements.

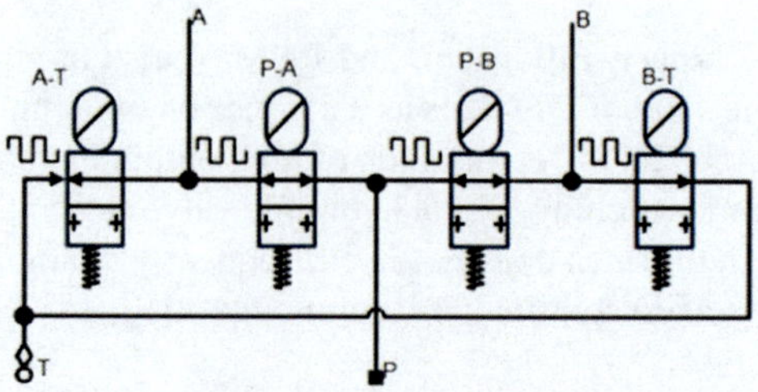

Fig. 3. High-speed switching valve.

4 Variable Output Control of Pump Flow

The main functioning of a high-speed switching digital hydraulic pump is illustrated in Fig. 4. This type of pump consists of a quantitative hydraulic pump and a high-speed switching valve. Through the use of a high-speed switching valve installed at the output of the quantitative pump, a high-speed switching digital hydraulic pump can regulate the flow of fluid entering the system. In addition, the valve switching frequency has a direct bearing on the control performance of the device. Figure 4 illustrates the operation of the parallel digital hydraulic pump, which is based on the idea that it is made up of many coaxial quantitative pumps operating in parallel.

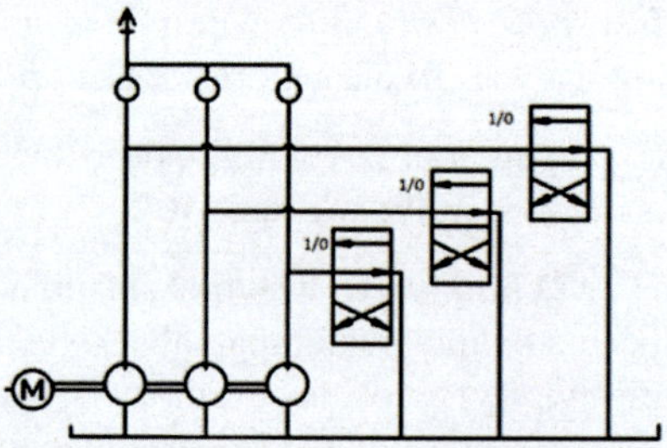

Fig. 4. Variable output control of pump flow.

A switch valve located at the outlet of each quantitative pump in a parallel digital hydraulic pump system provides independent control of the individual quantitative pumps. The difference between the maximum and the minimum flow ratio is determined by the encoding mode of the switching valves. The maximum flow ratio is equal to the sum of the flow ratios of all quantitative pumps operating in parallel. The minimum flow ratio is equal to the flow ratio of the minimum quantitative pump. A parallel digital hydraulic pump therefore has 2N different kinds of flow ratio (N is the number of quantitative pumps in parallel). The main difference between a parallel digital hydraulic pump and a high-speed switching digital hydraulic pump is that the former does not require the output of the system to be controlled by a single switching valve's frequent state change between on and off. This is the defining characteristic of the high-speed switching digital hydraulic pump. The only purpose served by switching valves' state changes is in the process of adjusting the combined form of parallel pumps [18–20].

4.1 Digital Control of a Variable Pump

The operating principle of the digital hydraulic pump of the combination cylinder control is shown in Fig. 5. When the controller outputs a different code in response to a digital input signal, not only does the amount of data flowing into the combined cylinder change, as well as the amount of opened switching valves. This, in effect, causes the extension length of the piston rod to change. In addition, the slope of a swash plate can be controlled by the piston rod, which enables intelligent regulation of the pump's displacement. The digital hydraulic pump is capable of reaching a number of pressures and flow rates, all of which are determined by the encoding combination used. There are 2N different types of pump displacement for every possible combination of N levels that the combination cylinder can offer [21–23]. However, this is just a possibility in theory because the forces acting on the swash plate are highly variable in practice, and controlling the forces acting on the swash plate alone is not sufficient to achieve the desired level of displacement control shown in Fig. 5.

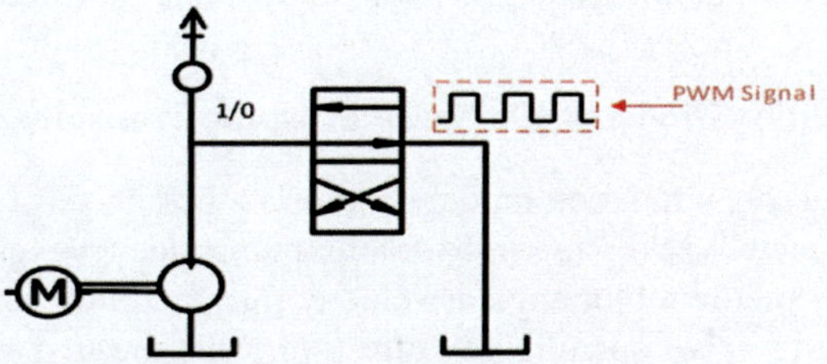

Fig. 5. Digital control of a variable pump

The high-speed switching valve control digital hydraulic pump is responsible for adjusting the extension length of the piston rod. This is accomplished by rapidly switching the state of the high-speed switching valve between the on and off positions. In order to implement intelligent control of pump displacement, the extension length of the piston rod can be employed to control the angle at which the swash plate is positioned. In Fig. 5 illustrates the basic operating principle of a pump like this one [21–24].

5 New Idea of Designing the Hydrostatic Transmission

Hydrostatic drives are systems that convert mechanical energy into fluid energy and turn it back into mechanical energy. Hydrostatic drives are classified into two types: fixed and variable displacement hydrostatic drives. In the whole industry, hydrostatic drive is used in various applications such as cranes, conveyors, mobile equipment and planes. The hydrostatic drive, also known as hydrostatic transmission, is usually composed of a combination of hydraulic pumps and motors.

The paper describes the first part of research answering the question: Is that possible to replace the proportional technique in multi-source hydrostatic drive systems with "digital hydraulics" ensuring a similar efficiency of energy recuperation while reducing the expenditure on the construction of multisource drive system?

By examining the above-described concept of replacing the classic hydraulic system with a system built according to the concept of "digital hydraulics", it will be possible to replace the expensive and complicated hydraulic system supply system (e.g. using a multi-piston axial hydraulic pump with the possibility of changing the flow rate) and control carried out by means of a proportional valve a set of gear pumps (simple and inexpensive) and a control system built using a set of simple and reliable valves (two-way, two-stage) with an appropriately selected, simple controller.

The benefits of the conducted research and analyzes will be:

- the possibility of reducing the costs of building the system and its maintenance.
- improvement of ecological parameters (reduction of energy consumption, noise, emission of harmful substances, reduction of the so-called carbon footprint, etc.)
- the use of less complex, and therefore more reliable and resistant to external and internal factors components (sensitivity to contamination, liquid parameters - viscosity, temperature, etc.).

To confirm this, it is necessary to conduct appropriate research and tests. First step is design of simulation model of classic hydrostatic transmission.

5.1 Simulation Model of Classic Hydrostatic Transmission

The goals of the simulation model creation is the analysis of hydrostatic transmission (Fig. 6.) especially: pressure in hydraulic pump side (p_p), pressure in hydraulic motor side (p_s), hydraulic motor angular velocity (ω_s), and flow through overflow valve (Q_z) with regard to valve opening pressure (p_o). This the first step to analyze if "digital" hydraulic is possible to use in multisource hydrostatic drive system.

a) Hydraulic motor torque Eq. (1):

$$q_s p_s = M_0 + R_0 \omega_s + I\frac{d\omega_s}{dt} \tag{1}$$

Initial condition: $\omega_s(0) = 0$, boundary conditions: for $q_s p_s \leq M0$ and $\omega_s = 0$, $\frac{d\omega_s}{dt} = 0$

b) Pump flow Eq. (2):

$$q_p\omega_p = G_p p_p + G_r(t)\sqrt{p_p - p_s} + Q_z + \frac{V_p}{E_c}\cdot\frac{dp_p}{dt} \tag{2}$$

Initial conditions: $p_p(0) = \left[\frac{q_p\omega_p}{G_r}\right]^2$

c) Distribution valve flow Eq. (3):

$$G_r(t)\sqrt{p_p - p_s} = q_s\omega_s + G_s p_s + \frac{V_s}{E_c}\cdot\frac{dp_s}{dt} \tag{3}$$

Initial conditions: $p_s(0) = 0$.

d) Overflow valve flow Q_z Eq. (4), (5):

Non inertial valve:

$$\text{For } p_p < p_o\text{: } Q_z = 0$$
$$\text{For } p_p > p_o,\ Q_z = K\left(p_p - p_o\right) \tag{4}$$

Inertial valve:

$$T_z \frac{dQ_z}{dt} = K_z(p_p - p_0) - Q_z \tag{5}$$

Nomenclature:

q_p, q_s- unit displacement of the hydrostatic motor (pump) [m^3/rad]; p_s, p_p pressure in motor and pump site [Pa]; Δp, dp- pressure loss, pressure difference [Pa]; t- time [s]; M_0 – torque of load [Nm]; R_0 – resistance coefficient; I- torque of inertia; ω_s- hydraulic motor angular velocity [rad/s]; G_p, G_r- flow ratio coefficient; V- volume of high pressure lines [m^3]; E_c- compressibility [N/m^2]; Q_z- flow through [m^3/s]; T_z- dynamic parameter of inertial valve; K- amplification.

e) Numerical data:

$q_p\omega_p = 723*10^{-6}[m^3/s]$, $E_c = 1.5*10^9[N/m^2]$, $p_o = 19*10^6[N/m^2]$, $q_s = 4.76*10^{-6}[m^3/rad]$, $R_o = 0.03[Nm/s]$, $G_p = G_s = 5.4*10^{-12}\ [m^4s/kg]$, $T_z = 0.2\ [s]$, $V_p = 150*10^{-6}\ [m^3]$, $V_s = 230*10^{-6}\ [m^3]$, $K_z = 0.52*10–9$ [m5/Ns], $I_0 = 54*10^{-4}[Nm/s^2]$, $M_0 = 60[Nm]$, $G_r = 7.23*10^{-7}\ [m^4/N^{0.5}]$.

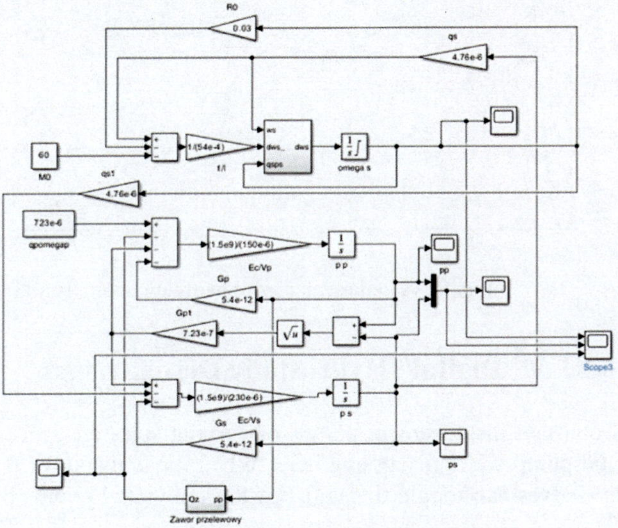

Fig. 6. Schema of simulation model.

An example of result obtained from simulation model of hydrostatic transmission is presented below on Figs. 7, 8, 9, 10 and 11.

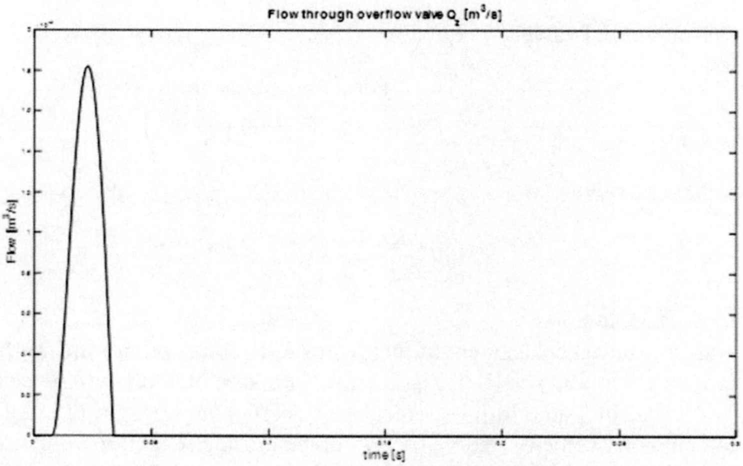

Fig. 7. Flow through overflow valve (Q_Z).

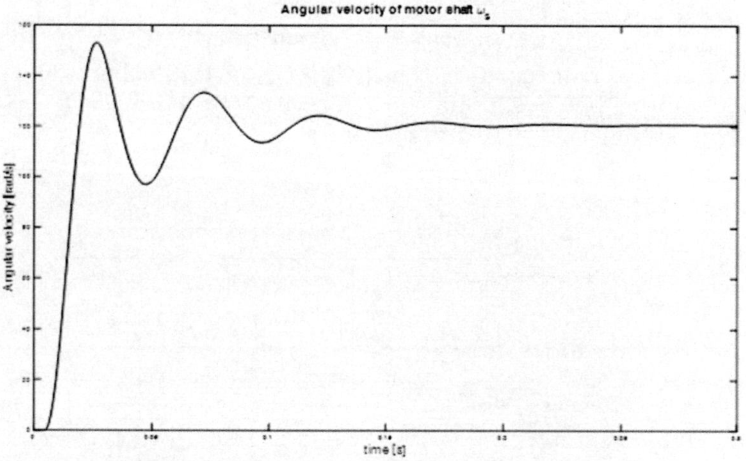

Fig. 8. Angular velocity of hydraulic motor shaft (ω_S).

6 Model of "Digital" Hydrostatic Drive System.

Several constant displacement pumps of varying sizes are arrayed in parallel to form the digital pump we're discussing here. When the demand for flow changes, a matrix of on-off valves can toggle the status of the individual pumps between load and idle (Fig. 12).

Any pump can construct a digital pump as long as a proper volumetric gradation is available. It is preferable to have designs that accommodate multiple units on the same shaft. As there are at least three to seven units in a digital pump, it makes sense to employ pump types with a small axial dimension, such as gear or vane pumps. Unfortunately,

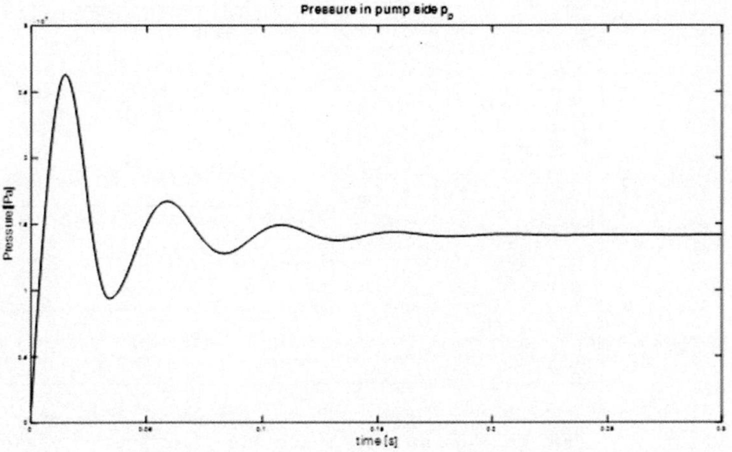

Fig. 9. Pressure in hydraulic pump side (p_P).

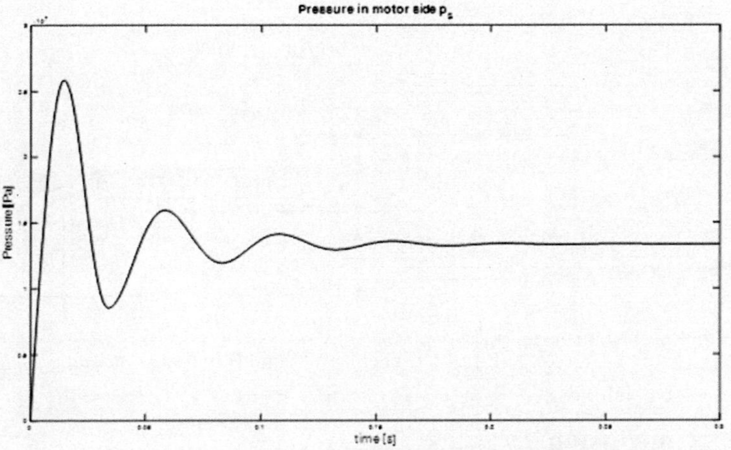

Fig. 10. Pressure in hydraulic motor side (p_S).

they have pressure constraints. High pressures can be achieved with axial or radial piston pumps (up to 70 MPa). The RAC radial piston pump, is a new radial design that is not yet in mass production but shows great promise for this application due to its compact size and the fact that its shaft is unaffected by any forces other than those associated with torque transmission. That's make it easy to stack multiple pumps on a single shaft [25–32].

This paper is described hydraulic drive system design from three two-way valves and three constant displacement pumps.

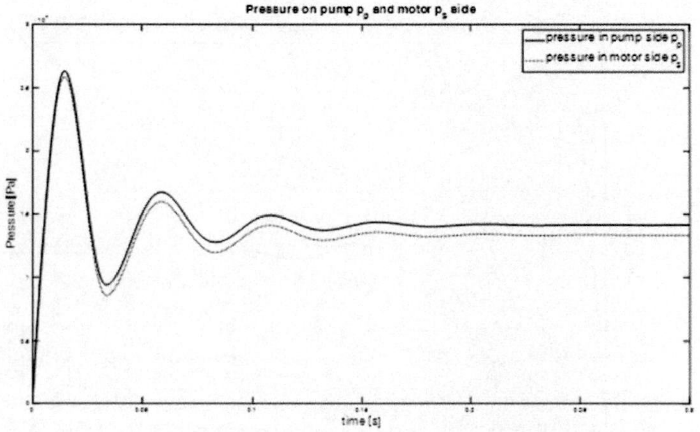

Fig. 11. Pressure in hydraulic motor (p_S) and pump (p_P) side

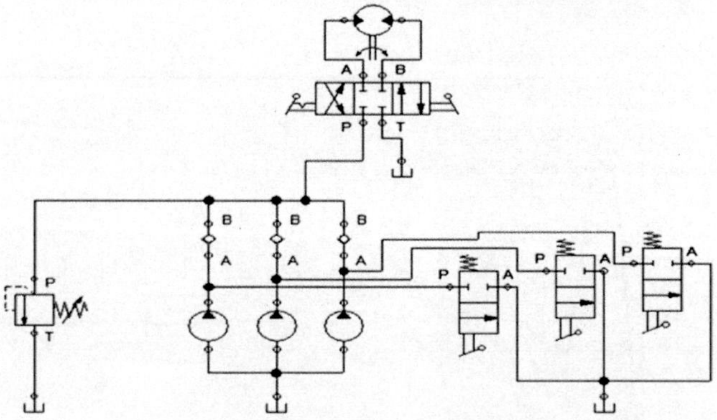

Fig. 12. Schema of "digital" hydraulic circuit

7 Conclusion

Digital flow controls with a separate or shared pump assignment are a promising concept to improve the energy efficiency of hydraulic motors. The energy savings potential of digital pumps relative to conventional variable displacement pumps is particularly high in working cycles with large quantities of partial loads or low speed movements. The challenges worth further research include energy losses, pressure peaks, and vibrations caused by valve changes between steps.

Digital Hydraulics' idea is to replace expensive and sensitive servo valves with a number of low-cost in/out valves. The basic concept is the parallel connection, allowing good flow control. This means that even slow response valves can achieve good control results, which reduces the price of valves and control electronics. The main advantages of Digital Hydraulics are: greater reliability due to simple valves and inherent redundancy. Total costs are also lower, especially in water hydraulic applications. Digital Hydraulics is therefore a real alternative to traditional servo and proportional systems, offering also unique advantages. However, some challenges must be resolved, the most important of which is valve technology and control methods. This technology is likely to fulfil the dreams of cheaper, better and more reliable hydraulics, but research and development is needed.

The next step to check the possibility of using digital pumps instead of multi-piston pumps is to perform tests to verify the correctness of the data obtained by simulation on the test stand.

References

1. The History of Hydraulics: From Ancient Times to Modern Day. https://hcsplating.com/history-of-hydraulics/
2. Merrill, K., Holland, M., Batdorff, M., Lumkes, J., Jr.: Comparative study of digital hydraulics and digital electronics. Int. J. Fluid Power **11**(3), 45–51 (2010)
3. Turner, S.B., Lakin, D.F.: Electrohydraulic proportional control valve assemblies. Patent (1997)
4. Laamanen, M.S.A., Vilenius, M.: Is it time for digital hydraulics. In: The Eighth Scandinavian International Conference on Fluid Power (2003)
5. Wang, F., Gu, L., Chen, Y.: A continuously variable hydraulic pressure converter based on high-speed on–off valves. Mechatronics **21**(8), 1298–1308 (2011)
6. Fundamentals of Digital Logic and Microcontrollers, 6th Edition by M. Rafiquzzaman. https://www.oreilly.com/library/view/fundamentals-ofdigital/9781118969304/9781118969304c01.xhtml
7. Rickenberg, F.: Valve. U.S. patent No. 1757059, 30 (1930)
8. Williamson, C., Ivantysynova, M.: Power optimization for multi-actuator pump-controlled systems. In: Proceedings of the 7th International Fluid Power Conference, RWTH Aachen, Germany (2010)
9. Long, G., Lumkes, J.: Comparative study of position control with 2-way and 3-way on/off electrohydraulic valves. Int. J. Fluid Power **11**(1), 21–32 (2010)
10. Schienbein, O., Kanter, M.: Hydraulic System with Variable Fluid Flow under Pressure to Fluid-operated Consumers, Patent US20030037545 (2003)
11. Linjama, M., Tammisto, J.: New alternative for digital pump-motor-transformer. In: The Second Workshop on Digital Fluid Power, Linz, Austria (2009)
12. Trostmann, E.: Water Hydraulics Control Technology. Danfoss, Nordborg, Denmark (1996)
13. Urata, E., Miyakawa, S., Yamashina, C., Nakao, Y., Usami, Y., Shinoda, M.: Development of a water hydraulic servovalve. JSME Int. J. Ser. B **41**, 286–294 (1998)
14. Zhang, Q. Research on the Static and Dynamic Characteristics of the 2D Digital Valve and Compensation of the Dead Zone Nonlinear. Master's Thesis, Zhejiang University of Technology, Hangzhou, China (2011).
15. Li, S.; Ruan, J.; Meng, B. Dither compensation technology for hysteresis of 2D digital valve. Trans. Chin. Soc. Agric. Mach. 42, 208–218 (2012)
16. Katakura, H., Yamane, R., Takenka, T.: Fundamental research on digital positioning by several hydraulic cylinders and a microcomputer. J. Jpn. Hydraul. Pneum. Soc. **22**, 63 (1991)
17. Ba, K., Yu, B., Gao, Z., Ma, G., Kong, X.: An improved force-based impedance control method for the legged robot HDU. ISA Trans. **84**, 187–205 (2019)
18. Lambeck, R.P.: Hydraulic Pumps and Motors: Selection and Application for Hydraulic Power Control Systems, Dekker, New York, NY, USA, p. 154 (1983)

19. Moorhead, J.R.: Saving energy with “digital” pump systems. Mach. Des. **56**, 40–44 (1984)
20. Rampen, W.H.S., Salter, S.H.: The digital displacement hydraulic piston pump. In: Proceedings of the 9th International Symposium on Fluid Power, Cambridge, UK, pp. 25–27. BHR Group, Cambridge, UK, pp. 33–46 (1990)
21. Ehsan, M., Rampen, W.H.S., Salter, S.H.: Modeling of digital-displacement pump-motors and their application as hydraulic drives for nonuniform loads. ASME J. Dyn. Syst. Meas. Control **122**, 210–215 (2000)
22. Payne, G.S., Kiprakis, A.E., Ehsan, M., Rampen, W.H.S., Chick, J.P., Wallace, A.R.: Efficiency and dynamic performance of digital displacementtm hydraulic transmission in tidal current energy converters. Proc. Inst. Mech. Eng. Part A **221**, 207–218 (2007)
23. Lumkes, J., Batdorff, M., Mahrenholz, J.: Characterization of losses in virtually variable displacement pumps. Int. J. Fluid Power **10**, 17–27 (2009)
24. Berbuer, J., Schulze Schenking, D.: Radial piston engine with cone valve plates. In: Proceedings of the 8th IFK (International Fluid Power Conference), Dresden University, Germany (2012)
25. Kędzia, K.: A method of determining optimal parameters for the secondary energy source of a multisource hydrostatic drive system in machines working in closed spaces. Energies **15**(14), 1–24 (2022) Article 5132
26. Śliwiński, P.: The influence of pressure drop on the working volume of a hydraulic motor. Eksploatacja I Niezawodność – Mainten. Reliab. **24**, 747–757 (2022). https://doi.org/10.17531/ein.2022.4.15
27. Osiński, P., Chruścielski, G., Korusiewicz, L.: Theoretical and experimental fatigue strength calculations of lips compensating circumferential backlash in gear pumps. Energies **14**(1), 1–14 (2021) Article 251
28. Stosiak, M., et al.: Analysis of the impact of vibrations on a micro-hydraulic valve using a modified induction algorithm. Machines **11**(2), 1–28 (2023). Article 184
29. Rădoi, R., Dumitrescu, C., David, I., Blejan, M., Şefu, S., Ionescu, C.: The effect of the main component ratios in the joint filling on the product quality
30. Joostberens, J., Pawlikowski, A., Prostański, D., Nieśpiałowski, K.: Method for assessment of operation of analog filters installed in the measuring lines for electrical quantities of a mining machine’s converter power supply system. Energies **14**(9), 2384 (2021)
31. Rădoi, R. I., Dumitrescu, C., David, I., Blejan, M., Sefu, S., Ionescu, C.: Utility vehicle with hydraulic transmission and hybrid energy source. Min. Mach. **41** (2023)
32. Antonchik, V., Zabolotnyi, K., Hankevich, V., Maltseva, V., Kuts, O., Dyczko, A.: Constructional changes of pneumopercussion machines for improving their efficiency. Min. Mach. **40** (2022)

Digital Control of a Vane Pump

Grzegorz Filo(✉) and Edward Lisowski

Faculty of Mechanical Engineering, Cracow University of Technology, Jana Pawla II 37, Cracow, Poland
{filo,lisowski}@mech.pk.edu.pl

Abstract. The progressive miniaturization in electronic system design allows us to implement digital controllers in an increasingly compact form, including installation directly in the hydraulic device. The control signals can be sent remotely from an external controller, or the built-in system can independently implement the control algorithm to some extent. The paper presents the results of numerical and experimental studies of a digital control system implemented in a variable displacement vane pump. It allows various control strategies to be implemented, including fixed value pressure control, constant power control, etc. The flow rate is set by controlling the position of the stator, while the pressure can be regulated using the built-in proportional valve. A mathematical pump and control system model was built as part of the work. Next, simulations were carried out in Matlab/Simulink environment, considering the dynamics of the system. To verify the mathematical model, a test bench was built for carrying out laboratory experiments with typical loads and changes in pump flow rate.

Keywords: vane pump · simulation · Matlab · experimental verification

1 Introduction

Hydraulic systems used to drive mechanisms or devices comprise actuators or hydraulic motors, control valve blocks, oil tanks, and equipment, including filters, sensors, pressure gauges, indicators, etc. However, one of the essential components that determine obtaining the required operational parameters with the lowest possible energy demand is the power supply unit. It usually consists of an electric motor, a pump and a reservoir. The best situation is when the pump characteristics are matched to the needs of the system. This is a difficult task since it is usually required for the system to operate at different loads and with variable speeds of actuators. Hence, it is especially convenient if the pump is able to adapt its parameters to the current demands. Variable displacement pumps, most often piston or vane pumps, have such possibilities. Piston pumps can operate at higher pressures; however, the main advantages of vane pumps include compact design, low flow rate pulsation, low noise, a significant reduction in leakages and good dynamic properties [1]. Particularly good results can be obtained using an IEHVP supply unit (Integrated Electro-Hydraulic Vane Pump), which contains a controller operating based on signals from sensors built directly inside the pump [2].

J. Stryczek and U. Warzyńska (Eds.): NSHP 2023, LNME, pp. 71–80, 2024.
https://doi.org/10.1007/978-3-031-43002-2_7

Current research on developing vane pumps is carried out in various directions, including innovative design solutions, modelling and simulation, analysis of phenomena occurring during operation, as well as study and optimization of operating parameters. In terms of new design solutions, Fatigati et al. proposed a low-speed sliding rotary vane pump for an engine cooling system [3], Bianchi et al. developed a vane pump as a power unit for waste heat to power conversion in energy recovery systems based on bottoming Organic Rankine Cycles [4, 5]. The authors analysed fluid dynamic filling and emptying processes, closed vane transformation, leakages as well as viscous and dry friction phenomena based on a comprehensive one-dimensional model built in GT-Suite package. In turn, M. Ciursys and W. Fiebig presented an innovative design solution of a balanced vane pump integrated with an electric motor [6]. In the area of vane vacuum pump development, Wang et al. designed and analyzed a three-chamber sliding vane vacuum pump [7] and Chen et al. provided a design methodology of an asymmetric cylinder profile for sliding vacuum pumps [8]. The second group includes publications focusing mainly on modelling vane pumps. Fornarelli et al. in [9] presented an analysis of a pressure-compensated vane pump in AMESim and Matlab/Simulink environments to estimate the friction forces and volumetric efficiency loss. Lobsinger et al. provided CFD studies of a fixed-type balanced vane pump 2D model [10], including the leakage and multiphase flow analyses using TwinMesh and Ansys CFX software. Similarly, Guerra et al. conducted a detailed numerical analysis of the kinematics of twin lip vanes in balanced vane pumps [11]. Moreover, Sun et al. conducted stator curvature optimization and dynamic performance simulation of an axial vane pump [12] with the help of particle swarm algorithms, as well as Matlab and ADAMS environments. Important issues are also the problems of wear and reliability of vane pumps, such as analysis of damage factors for pump wear progression [13], methodology of fast and efficient ranking of hydraulic fluids for vane pumps [14], and analysis of the lubrication regime and wear that occurs between vanes and pressure ring in variable displacement vane pumps [15, 16]. Furthermore, matrix methods can be used for a classification of variable delivery vane pump potential failures [17]. Publications related to the research on vane pump operating parameters include flow ripple analysis [18], estimation of unbalanced radial forces [19], examination of pressures in the internal displacement chambers [20], analysis of the vane-cam ring mechanism kinematics [21], or assessment of variable radial loads applied to the vanes [22].

The article presents a numerical model of a vane pump developed in the Matlab/Simulink program. The model allows the simulation of pump flow rate and pressure control based on current demand information as well as the evaluation of the dynamic operational parameters. The results of numerical simulations are verified on a laboratory test bench containing a variable-displacement vane pump with an integrated control system.

2 Case Study: Working Principle of a Vane Pump

Figure 1 shows a schematic diagram of the modelled vane pump. The working fluid supply to the pressure space through each of the eleven vanes begins when the vane enters the sealed area. The pump displacement is controlled by the spool (8), which position determines the value of the eccentricity e_p.

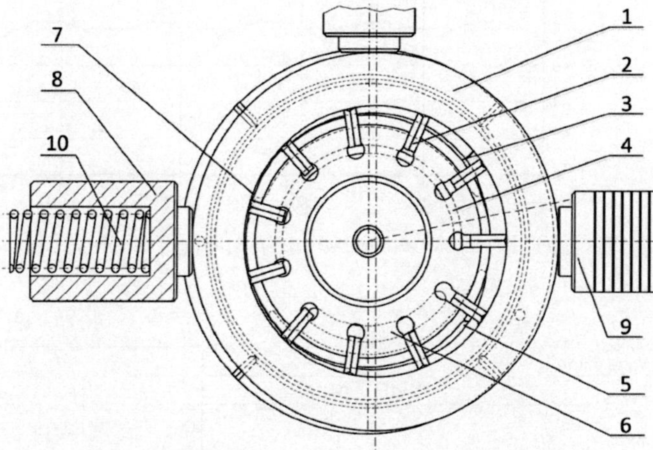

Fig. 1. Vane pump schematic view; 1 – stator, 2 – vane, 3, 4 – output channels, 5, 6 – suction channel, 7 – damping gaps, 8 – control spool, 9 – return spool, 10 – spring.

The instantaneous active height of the vane h against the rotation angle ϕ can be determined from the formula [23]:

$$h = e_p(1 - \cos(\phi)) \tag{1}$$

Hence, the circumferential v_c and radial v_r velocity components of the vane's center of gravity, assuming the inner stator radius r and angular velocity ω are equal to [23]:

$$\begin{cases} v_c = \left(r - \frac{1}{2}e_p(2 + \cos(\phi))\right) \cdot \omega \\ v_r = e_p \cdot \omega \cdot \sin(\phi) \end{cases} \tag{2}$$

The total theoretical pump flow rate Q_0 can be calculated as the sum of the flows due to the circumferential and radial movement of the vanes Q_ϕ and Q_r, respectively, decreased by the losses due to the non-zero vane thickness Q_s. For the considered construction of the pump, it is possible to assume approximately $Q_r = Q_s$, hence:

$$Q_0 = Q_\phi + Q_r - Q_s = Q_\phi = e_p\left(r - \frac{1}{2}e_p(1 + \cos(\phi))\right) \cdot \omega \cdot b \cdot (1 - \cos(\phi)). \tag{3}$$

The geometrical parameters have the following values: inner stator radius r = 43 mm, vane height b = 20 mm, maximum eccentricity e_p = 2.5 mm. The maximum theoretical flow rate against time for the rotational speed ω = 157 rad/s is shown in Fig. 2. As can be seen from the graph, the coefficient of non-uniformity of the pump flow rate does not exceed 2.14%. This is a relatively low value, comparable to internal gear pumps and multi-piston pumps with nine or more pistons.

To analyze the operational parameters of the vane pump, in the first step, a simplified mathematical model of the pump was built according to the diagram shown in Fig. 3. It was assumed that the pulsation of the pump would be omitted due to the negligible impact on the unevenness of the flow. In addition, the delays in the electronic system

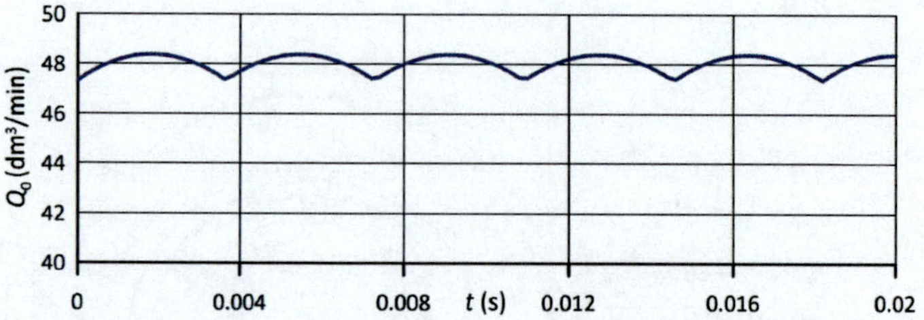

Fig. 2. Theoretical flow rate of the studied vane pump for $e_p = 2.5$ mm and $\omega = 157$ rad/s.

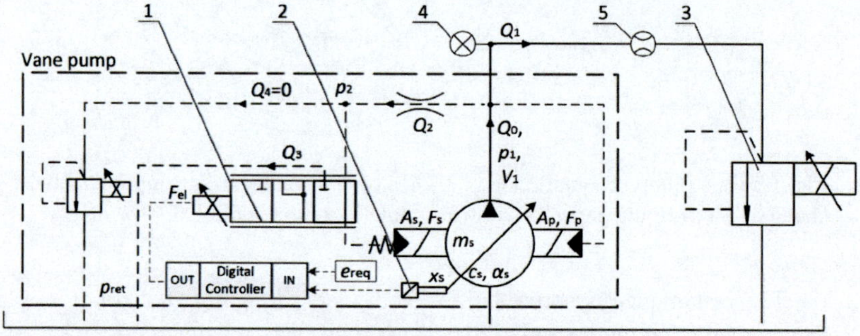

Fig. 3. Diagram of the studied pump with the controller: 1 - flow control valve, 2 - stator position transducer, 3 - relief valve, 4 - pressure transducer, 5 - flowmeter.

and the proportional control valve operation will not be considered, as these elements have a significantly shorter response time than the stator of the pump.

Theoretical pump flow rate can be determined based on the geometric unit capacity per revolution $\dot{V}_g = 32$ cm³/rev, normalized eccentricity $e_p[0, 1]$, rotational speed. $n_p = \omega/(2\pi)$ and volumetric efficiency η_p:

$$Q_0(t) = \dot{V}_g \cdot e_p(t) \cdot n_p(t) \cdot \eta_p. \tag{4}$$

The Q_2 and Q_3 flow rates can be determined from the following flow equations, where ρ is fluid density, μ_2, μ_3 – discharge coefficients and A_2, A_3 – flow areas, respectively:

$$Q_2(t) = \mu_2 \cdot A_2 \cdot \sqrt{\frac{2}{\rho} \cdot |p_1(t) - p_2(t)|}, \tag{5}$$

$$Q_3(t) = \mu_3 \cdot A_3 \cdot \sqrt{\frac{2}{\rho} \cdot |p_2(t) - p_{ret}|}, \tag{6}$$

In the steady state, it can be assumed that $Q_2 = Q_3$, hence the comparison of formulas (5) and (6) makes it possible to determine the p_2 pressure:

$$p_2(t) = \frac{(\mu_2 A_2)^2 \cdot p_1(t) + (\mu_3 A_3)^2 \cdot p_{ret}}{(\mu_2 A_2)^2 + (\mu_3 A_3)^2} \tag{7}$$

The mass conservation equation for pump supply line volume V_0, assuming the fluid bulk modulus B has the following form:

$$(Q_0(t) - Q_1(t) - Q_2(t)) = \frac{V_0}{B}\frac{dp_1(t)}{dt} \tag{8}$$

The stator motion equation includes the viscous friction force $F_\alpha = \alpha\frac{dx_s}{dt}$, spring force $F_{spr} = c_s \cdot x_s$, the hydrostatic force of the control spool $F_s = A_s \cdot p_2$, the hydrostatic force of the return spool $F_p = A_p \cdot p_1$ and the force resulting from the pressure difference in the discharge and suction space $F_t = h \cdot b \cdot p_1$:

$$m_s\frac{d^2x_s}{dt^2} = F_s - F_p - F_t - F_\alpha - F_{spr}. \tag{9}$$

Finally, the relief valve flow rate $Q_1(t)$ was defined by a simplified equation based on its nominal operational characteristics:

$$Q_1(t) = Q_{1,nom} \cdot \frac{p_1(t) - p_{ret}}{\Delta p_{nom}}. \tag{10}$$

3 Numerical Simulations and Laboratory Experiments

Numerical simulations were conducted in Matlab/Simulink. Next, laboratory experiments were carried out on a test bench.

3.1 Simulation Model and Numerical Studies

Figure 4 shows a block diagram of the simulation model built in the Simulink environment. The main components are the pump and two overflow valves, including one electromagnetically-controlled. The adjustable parameters are the voltage signal controlling the pump displacement `U_pump` [0.0, 5.0] V, which corresponds to the value of the required eccentricity in the range [0.0, 2.5] mm and the setting of the overflow valve `U_rel` setting the pressure in the supply line in the range [0, 16] MPa.

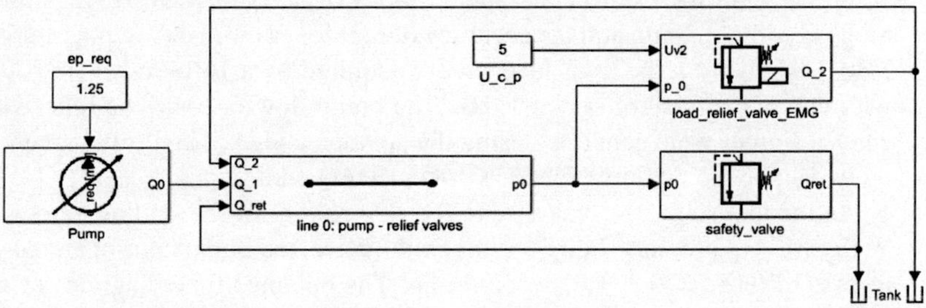

Fig. 4. Simulink model; `ep_req` – required eccentricity, `U_c_p` – pressure control signal

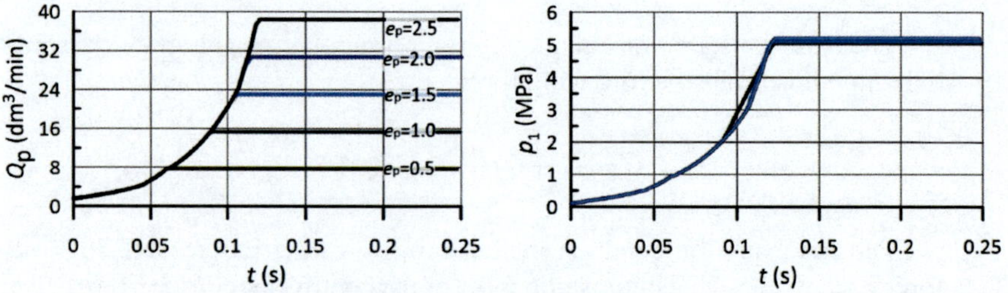

Fig. 5. Time diagrams of flow rate and supply line pressure for $p_{1,req1} = 5.0$ MPa

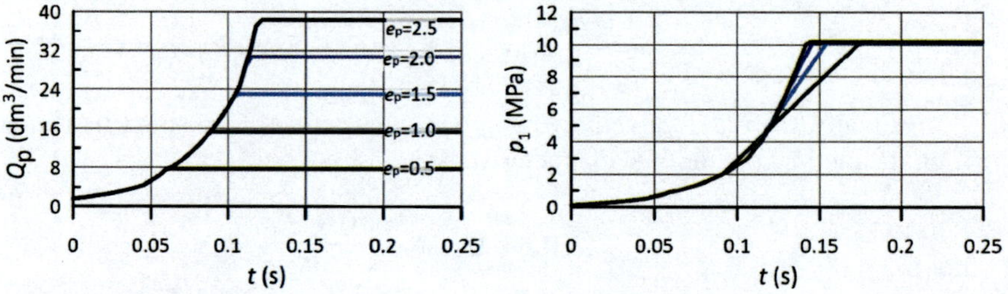

Fig. 6. Time diagrams of flow rate and supply line pressure for $p_{1,req2} = 10.0$ MPa

The pump response to the input step signal $e_p = 0.5, 1.0, \ldots, 2.5$ mm was tested at the set supply line pressures $p_{1,set1} = 5.0$ MPa and $p_{1,set2} = 10.0$ MPa. The obtained results in the form of time charts for $p_{1,set1}$ and $p_{1,set2}$ pressures are shown in Fig. 5 and Fig. 6, respectively.

As can be seen from Fig. 5 and Fig. 6, change in the eccentricity setting causes a proportional change in the flow rate. The time to reach the flow rate fixed value fits in the range 0.07 – 0.12 s. The pressure in the supply line stabilizes in 0.12 to 0.15 s.

3.2 Test Bench and Laboratory Experiments

The test bench was built according to the diagram in Fig. 3. In turn, Fig. 7 shows the view of the tested pump and the computer system for generating the flow rate and pressure control signals. The measuring equipment included a pressure transducer Trafag NAT with a range of [0, 25] MPa and the accuracy of $\pm$ 0.2%, a KEM HM flowmeter of the range [1, 40] dm^3/min and the accuracy of $\pm$ 0.5%, as well as a temperature meter Introl FH0. Signals from the transducers were acquired by a 16-bit NI PCIe-6321 DAQ card with the SCB-68A terminal at 1kHz. The pump flow rate and the relief valve pressure control signals were generated using the Advantech DAQNavi software and transmitted by two analogue outputs of the USB-4711 DAQ card.

In the first stage, the response to the step change of the set flow rate was examined. With different pressure values on the relief valve, the pump control signal was changed from zero to $e_p = 0.5, 1.0, \ldots, 2.5$ mm. The obtained time diagrams of flow rate and pressure are presented in Fig. 8 a) and b), respectively. The pressure increase against the time during the pump eccentricity adjustment for the three highest e_p values is shown in

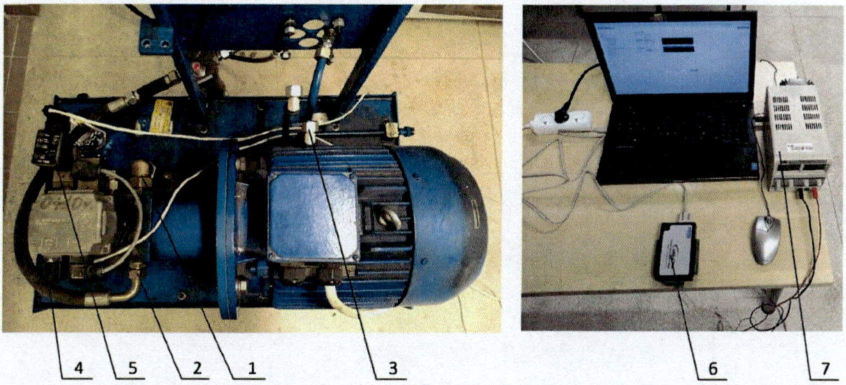

Fig. 7. Test bench: 1 – suction channel, 2 – pump output, 3 – return line, 4 – flow rate control, 5 – internal relief valve pressure control, 6 – Advantech DAQ card, 7 – power supply 24 V

Fig. 9. It can be observed, that the pressure adjustment time is within the range of 0.11 s to 0.13 s, which fits well with the results of simulations.

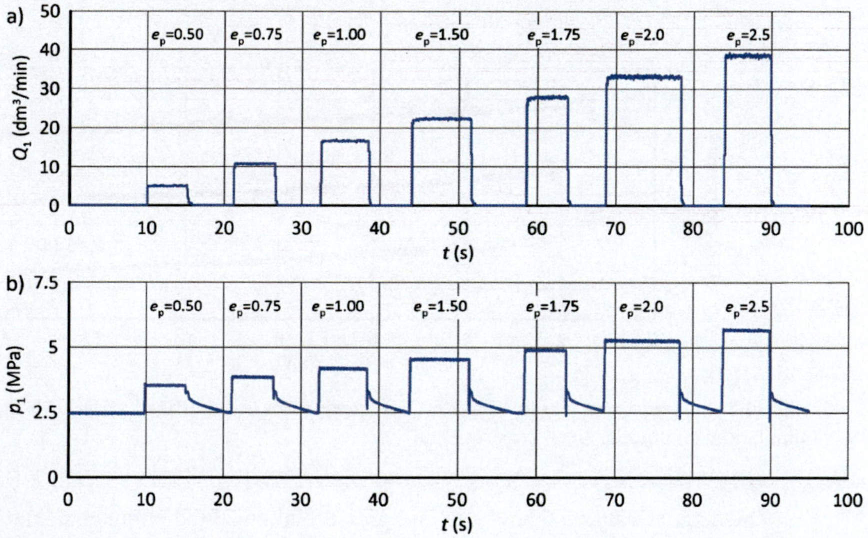

Fig. 8. Flow rate and pressure against time for different pump setting; a) flow rate, b) pressure

In the second stage, the dependence of the flow rate on the pressure in the supply line was examined. With the fixed ep value, the pressure was smoothly increased from the minimum to about 15 MPa and then decreased to the initial value. Based on the results, the $Q(p)$ function was obtained. The results of the flow rate and pressure measurements acquired for $e_p = 1.0$, 1.5 and 2.0 mm are shown in Fig. 10, while the obtained p-Q characteristic is shown in Fig. 11, respectively.

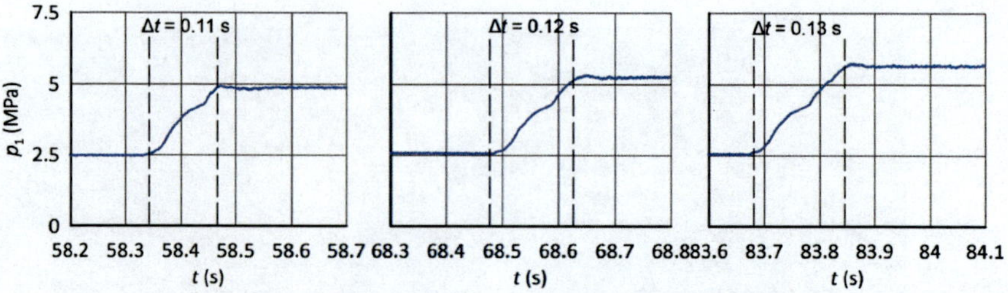

Fig. 9. Pressure increase during the pump flow rate adjustment for $e_p = 1.75, 2.0, 2.5$ mm.

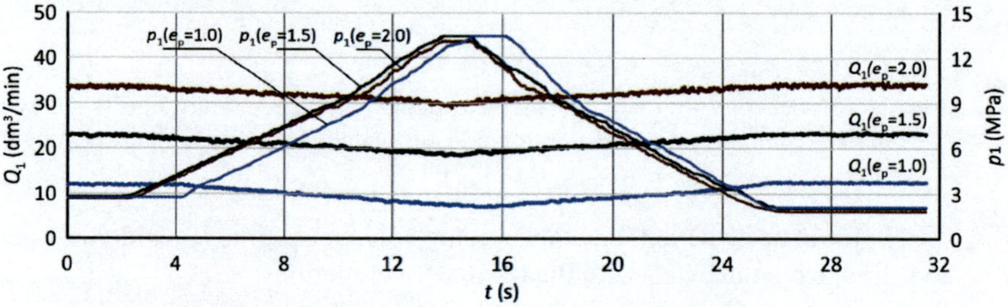

Fig. 10. The measurement results for the fixed e_p values and variable pressures.

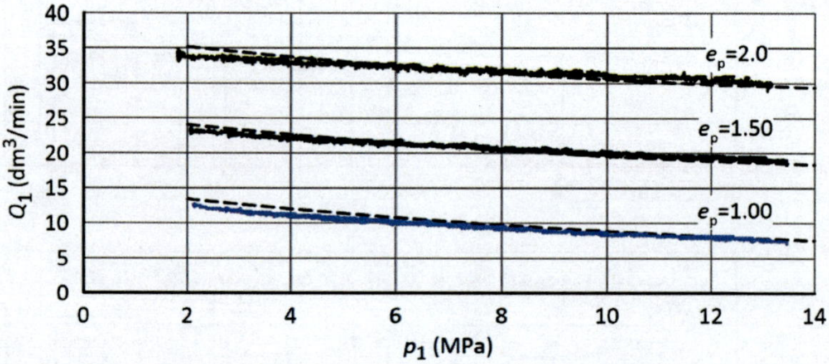

Fig. 11. Comparison of the obtained *p*-*Q* characteristics for the fixed e_p values: dotted line - simulation, continuous line - experimental.

The acquire values of flow rate and pressure allowed the pump characteristic to be formulated. The results show that, regardless of the pump setting, increase in the pressure from 2 to 14 MPa causes a linear drop in flow rate by approximately 4 dm^3/min.

4 Conclusions

The article presents simulations and bench tests of a vane pump, whose flow rate and pressure are controlled by means of electronic controllers built into its body. Pressure control is carried out using a pressure relief valve, while the flow rate is adjusted by changing the pump eccentricity. The eccentricity is electro-hydraulically controlled in a feedback loop through an electronic displacement transducer, since in the current state-of-art accurate measurement of the fast-changing dynamic flow rate is a difficult task. The carried out experiments indicate that the applied solution is characterized by high accuracy of the pump displacement setting, repeatability and the possibility of throttling-free control of the system start-up. This is especially useful in systems where a smooth start is required. With rapid changes in the flow rate setting, the shift time from zero to maximum eccentricity is relatively short compared to other pump types.

The presented test bench, including the control and data acquisition systems, can be used to test various hydraulic components and subsystems. The built-in electronic control system allows pressure losses to be minimized when the system is unloaded. Both flow rate and pressure can be quickly adjusted to the current needs of the receivers. Moreover, a quick pressure discharge in the supply line is also provided.

References

1. Blackmer, R., Pyk, M.: Sliding vane pumps. The benefits of sliding vane pump technology. World Pumps 2019 (3), 38–42 (2019). https://doi.org/10.1016/S0262-1762(19)30039-2
2. Yang, W., Wang, Y., Yang, Y., Zhang, W.: Design and experimental verification on performance of a novel integrated electro-hydraulic vane pump. Appl. Sci. **12**, 5006 (2022). https://doi.org/10.3390/app12105006
3. Fatigati, F., Di Bartolomeo, M., Cipollone, R.: Development and experimental assessment of a low speed sliding rotary vane pump for heavy duty engine cooling systems. Appl. Energy **327**, 120126 (2022). https://doi.org/10.1016/j.apenergy.2022.120126
4. Bianchi, G., Fatigati, F., Murgia, S., Cipollone, R.: Design and analysis of a sliding vane pump for waste heat to power conversion systems using organic fluids. Appl. Therm. Eng. **124**, 1038–1048 (2017). https://doi.org/10.1016/j.applthermaleng.2017.06.083
5. Bianchi, G., Fatigati, F., Murgia, S., Cipollone, R., Contaldi, G.: Modeling and experimental activities on a small-scale sliding vane pump for ORC-based waste heat recovery applications. Energy Procedia **101**, 1240–1247 (2016). https://doi.org/10.1016/j.egypro.2016.11.139
6. Ciurys, M., Fiebig, W.: Experimental investigation of a double-acting vane pump with integrated electric drive. Energies **14**, 5949 (2021). https://doi.org/10.3390/en14185949
7. Wang, J., Chen, Z., Yang, S., Li, H., Cui, S.: Geometric design and analysis of a novel sliding vane vacuum pump with three chambers. Mech. Mach. Theory **141**, 52–66 (2019). https://doi.org/10.1016/j.mechmachtheory.2019.07.003
8. Chen, Z., Wang, J., Cui, S., Haozhi Feng, H., Sha, R.: Numerical simulation and design methodology of a novel asymmetric cylinder profile for sliding vane vacuum pumps. Vacuum **169**, 108945 (2019). https://doi.org/10.1016/j.vacuum.2019.108945
9. Fornarelli, F., Lippolis, A., Oresta, P., Posa, A.: Investigation of a pressure compensated vane pump. Energy Procedia **148**, 194–201 (2018). https://doi.org/10.1016/j.egypro.2018.08.068
10. Lobsinger, T., Hieronymus, T., Brenner, G.: A CFD investigation of a 2D balanced vane pump focusing on leakage flows and multiphase flow characteristics. Energies **13**, 3314 (2020). https://doi.org/10.3390/en13133314
11. Guerra, D., Battarra, M., Mucchi, E.: Kinematics and geometrical admissibility conditions of balanced vane pumps with twin lip vanes. Mech. Mach. Theory **167**, 104534 (2022). https://doi.org/10.1016/j.mechmachtheory.2021.104534
12. Sun, Y., Xue, D., Liu, S., Wu, J., Bai, X.: Stator curvature optimization and analysis of axial hydraulic vane pumps. Energies **15**, 6229 (2022). https://doi.org/10.3390/en15176229

13. Doikin, A., Zadeh, E., Campean, F., Priest, M., Brown, A., Sherratt, A.: Impact of duty cycle on wear progression in variable-displacement vane oil pumps. Procedia Manuf. **16**, 115–122 (2018). https://doi.org/10.1016/j.promfg.2018.10.170
14. Georgiou, E.P., Drees, D., De Bilde, M., Anderson, M.: Pre-screening of hydraulic fluids for vane pumps: An alternative to vickers vane pump tests. Wear **404–405**, 31–37 (2018). https://doi.org/10.1016/j.wear.2018.02.022
15. Mucchi, E., Agazzi, A., D'Elia, G., Dalpiaz, G.: On the wear and lubrication regime in variable displacement vane pumps. Wear **306**, 36–46 (2013). https://doi.org/10.1016/j.wear.2013.06.025
16. Summer, F., Pusterhofer, M., Grün, F., Gódor, I.: Tribological investigations with near eutectic AlSi alloys found in engine vane pumps – characterization of the material tribo-functionalities. Tribol. Int. **146**, 106236 (2020). https://doi.org/10.1016/j.triboint.2020.106236
17. Fabis-Domagala, J., Domagala, M., Momeni, H.: A matrix FMEA analysis of variable delivery vane pumps. Energies **14**, 1741 (2021). https://doi.org/10.3390/en14061741
18. Battarra, M., Mucchi, E.: On the relation between vane geometry and theoretical flow ripple in balanced vane pumps. Mech. Mach. Theory **146**, 103736 (2020). https://doi.org/10.1016/j.mechmachtheory.2019.103736
19. Wang, S., et al.: Mechanical analysis of radial hydraulic force in oil recovery by sliding vane pump. J. Petrol. Sci. Eng. **213**, 110365 (2022). https://doi.org/10.1016/j.petrol.2022.110365
20. Hieronymus, T., Lobsinger, T., Brenner, G.: Investigation of the internal displacement chamber pressure of a rotary vane pump. Energies **13**, 3341 (2020). https://doi.org/10.3390/en13133341
21. Battarra, M., Blum, A., Mucchi, E.: Kinematics of a balanced vane pump with circular tip vanes. Mech. Mach. Theory **137**, 355–373 (2019). https://doi.org/10.1016/j.mechmachtheory.2019.03.034
22. Battarra, M., Mucchi, E.: Analytical determination of the vane radial loads in balanced vane pumps. Mech. Mach. Theory **154**, 104037 (2020). https://doi.org/10.1016/j.mechmachtheory.2020.104037
23. Stryczek, S.: Napęd hydrostatyczny, tom I, element, 4th edition. WNT, Warsaw. ISBN: 978–837–9261–444 (2013)

Hydraulic Systems

Volumetric Control of Hydrostatic Drives System According to the Principle of the Forced Flow

Piotr Woś(✉), Ryszard Dindorf, Jakub Takosoglu, and Łukasz Chłopek

Kielce University of Technology, Kielce, Poland
{wos,dindorf,qba,lchlopek}@tu.kielce.pl

Abstract. The article presents considerations of controlling the hydraulic power source according to the principle of forced flow. In systems with forced flow, the pump generates a flow of liquid regardless of the load level. A pump with a constant geometric capacity produces an appropriate liquid stream by changing the rotational speed of its drive. The use of a pump with constant capacity in the hydraulic system, driven at variable speed employing an adjustable electric drive powered by a variable frequency drive (VFD), ensures high dynamics of fluid stream changes, good control quality of the hydrostatic drive, energy efficiency and low noise emission. The use of a dedicated measurement system in the hydrostatic power system enables the development of control methods for hydraulic drive systems that maintain constant efficiency, pressure and power output. The conducted simulation and experimental studies show the legitimacy of using new control methods in hydrostatic power supply systems.

Keywords: volumetric control system · electro-hydraulic system · hydrostatic drive and control

1 Introduction

Volumetric capacity control, using a hydraulic setting device, can be divided into volumetric and pressure control. In the former case, the pump capacity setting is proportional to the volume of fluid supplied to the setting device, while in the latter it is proportional to the pressure [1, 2]. Hydraulic actuators in combination with various types of valves and manifolds are used as setting devices to control the pump output. Volumetric control of the operating speed of the motor or actuator is achieved by changing the flow rate of the stream feeding the consumer or by changing the absorptive capacity of the consumer. An example of such control is encountered in hydrostatic transmission. Hydrostatic transmissions can be built, as compact or loose units. Compact units mean that both the generating and the receiving volumes of the operating medium are located in a common body. Loose units mean that the pump and hydraulic motor are structurally separate. Loose hydrostatic transmissions can operate in an open or closed system. By open system, he means such a system in which the power pump in the hydrostatic transmission draws the medium from the tank, and the hydraulic motor discharges it into the tank.

J. Stryczek and U. Warzyńska (Eds.): NSHP 2023, LNME, pp. 83–92, 2024.
https://doi.org/10.1007/978-3-031-43002-2_8

In a closed system, the operating medium is in circulation between the pump and the motor. In such a system, there must be replenishment of volume losses and cooling of the working medium. Closed hydrostatic transmission can be built with adjustable and non-adjustable parameters and with one and two directions of rotation [3, 4]. Controlling the speed of a hydraulic actuator requires a valve in the system that allows a continuous change in the flow rate of the working fluid or a variable displacement pump. The change in capacity can be done by changing the pump's unit capacity or by changing its speed [5].

Hydraulic pumps and motors have some limitations when it comes to low angular velocity. Both motors and hydraulic pumps should be operated above their minimum stable angular velocity. When selecting hydraulic power systems, it is important to remember that different models of hydraulic pumps may have different specifications and performance. Therefore, it is important to understand the requirements and limitations of a specific hydraulic pump and motor before using it in your application.

The basic assumption of this work is the use of a constant displacement pump in the hydraulic system, driven at variable speed by an adjustable AC electric drive, instead of the classic solution with a variable displacement pump. Which will ensure sufficient or even high dynamics of changes in the liquid stream, good quality control of the hydrostatic drive and low noise emission.

2 Operating Principle and Mathematical Modelling

The control parameter of a hydraulic pump or motor is the quantity that determines the value of the pump capacity or motor absorption. Volumetric control of the speed of receivers (motors) is possible by changing the capacity of the pump or changing the absorption of the motor. In the case of a fixed-volume pump and motor station, the only way to change the pump capacity is by changing the speed of the electric drive motor using a frequency converter. So far, mainly two types of motors have been used in electric drives, i.e. induction (AC) or classic DC [6]. Induction motors have good operating and control characteristics. The great popularity of three-phase induction motors is also influenced by design and technological advances in the construction of solid-state power components. In evaluating an electric pump drive, the following parameters are important: angular speed control time, torque overload resistance, low value of torque and angular speed pulsation, measurement requirements in implementing control systems and availability of ready-made control system applications.

A schematic of a volumetric control system with units (pump and motor) with a fixed working volume is shown in Fig. 1. The drive of a constant geometric volume pump consists of an inverter drive (frequency converter) (1) and an AC electric motor (3). The voltage analogue signal at the input is proportional to the value of the output frequency f of the VFD. The drive unit consists of a hydraulic motor (4) with a coupled load pump. Flow measurement in the hydraulic system is carried out by a flow meter (6) equipped with a voltage transducer (7).

Using the flow characteristics of the pump ($Q_{p\text{-}n}$), it is possible to control the volumetric flow rate Q_p. The operation of the frequency converter in torque control mode allows quasi-control of the supply pressure of the hydrostatic system. In addition, the

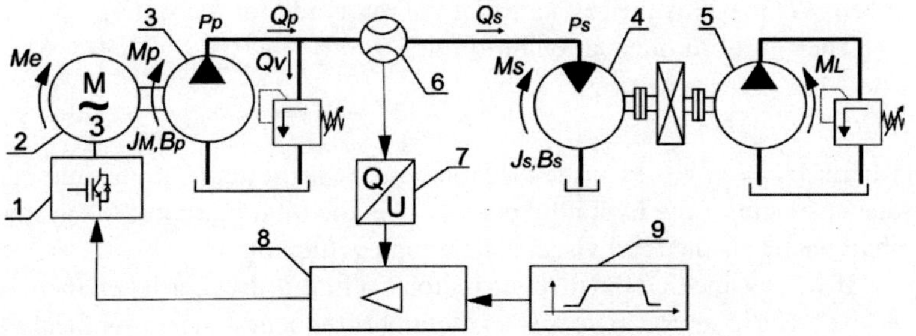

Fig. 1. Schematic diagram of the hydrostatic system: 1 - Variable-frequency drive (VFD), 2 - AC motor, 3 - Fixed displacement pump; 4 - Hydraulic motor, 5 - Motor load pump, 6 - Flow meter, 7 - Analog flow transducer, 8 - Controller, 9 - Reference signal generator.

introduction of compensation for changes in the mechanical-hydraulic efficiency of the pump as a function of load can improve the static accuracy of the set pressure. The application of flow feedback Q_p enables the regulation of the volumetric flow rate. On the other hand, using the properties of the inverter drive (VFD), it is possible to realize the function of the power limiter or operation according to the principle of constant power [9].

2.1 Variable Speed Pump Link

In practical applications of AC motor angular velocity or torque control systems, the following methods are most commonly used: scalar, direct torque control DTC, (Direct Torque Control) and field-oriented control FOC (Field Oriented Control) [7]. In the previous considerations, it was assumed that the rotation reference signal was input using a voltage signal U.

On the other hand, the characteristic $U/f = const.$ is the relationship between the output voltage and the output frequency [8]. It is assumed that the inverter is a proportional element, that is:

$$f_p = k_u u_p, \tag{1}$$

where: k_u is the gain factor of the frequency converter, u_p is the input voltage of the frequency converter (VFD), and f_p the output frequency of the VFD.

The electromagnetic torque of the electric motor is:

$$M_e = k_{e1} U - k_{e2} \omega_s, \tag{2}$$

where: $U = k_f f_p$ is the stator phase voltage, k_f is the gain factor for the frequency/voltage transformation, and ω_s is the angular velocity of the electric motor. The relationships: k_{e1}, k_{e2} are described in (3) and (4),

$$k_{e1} = \frac{3N}{2\pi R} k_f, \tag{3}$$

$$k_{e2} = \frac{3N^2}{2\pi R^2} k_f^2, \tag{4}$$

where: $N = 1$, R is the resistance of the rotor side for the stator.

The equilibrium equation of torque is:

$$J_M \dot{\omega}_p = M_e - M_p - B_p \omega_p, \tag{5}$$

where: $M_p = p_p V_p / 2\pi$ is the load moment, p_p is the actual pump output pressure, V_p the displacement of the hydraulic pump, J_M is the total inertia of the system for the motor shaft and B_p is the total viscous damping coefficient.

If we assume that the inverter-motor is a first-order inertial element, and the velocity of the electric motor increases as the input frequency increases then we can transform Eqs. (1), (2) and (3) and get:

$$\omega_p = \frac{\frac{k_f k_{e1}}{k_{e2}+B_p} f_P - \frac{V_P/2\pi}{k_{e2}+B_p} p_P}{\frac{J_M}{k_{e2}+B_p} s + 1}, \tag{6}$$

On the other hand, the flow rate of a variable speed pump is:

$$Q_p = V_p \omega_p - C_p p_p, \tag{7}$$

where: C_p - the total leakage rate of the pump (Table 1).

Table 1. Parameters for the electric motor-pump system (speed pump link)

Symbol	Value	Unit
up	0–10	V
kf	4.4	
ke1	4.82	
ke2	3.36	
R	0.437	Ω
JM	0.0015	kg/m2
B_p	0.0128	N m s/rad
Cp	0.002	l/min bar
Vp	8.4	cm^3/rev

2.2 Hydraulic Motor Link

In leaky control mode, the system flow balance equation is:

$$Q_p - Q_v = Q_s = C_s p_s + \omega_s V_s \tag{8}$$

where: C_s is the leakage rate of the hydraulic motor, p_s is pressure in the motor input line, Q_v is the flow rate through the pressure valve, ω_s is the angular velocity of the hydraulic motor and V_s is the displacement of the hydraulic motor.

The equilibrium torque equation of a hydraulic motor is:

$$M_s = J_s\dot{\omega}_s + B_s\omega_s + M_l \tag{9}$$

where: J_s is total inertia related to the shaft of the hydraulic motor, B_s is the coefficient of viscous damping, M_l external torque (Table 2).

Table 2. Parameters for the pump - hydraulic motor link system.

Symbol	Value	Unit
Js	0.0003	kg/m2
Bs	0.0128	N m s/rad
Cs	0.002	l/min bar
Vs	5	cm3/rev
Ml	13.6	Nm

An analysis of changes in the operating parameters of a hydrostatic transmission (Fig. 2 and Fig. 3) with adjustable pump parameters (primary control) was carried out.

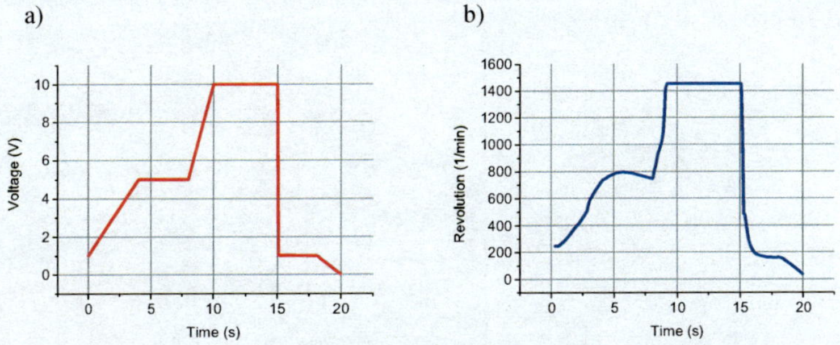

Fig. 2. Graphs of input voltage signals of the frequency converter (a) and rotation speed $n_s = \omega_s 30/\pi$ of the electric motor.

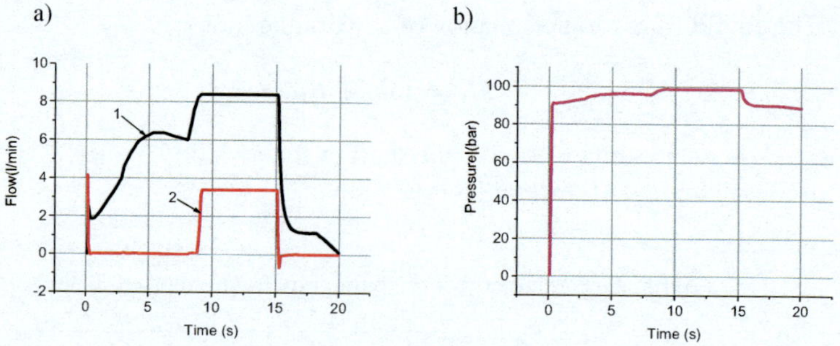

Fig. 3. Simulation diagrams of flow rate signal (a) of hydraulic motor Q_p (1) and hydraulic pressure relief valve Q_v (2), p_s hydraulic motor input pressure (b).

3 Experimental Studies

In the Department of Mechatronics and Armament, Kielce University of Technology, an experimental hydraulic power pack has been built, consisting of a gear pump with a unit displacement of 8.4 cm^3/rev driven by a 4 kW AC electric motor.

($n_s = 1450$ 1/min) powered by an inverter (Astraada DRV-28A). The control system of the hydraulic power pack is implemented using a PLC (Astraada One).

Figure 4 shows views of the hydraulic power pack (1) with the attached motor-hydraulic pump drive (3), load gear pump (4), rotation sensor (5), measurement unit (2) and control system (6).

Fig. 4. View of the electrohydraulic drive laboratory test stand: 1 - Hydraulic power pack with *AC* electric motor ($n_s = 1450$ 1/min) and hydraulic pump ($V_p = 8.4$ cm^3/rev), 2 - Flow rate and pressure measurement module (Hydac HMG 3000), 3 - Hydraulic axial piston motor ($V_s =$ 6.4 cm^3/rev), 4 - Load gear pump ($V_l = 6.4$ cm^3/rev), 5 - Rotation sensor, 6 - PLC control system (Astraada).

The hydraulic power unit pump is a gear pump with constant capacity and internal gearing, where the minimum stable rotational speed given by the manufacturer is 500 rpm. The hydraulic motor (3) is a low initial speed of rotation 90 rpm axial piston motor. The hydraulic load pump (4) is also an internal gear pump.

The frequency converter used in the system to power the AC motor is a modern vector inverter with a maximum power of 4kW. It has the ability to use one of the many communication options in the form of cards such as Profinet, EtherCAT, Profibus DP, WiFi, Modbus TCP, Ethernet/IP, and facilitates the integration of inverters with the control system (PLC controller) of the test stand.

In addition, the software used in the controller provides a real-time display of measured values and processed parameters mainly pressure, flow rate and speed. The main screen, in addition to graphic and digital displays, includes bars: a menu, function buttons and alarm indicators. Based on the measurement of basic parameters, such as pressure, flow rate and rotational speed, it is possible to calculate indicators of the condition of equipment and hydraulic system. Implemented calculation functions make it possible, for example, to calculate the efficiency of pumps, actuators and motors, flow parameters or pressure losses in hydraulic systems.

Figure 5 shows the time waveforms of flow rates Q_s and pressure p_s at the inlet of a loaded hydraulic motor with a pump operating at 30 bar.

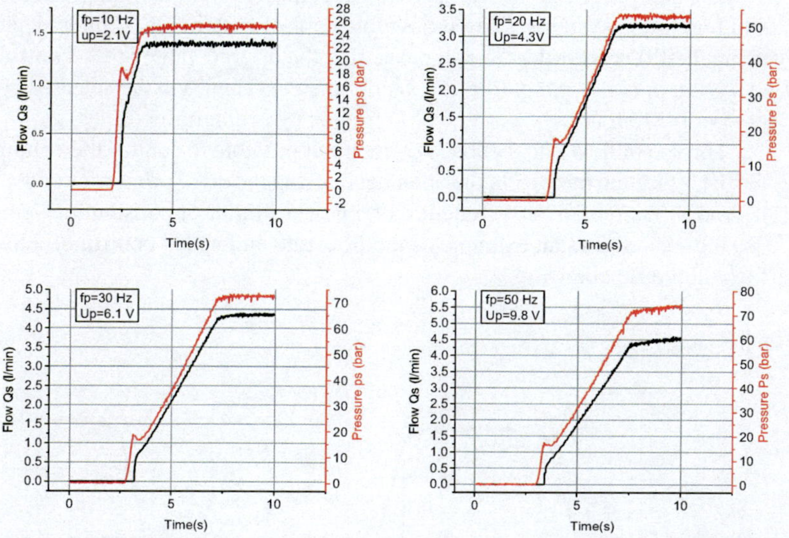

Fig. 5. Step characteristics of flow rate and pressure for a loaded hydraulic system.

In laboratory tests, the drive's time waveforms were analyzed when the inverter's frequency control voltage set point was pulsed, respectively, in the range of $u_p = 2.1$–9.8 V, which corresponded to the frequency range VFD $f_p = 10$–50 Hz.

3.1 Controlling the Flow of the Hydraulic System

To control the flow rate in the presented hydraulic system, control systems are used. This system maintains a constant flow rate by appropriately controlling the rotational speed of the pump.

Direct control of the flow rate Q_S (volumetric control) is carried out in the system, the diagram of which is shown in Fig. 6.

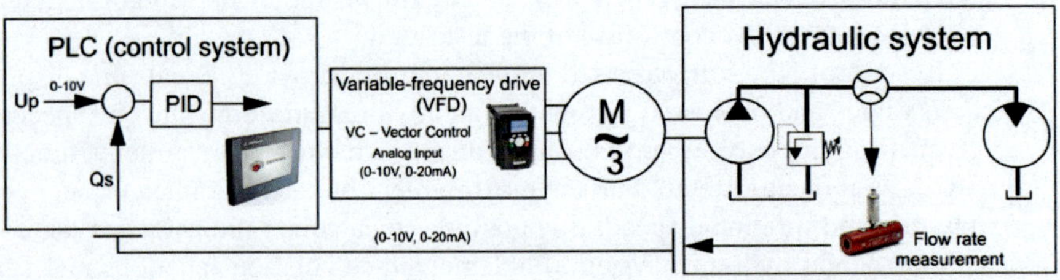

Fig. 6. Block diagram of the hydraulic flow rate control system (volumetric control).

The desired value of the controlled angular velocity was obtained by applying a voltage in the range of 0–10 V to the input of the VFC controller.

The controller was used as an external control system to implement the basic control and regulation systems. Closed flow rate control system Q_s is implemented in the PLC using the PID controller. The signal setting the desired value of the control voltage is set using the operator panel (digital potentiometer). Then, via the selected communication protocol (Modbus), the correct VFD frequency is determined.

The use of flow rate feedback Q_s makes it possible to control the volumetric flow rate for the hydraulic motor. On the other hand, using the drive properties of the VFD inverter, it is possible to realize the function of a power limiter or constant power operation.

Figure 7 shows an example of the flow rate and speed control of a hydraulic motor for volumetric control.

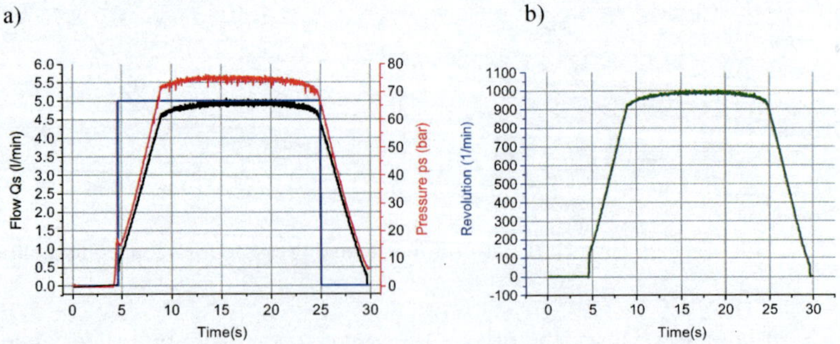

Fig. 7. Flow rate control of a loaded hydraulic system: a) flow rate Q_S and pressure p_S waveforms of the measurement signal, b) rotational speed signal of the hydraulic motor.

The control system uses a variable-frequency drive (VFD) with a built-in application of the AC motor angular velocity and torque control system. In the application, the control settings can be entered manually or use the autodetection of the dynamic parameters of the drive.

For this hydraulic station with a fixed displacement pump and motor, the only way to change the pump capacity is by changing the rotational speed of the electric drive motor using a frequency converter. The analogue signal at the output (u_p the voltage of the control current and its rotational speed n_s) is proportional to the value of the output frequency f_p of the frequency converter (VFD).

The analyzes carried out show that there is an approximately proportional relationship between the angular speed of the hydraulic motor and its absorption capacity. In addition, there are approximately proportional relationships between the pump capacity and the angular speed of the AC motor. This relationship can be used to appropriately control the speed of the electric motor driving a fixed-volume positive displacement pump, which in many cases will replace inefficient throttle control. More so that the development of the construction of electric motors, power electronic converters (VFD), control methods and microprocessor systems enables effective and energy-saving control of the operating parameters of hydraulic systems.

The presented control system is a relatively simple solution, but taking into account the limitations related to the drives, i.e. the minimum stable rotational speed and generated volumetric losses, they can be successfully used in industrial applications. It should also be noted that flow rate control generates power losses, and the flow rate depends on the differential pressure in the branch of the hydraulic system and the viscosity of the hydraulic fluid, which decreases with increasing temperature. Then it is reasonable to take into account these changing operating conditions in the control systems of hydrostatic drives.

4 Summary

This paper presents the results of an analysis of flow control in a hydraulic system, with a pump of constant geometric volume, by controlling the speed of an induction motor. The induction motor was powered by a voltage inverter. The study was aimed at evaluating the possibility of effectively controlling the flow or speed of a hydraulic system without using a proportional throttling valve. The possibility of controlling the speed of a hydraulic pump by measuring the flow conditions at its output was analyzed. Flow control by controlling the speed of the motor driving the pump has good control characteristics and can be fully competitive with flow control by using a proportional valve. In hydrostatic stationary drives, where variable unit capacity pumps are used for some reason, a power supply system with an adjustable electric drive can be used conditionally. This results in a more energy-efficient and quiet drive, as well as greater possibilities for machine automation and the use of a PLC. In a hydrostatic drive with an adjustable electric drive of the pump, there is a combination of the advantages of both types of drive in a system usually equipped with a digital interface, with the possibility of communication over a standard industrial bus. This advantage becomes particularly important when the drive is integrated into the machine's overarching control strategy. A non-adjustable pump system with an adjustable electric drive requires the development of a suitable way to control the applied hydraulic system.

References

1. Akkaya, A.V.: Effect of bulk modulus on performance of a hydrostatic transmission control system. Sadhana - Acad. Proc. Eng. Sci. **31**, 543–556 (2006)
2. Danh, D.N., Aschemann, H.: Discrete-time state-dependent proportional-integral control for torque tracking of hydrostatic transmissions. IFAC-PapersOnLine **53**, 6591–6596 (2020)
3. Dindorf, R., Woś, P.: Development of energy efficient hydrostatic drives with energy recovery. Mechanik **90**, 776–782 (2017)
4. Prabel, R., Aschemann, H.: Torque control of a hydrostatic transmission using extended linearisation techniques. In: Proceedings of the 15th Scandinavian International Conference on Fluid Power, SICFP 2017, 7–9 June 2017 - Linköping, Sweden, vol. 144, pp. 352–359 (2017)
5. Schulte, H., Gerland, P.: Observer-based estimation of pressure signals in hydrostatic transmissions. IFAC Proc. **43**, 425–430 (2010)
6. Skorek, G.: Study of losses and energy efficiency of hydrostatic drives with hydraulic cylinder. Polish Marit. Res. **25**, 114–129 (2018)
7. Stefanski, T., Zawarczynski, L.: Analysis of inverter-fed drive of hydraulic pump in volumetric control system. In: 2016 21st International Conference Methods Models in Automation and Robotics, MMAR 2016, pp. 146–151 (2016)
8. Tarasov, O., Scherbachev, P.: Development of BLDC electric motor control system in hydraulic servo drive based on variable hydrostatic transmission. Sci. Educ. Bauman MSTU. **14** (2014)
9. Yanishevskyy, V.: Basic principles of operation and energy balance of volume hydraulic drive. Her. Khmelnytskyi Natl. Univ. Tech. Sci. **309**, 115–118 (2022)

Control Strategy for a Hardware in the Loop Test Bench

Christian Haas(✉) and Katharina Schmitz

Institute for Fluid Power Drives and Systems (ifas), RWTH Aachen University, Aachen, Germany

Christian.haas@ifas.rwth-aachen.de

Abstract. This paper presents a control concept for a hardware-in-the-loop test rig. The aim is to develop optimized control concepts for aircraft fatigue tests, without risking component damage. The test rig simulates a mechanical structure´s response on a force-controlled axis through two servohydraulic axes. Since speed and position have a significant influence on the force control, both must be controlled as precisely as possible and independently of the force control. Classic single-loop PID controllers are not sufficient in this case. Therefore, a control concept consisting of a position controller and a velocity feedforward control using a characteristic map for partial linearization of the controlled system is proposed.

The performance of the controller is compared with a single-loop PID control through step tests, and its influence on system dynamics is analyzed. The results show that the proposed control concept leads to better control behavior without noticeable overshooting compared to the PID control. The use of a characteristic map for load pressure correction also simplifies the mapping of non-linear behavior of hydraulic valves and determination of suitable parameters for feedback only considering information usually available from data-sheets.

Keywords: aircraft fatigue testing · hardware in the loop · hydraulic control

Nomenclature

A	Piston area
B	Valve coefficient
F_L	Load force
K_F	Force feedback gain
$K_{\mathrm{Li}\dot{\mathrm{x}}}$	Velocity proportional leakage
$K_{\dot{\mathrm{x}}}$	Velocity feed forward gain
p_0	Supply pressure
p_A	Pressure chamber A
p_B	Pressure chamber B
p_L	Load pressure
p_{ref}	Reference pressure
p_T	Tank pressure
Q_{ref}	Reference volume flow

J. Stryczek and U. Warzyńska (Eds.): NSHP 2023, LNME, pp. 93–103, 2024.
https://doi.org/10.1007/978-3-031-43002-2_9

T_H Hydraulic time constant
T_M Mechanical time constant
V_H Hydraulic amplification
V_M Mechanical amplification
V_{Qy} Volume flow pressure gain
x Displacement
x_{set} Displacement set point
y Valve opening
y_{max} Maximum valve opening

1 Introduction

Applying load forces using hydraulic axes can be challenging. Especially when applied to systems with low stiffnesses and therefore large motions, with low natural frequencies, or a strong coupling between different controlled axes. An example of this is fatigue testing of aircraft structural components, which are a critical part of the certification process for new aircraft series. In these tests, the aircraft structure is subjected to load cycles through servo-hydraulic axes, simulating approximately three times the service life. Due to influences such as mutual interference of the actuators or natural oscillations of the components, the force control concepts currently used only allow slow load application, which increases the effort and cost of these tests and can significantly delay the certification process.

To investigate optimized load impingement concepts without the risk of component damage, while at the same time reproducing the behavior of the hydraulic drives as realistically as possible, a hardware-in-the-loop test rig is set up at the Institute for Fluid Power Drives and Systems (ifas). This simulates the behavior of corresponding tests and imprints the component response on a force-controlled axis by means of further servohydraulic axes.

Speed and position of the force-controlled axis represent a decisive influencing variable and must therefore be controlled as precisely as possible and independently of the force control. Classic single-loop PID controllers are to some extend not sufficient here. The use of model-based controllers, such as flatness-based controllers, or model predictive controllers, might be a solution. But such control schemes require complex modeling, especially for hydraulic drives due to the numerous non-linearities and parameter uncertainties and therefore might lead to unsatisfying control behaviors [1]. Using artificial intelligence to reduce the modelling effort might solve this problem, but requires system data, extensive training of the algorithm and proving stability is challenging. [2, 3, 4].

Therefore, this paper presents a control concept consisting of a position controller and a velocity feedforward control using a characteristic map for partial linearization of the controlled system. The performance of the controller is validated against a single-loop PID control and analyzed with respect to the influence of the control concept on system dynamics.

2 Test Rig

To optimize force control concepts for aircraft fatigue tests, a hydraulic hardware-in-the-loop test rig is set up at ifas. The test rig consists of three servo-hydraulic axes as shown in Fig. 1.

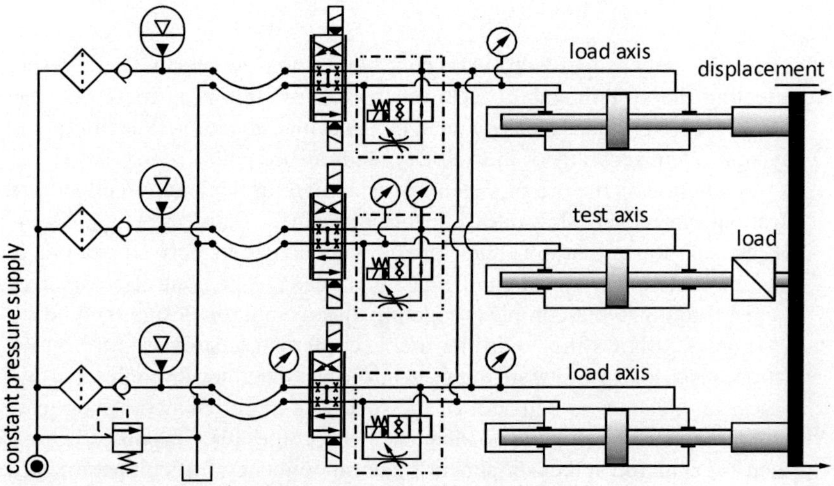

Fig. 1. Schematic diagram of the hydraulic circuit

The outer two axes imprint the response of any model on the middle axis in a path-controlled manner. The middle axis represents the force-controlled test axis. All three axes are fixed on one side in a machine bed and connected on the other side via a crossbeam that can move freely in the direction of the rod. The piston surfaces of the outer two cylinders correspond to the piston surface of the middle one to prevent the risk of cavitation or exceeding the maximum operating pressure. To avoid coupling of the axes via the supply line, all axes also have accumulators upstream of the servo-valves.

The measured state variables from the system include the position of the crossbeam, the force between the crossbeam and the test axis, the supply and tank pressures, and the chamber pressures of the load axes and the test axis.

3 Control Architecture

As depicted in Fig. 2, there exists a correlation between valve opening, load force, and speed for the control system of a servo-valve and a constant velocity cylinder. This implies that the force control of hydraulic actuators is substantially affected by the speed generated by the excitation of the loaded system. As force control is being examined on the hardware-in-the-loop test rig, it is critical to provide the test axis with an as precise as possible velocity profile.

While a multi-loop control with speed path is feasible, implementing closed-loop velocity control may trigger additional instabilities or result in poor system dynamics.

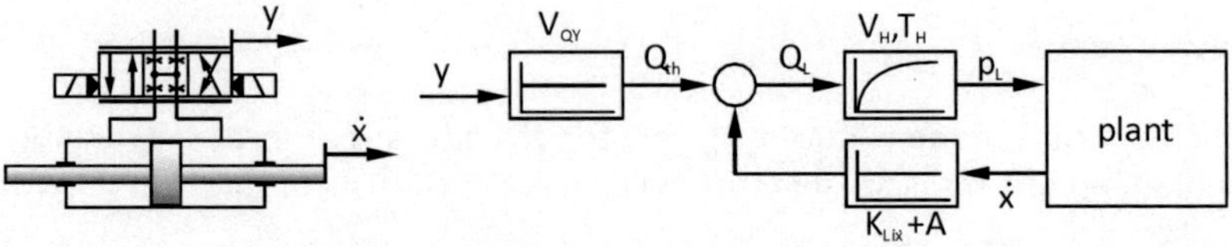

Fig. 2. Control system valve linear motor

Therefore, only using feedforward control is optimal, due to the direct response, without affecting the system stability. However, the flow rate is restricted by the differential pressure across the valve, as dictated by the orifice equation, thus limiting the attainable dynamics and accuracy of the velocity feedforward path.

A solution, is the use of a characteristic map for the load force/load pressure, valve opening and velocity/flow rate triplet occurring in the static case [1, 5]. A major advantage of this solution is that the characteristic diagram can be derived from data sheet information, is descriptive and can be easily validated. The control parameters themselves can be found without complex modeling, using common design methods.

Figure 3 shows the controller architecture of the hardware-in-the-loop test rig. In fatique tests, load profiles are imposed on the test specimen through servo-hydraulic axes. Since the mechanical behavior of the specimen might change during testing, the load profiles consist in force and not displacement profiles [6]. This poses various challenges, such as excitation at the component's natural frequency, difficult parameterization of the control system, or strong interactions between the axes. These properties are represented by a model simulated in real-time. For most investigations a state space model is sufficient here.

The principle of the hardware in the loop test bench is that one of the hydraulic axes runs as a real axis in a real force control loop, while the other axes and the tested mechanical structure are simulated. The model specifies set values for speed and position, and the measured force at the test axis is in turn incorporated into the model.

The actual position control structure consists of a velocity feedforward via the characteristic map, as well as an outer PI position controller for correcting the model error. The characteristic map gets the velocity setpoint and the load pressure, calculated from pressure measurements in each cylinder chamber as input.

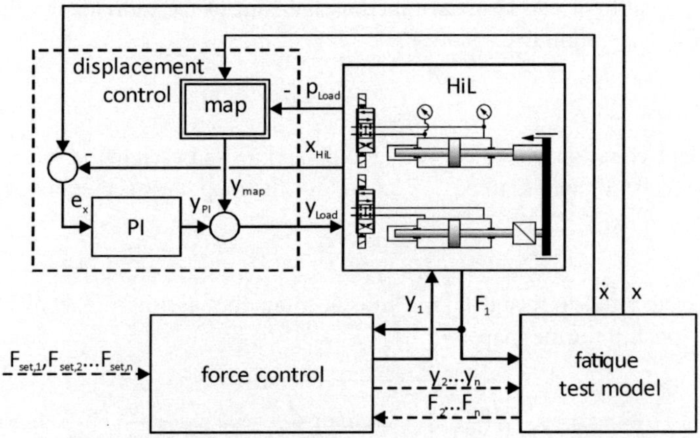

Fig. 3. Control Architecture of the Hardware in the Loop test rig

3.1 Derivation of the Characteristic Map

According to [7] and as shown in Fig. 4, a flow rate can be described for the control system linear-motor-valve for each control port as a function of the valve opening and the differential pressure across the control port. For the following derivation, non-linearities or influences due to temperature, for example, are neglected and the valve behavior is combined in the valve coefficient B. In addition, the valve is idealized and assumed to be zero-covered.

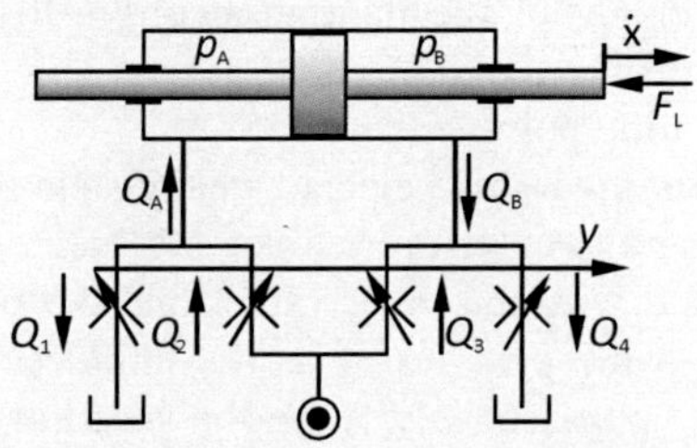

Fig. 4. Control system linear-motor-valve

The volume flow rates over the control edges for $y > 0$ result in:

$$Q_1 = 0; Q_2 = B \cdot y \cdot \sqrt{p_0 - p_A}; Q_3 = 0; Q_4 = B \cdot y \cdot \sqrt{p_B - p_T} \tag{1}$$

As well as for $y < 0$ to:

$$Q_1 = -B \cdot y \cdot \sqrt{p_A - p_T}; Q_2 = 0; Q_3 = -B \cdot y \cdot \sqrt{p_B - p_0}; Q_4 = 0 \tag{2}$$

With the valve coefficient B, assuming a linear valve characteristic:

$$B = \frac{Q_{ref}}{y_{max} \cdot \sqrt{p_{ref}}} \tag{3}$$

Q_A and Q_B can be determined as the sum of the individual volume flows and are used in the continuity equation.

$$\dot{x}A = Q_A = Q_B \tag{4}$$

The equation can be converted to p_A and p_B and used in the force equilibrium. Since only the static part is to be considered for the map, the acceleration can be set to 0.

$$\ddot{x}m = 0 = A(p_A - p_B) - F_L \tag{5}$$

Converted to $\dot{x}$ and taking into account the sign of y, the formula results in the relationship for the map.

$$\dot{x} = y\frac{B}{A}\sqrt{\frac{p_0}{2}}\sqrt{1 - sign(y)\frac{F_L}{A \cdot p_0}} = y\frac{B}{A}\sqrt{\frac{p_0}{2}}\sqrt{1 - sign(y)\frac{p_L}{p_0}} \tag{6}$$

In addition to the supply pressure p_0 and the piston area A, the entire map can be described via the valve coefficient B and can therefore be described purely in terms of generally available data. However, the characteristic field equation is only valid for servo valves with an approximately linear characteristic curve. In other cases creating the map from measurements might be required.

3.2 Map Optimization

The valve coefficient B is determined from the valve's data sheet information and therefore does not contain any information on pressure drops downstream of the valve. Figure 5 shows the characteristic diagram according to the data sheet. In addition, steady-state operating points measured on the test rig are plotted, as well as an optimized map fitted to these operating points.

For this purpose, randomly selected steady-state operating points were determined for valve opening and counterforce. Since all three cylinders in the test setup considered here were fitted with the same valves and the two outer cylinders together have approximately the same piston area as the middle cylinder, the achievable speed of load application is approximately twice the speed of the tested cylinder for the same pressure difference across the valves. Therefore, the adjustment range of the valves for load impingement was limited accordingly. Since the clean adjustment of the operating points is challenging, especially at high speeds and high forces for force control, operating points at which the steady-state operating point was reached for at least 100ms were considered. Since controlling forces at high speeds is challenging a backstepping controller for force control was used, which is not focus of this contribution [8]. Using the MATLAB MSB Toolbox, the map can be fitted to the measured values via Gaussian regression for optimizing the valve coefficient B.

As can be seen in Fig. 5, there is a deviation between the map determined from the data sheet and the fitted map, of about 20%. This is meanly due to the fact, that with only the valve datasheet effects like, pressure losses, or friction in the cylinder are not covered. So by only using the datasheet data, the controller might overestimate

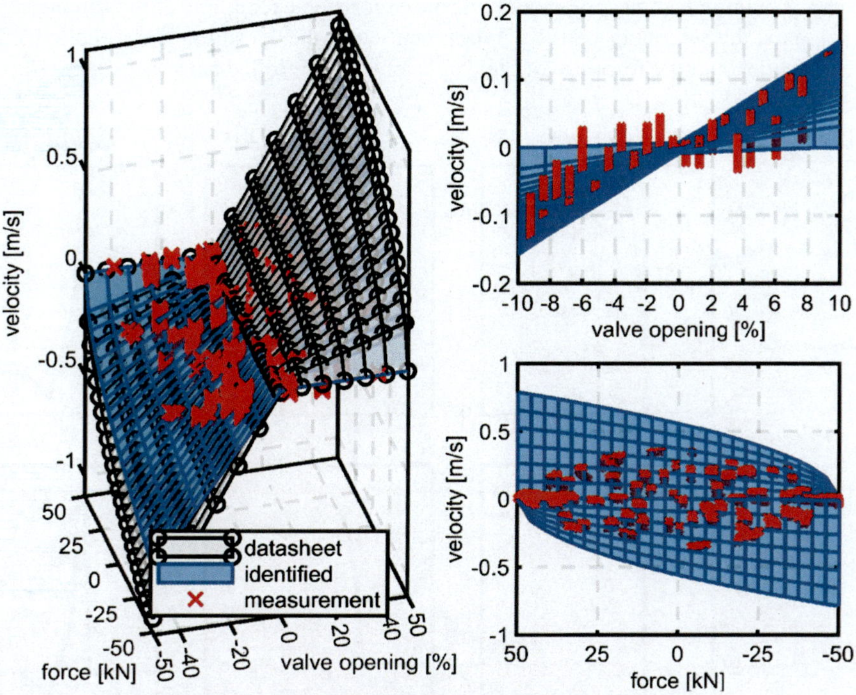

Fig. 5. Calculated and identified characteristic map

the cylinder speed. Also the measured data deviate noticeably from the identified map, especially around the center position of the valve. The reason for this is in particular the map´s assumption of zero overlap, which does not apply in reality and stribec friction for low cylinder velocities. Nevertheless, especially for low velocities the controller can tolerate the error, since only a small controller gain is required. A promising method for optimizing characteristic maps of hydraulic systems by using AI and measurement data is presented at [5]. Both maps can be used for the realization of a control system. Even a map with moderate deviations from reality represents an informative added value for the control. In the following the identified map is used.

4 Experimental Comparison

In order to evaluate the controller, a single-mass oscillator according to Eqs. (7) and (8) is used as the structure model.

$$\begin{pmatrix} \dot{x} \\ \ddot{x} \end{pmatrix} = \begin{pmatrix} 0 & 1 \\ -\frac{k_c}{m} & -\frac{k_D}{m} \end{pmatrix} \begin{pmatrix} x \\ \dot{x} \end{pmatrix} + \frac{1}{m} F \tag{7}$$

$$\begin{pmatrix} x \\ \dot{x} \end{pmatrix} = \begin{pmatrix} 1 & 0 \\ 0 & 1 \end{pmatrix} \begin{pmatrix} x \\ \dot{x} \end{pmatrix} \tag{8}$$

Figure 6 shows the model responses for step excitations of the force-controlled test axis, the set values of the displacement controller, resulting from the loaded structure model, and the response of the displacement-controlled axes.

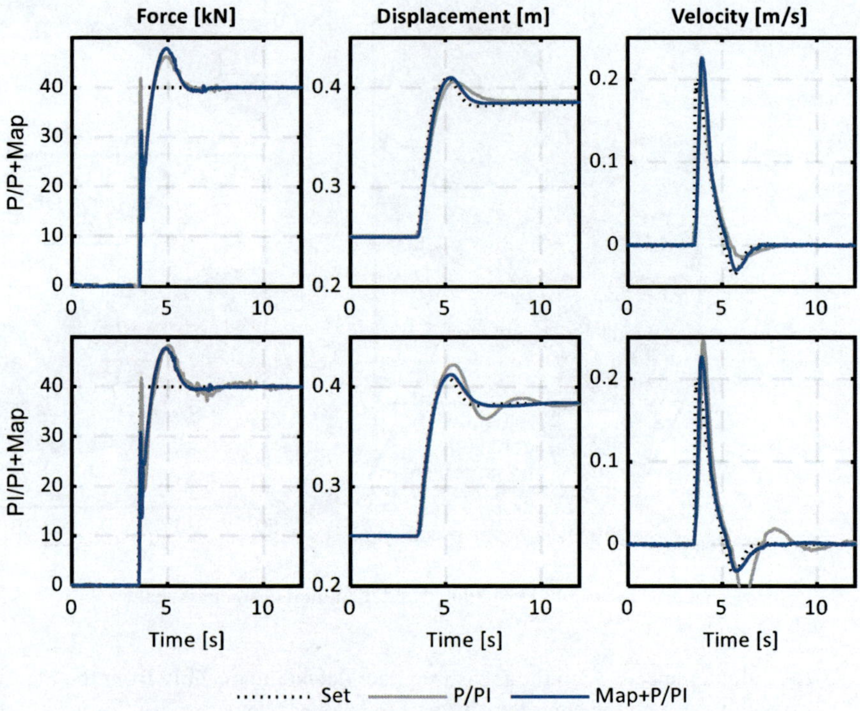

Fig. 6. Controller comparison

The spring stiffness k_C was selected so that 80% of the stroke is covered when the maximum cylinder force is applied to the unloaded system. Mass and damping were selected to give a natural frequency of 0.1 Hz with a decay rate of 0.7. Due to the low natural frequency, the velocities are limited to an acceptable level for the force control. As a reference for the map controller, curves for a P or PI control are also shown.

As shown in Fig. 6, the feedforward control can follow the setpoint signal well, both in terms of displacement and speed, especially in conjunction with a superimposed PI position controller. Especially in direct comparison with the PI controller, hardly any overshoot occurs. It has to be stated, that the overshoot of the PI controller is strongly dependent on the control parameters, but a reasonable compensation of the control error could not be achieved without overshooting. For further evaluating the controller performance Table 1 shows the ITAE values of the four cases, calculated according to 9, in relation to the speed and the position control.

$$\text{ITAE} = \int_0^\infty |e| \cdot t \cdot dt \tag{9}$$

Table 1. ITAE comparison

	P	PI	Map + P	Map + PI
Displacement	171.64	211.13	150.82	59.01
Velocity	161.15	419.26	189.79	93.21

The utilization of characteristic maps in control schemes yields improved performance, as evidenced by lower ITAE values. Specifically, employing the PI-controller in conjunction with the map results in an ITAE value approximately four times lower than that achieved with PI position control.

5 Influence on System Dynamics

The use of a pure speed feed-forward control does not influence on the system dynamics due to the principle of the non-existing feedback loop. In the case of the control concept presented here, there is no feedback of the speed, but there is feedback of the force or the load pressure, which is why there is an influence on the system dynamics. For the evaluation of the influence, the substitute system shown in Fig. 7 is considered in the following. The characteristic map is represented in the operating point by $K_{\dot{x}}$ and K_F.

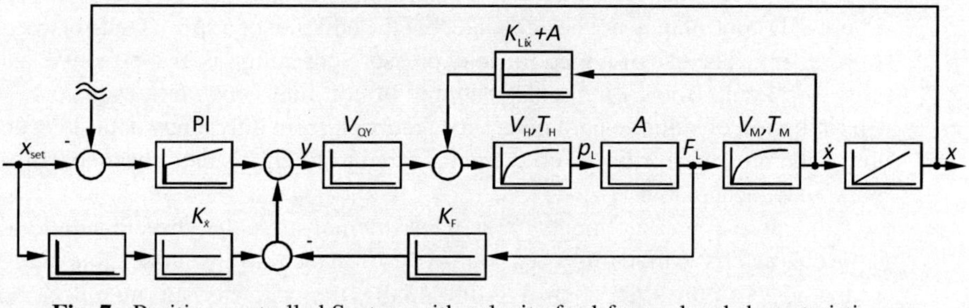

Fig. 7. Position controlled System with velocity feed-forward and characteristic map

It should be noted that the derivation of the velocity from the displacement in real applications is often done through a DT1 element, since the realization of control elements with more zeros than poles is not possible. This leads to additional poles in the natural frequency of the filter. Since velocity setpoints can be determined directly, as shown in the previous, or without phase delay by means of an observer, this is neglected in the following. The transfer function for the open loop is thus given by Eq. 10.

$$\frac{x}{x_{\mathrm{set}}} = \frac{s^2 \cdot \frac{V_{\mathrm{Qy}} K_{\dot{x}} V_{\mathrm{M}} V_{\mathrm{H}} A}{T_{\mathrm{M}} T_{\mathrm{H}}} + s \cdot \frac{K_{\mathrm{P}} V_{\mathrm{Qy}} V_{\mathrm{M}} V_{\mathrm{H}} A}{T_{\mathrm{M}} T_{\mathrm{H}}} + \frac{K_{\mathrm{I}} V_{\mathrm{Qy}} V_{\mathrm{M}} V_{\mathrm{H}} A}{T_{\mathrm{M}} T_{\mathrm{H}}}}{s^2 \cdot \left(s^2 + \mathrm{s} \cdot \frac{T_{\mathrm{M}} + T_{\mathrm{H}} + T_{\mathrm{M}} K_{\mathrm{F}} V_{\mathrm{Qy}} V_{\mathrm{M}} V_{\mathrm{H}} A}{T_{\mathrm{M}} T_{\mathrm{H}}} + \frac{1 + A V_{\mathrm{M}} V_{\mathrm{H}} (K_{\mathrm{Li}\dot{x}} + A + K_{\mathrm{F}} V_{\mathrm{Qy}})}{T_{\mathrm{M}} T_{\mathrm{H}}} \right)} \tag{10}$$

Neglecting the double integrating pole, which arises from the drifting of the path in the open loop case, as well as the integrating part of the controller, the natural frequency and the damping of the open loop can be described according to Eqs. 11 and 12.

$$\omega_0 = \sqrt{\left(\frac{1 + AV_{\mathrm{M}}V_{\mathrm{H}}\left(K_{\mathrm{Li}\dot{x}} + A + K_{\mathrm{F}}V_{\mathrm{Qy}}\right)}{T_{\mathrm{M}}T_{\mathrm{H}}}\right)} \tag{11}$$

$$D = \frac{T_{\mathrm{M}} + T_{\mathrm{H}} + T_{\mathrm{M}}K_{\mathrm{F}}V_{\mathrm{Qy}}V_{\mathrm{M}}V_{\mathrm{H}}A}{2 \cdot T_{\mathrm{M}}T_{\mathrm{H}} \cdot \sqrt{\left(\frac{1+AV_{\mathrm{M}}V_{\mathrm{H}}\left(K_{\mathrm{Li}\dot{x}}+A+K_{\mathrm{F}}V_{\mathrm{Qy}}\right)}{T_{\mathrm{M}}T_{\mathrm{H}}}\right)}} \tag{12}$$

The force feedback increases the damping and the natural frequency of the system. On the one hand, this can lead to better system behavior, but on the other hand, it can also overdampen the system and slow it down if the system properties are unfavorable. In this case, a pure velocity pre-control makes sense. The characteristic diagram thus simplifies to the characteristic curve at $p_{\mathrm{L}} = 0$. However, according to Eq. 10, neither velocity feedback nor force feedback have a negative influence on stability.

6 Conclusion

This paper presents a hardware-in-the-loop test rig and its associated control concept aiming to develop optimized control strategies for aircraft fatigue tests by imprinting the displacement and velocity response of a simulated test onto a hydraulic axis. The accuracy of both position and velocity imprinting, independent of load force, is crucial. A pure PID control may not be adequate, but the addition of a speed feed-forward control is promising. The use of a map for load pressure correction is also effective, as characteristic diagrams allow for simple mapping of non-linear hydraulic valve behavior and determination of suitable parameters for feedback from only a few data sheet specifications. The model error based on the considered test bench's data sheet specifications is found to be acceptable.

Comparative measurements for step tests demonstrate that the sub-supplementation of the characteristic map improves control performance without noticeable overshooting compared to a pure P/PI control. Additionally, the characteristic map does not negatively affect system stability but can lead to lower dynamics with unfavorable system parameters.

References

1. Anders, P., Ströbel, S.: User-oriented systematic of control concepts for fluidmechatronic servo drives. In: Proceedings of the 12th IFK, Dresden (2020)
2. Brumand-Poor, F., Matthiesen, G., Schmitz, K.: Control of a Hydromechanical Pendulum with a Reinforcement Learning Agent. In: Proceedings of the 13th IFK, Aachen (2022)
3. Brinkschulte, L., Graf, M., Geimer, M.: Reinforcement learning: a control approach for reducing component damage in mobile machines. In: Proceedings of the 12th IFK, Dresden (2020)
4. Demir, O., Ehlers, B., Bender, F., Trachte, A.: Learning based feed-forward control for advanced excavator assistance functions. In: Proceedings of the 13th IFK, Aachen (2022)

5. Helmke, M., Ströbel, S., Anders, P., Schulze, T.: Computer-assisted modeling and automatic controller adjustment for hydraulic drives based on an innovative nonparametric identification method. In: Proceedings of the 11th IFK, Aachen (2018)
6. Rossow, C., Wolf, K., Horst, P.: Handbuch der Luftfahrzeugtechnik. Carl Hanser Verlag GmbH (2014). ISBN 978–3–446–43604–6
7. Murrenhoff, H.: Servohydraulik – Geregelte hydraulische Antriebe. Shaker Verlag GmbH (2012). ISBN 978–3–8440–0947–7
8. Jeladi, M., Kroll, A.: Hydraulic Servo-Systems: Modelling, Identification, and Control, Springer, London (2003). https://doi.org/10.1007/978-1-4471-0099-7 ISBN 978–1–4471–1123–8

Model Based Design of the Hydraulic Actuator

Petr Noskievič[1(✉)] and Jan Milata[2]

[1] Faculty of Mechanical Engineering, VŠB-Technical University of Ostrava, 17. Listopadu 2172/15, Ostrava–Poruba, Czech Republic
petr.noskievic@vsb.cz

[2] Faculty of Electrical Engineering and Computer Science, VŠB-Technical University of Ostrava, 17. Listopadu 2172/15, Ostrava–Poruba, Czech Republic

Abstract. The paper is focused on the Model Based Design of the closed loop control of the linear hydraulic drive. Model-Based-Design is progressive methodology for design of control systems, embedded control systems and for testing of the implemented software. The hydraulic servo drives represent very important parts of the modern machines. The demands on their accuracy and dynamics require the use of the modern controller design approaches based on the computer simulation and comprehensive tools of their implementation. The novel created MATLAB-application named Hydraulic System Control–Model Based Design is introduced. The application concentrates the knowledge of the possible solutions of the closed loop control systems suitable for hydraulic drives and supports the use and tuning of the controllers different structures. The parameterization of the drive can be done using the known data from the data sheets of the servo valve and cylinder or can be based on the experimentally obtained characteristics which can be evaluated using the identification methods. The application is a user-friendly environment and creates also comfortable access to the MATLAB-toolboxes and their use. It simplifies the process of the controller design based on the implemented non-linear, linear respectively linearized mathematical models of the hydraulic drive, because the tools for the linearization of the non-linear model are available. The parameters of the controller can be verified by the simulation of the linear or non-linear model. The created MATLAB-application represents the Model-in-the-Loop (MIL) simulation of the control of the hydraulic drive and supports very important part of the development cycle called V-Model. Practical use of the application is demonstrated on the design of the closed loop position control of the hydraulic drive installed on the test rig. The simulated and measured data are presented as well as the results of the Hardware-in-the-Loop simulation of the control of the laboratory hydraulic drive.

Keywords: Hydraulic Servodrive · Closed Loop Control Design · Model Based Design · Model in the Loop simulation (MIL) · Hardware in the Loop simulation (HIL) · MATLAB Application Design · PID Controller · State Controller

J. Stryczek and U. Warzyńska (Eds.): NSHP 2023, LNME, pp. 104–113, 2024.
https://doi.org/10.1007/978-3-031-43002-2_10

1 Introduction

Model-Based-Design methodology is based on the use of the simulation using mathematical models of the controlled systems. Typically, the design starts with the simulation of the controlled system and controller in common software at one computer as a Model in the Loop simulation. Next step can be the generation of the controller software and simulation of the control as a Software in the Loop simulation, or Processor in the Loop simulation, if the code of the controller is running on the given part of the processor capacity. Next step can be the Hardware in the Loop simulation. The plant is typically simulated in real time on the external special computer for real time simulations and the real controller is connected using the physical signal to the real time model. Important advantage of the Model-Based Design is the verification of the control algorithms within a simulation [10].

Model-based design workflow is very often used by the development of the embedded control systems and by testing of the implemented software application. It accelerates product development, improves performance, increases reliability and reduces engineering costs [8]. The development of the embedded hardware based on the application of the Model-Based Design is introduced in [13]. The design of the new Brake-By-Wire-Controller, which is a part of the motion controller of the autonomous vehicles, using this methodology is described in [15] as well as the design of the control of the dynamics of the omnidirectional automated guided vehicle presented in [14]. The use of the Model-Based Design by testing of cyber-physical systems like cooperative mobile robots is a topic of the paper [12].

In [9] the term Model Driven Engineering is introduced as a framework for implementation of the control software, e.g. using the V-Model and the importance of the HiL simulation in this process is deeply described. Also, in power electronics these methods are used. The design of the power electronic system with the implementation of the efficient direct model predictive control algorithms was managed with the use of Model-Based Design [11].

1.1 Introduction to the MATLAB Application for Hydraulic System Control

The design of the control systems of the hydraulic drives is very important part of the modern machine design which are driven by the hydraulic drives and must perform high quality movements and technological functions. The specification of the suitable components of the hydraulic circuit is the core decision, which should be done by the project designer, especially the specification of the control valve, which predetermines the dynamic behaviour of the controlled drive through their static and dynamic properties and in connection with the hydraulic cylinder of the whole controlled mechanism. The use of the Model-Based Design methodology, which makes possible to do in short time variant design of the hydraulic drive and closed loop control system, was the main idea by the development of the described application [6]. The dynamic behaviour of the drive can be analysed using the simulation models, the suitable controller can be set and tuned using the Model-in-the-Loop simulation and finally the behaviour of the whole closed loop control system can be tested using the Hardware-in-the Loop simulation.

MATLAB and his extension using the toolbox Simulink is a great environment for the simulation of the hydraulic circuits different structure. Comfortable access to the predefined simulation models of the mostly used hydraulic circuits and their use by the closed loop control design was the goal of the development and presented MATLAB-Application Hydraulic System Control, Fig. 1.

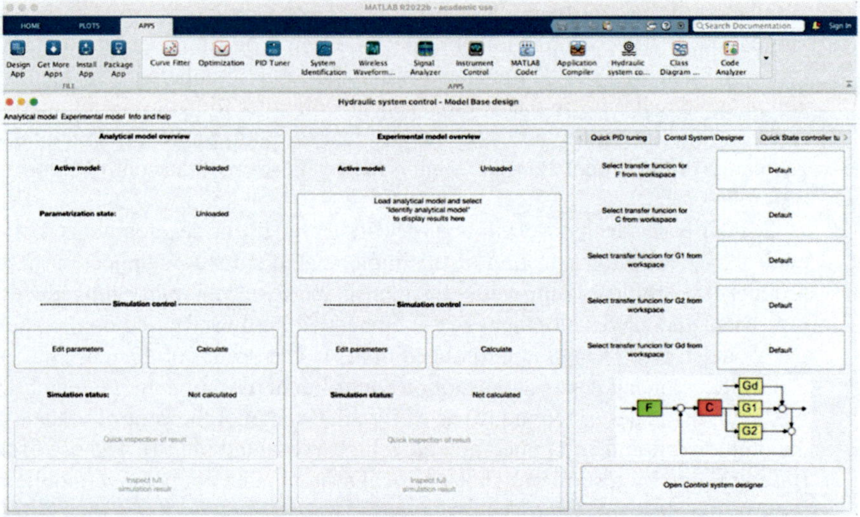

Fig. 1. MATLAB window with the opened Application for Hydraulic System Control.

The available simulation models of the hydraulic circuits consisting of hydraulic cylinder and control valve can be parameterized using the values of their parameters obtained from the documentation and data sheets or using the measured data evaluated from the experimental identification. The detailed description of the components of the hydraulic circuit is done in [2], control strategies used by the servo drives can be found in [3].

The simulation of the open loop system can be done after choosing the designed drive configuration and entering the actual parameters values and the simulation of the closed loop control system after the controller choosing. The suitable components of the hydraulic drive and the appropriate tuning of the controller adjusted to the demands on the drive can be found after repeating of this process. The methodology for the fluid power system analysis, modelling and control is deeply described in [1]. The used methods of control theory and their application and support in MATLAB can be found in [5]. The modern control approaches used by the control system design also with respect to different industrial application are described in [4].

2 Design of the Hydraulic System Control

At the beginning of the design process, it must be decided, if the design is based on the data sheet parameters and based on the analytical mathematical model of the drive implemented in the application or on the data obtained from the experimental identification.

2.1 Design Using Analytical Models

The design process using the analytical model begins with the determination of the type of the controller–PID type controller or state feedback controller. Afterwards the setting of the model parameters is required. The parameters of the hydraulic cylinder and corresponding equations of the used mathematical model are shown in Fig. 2.

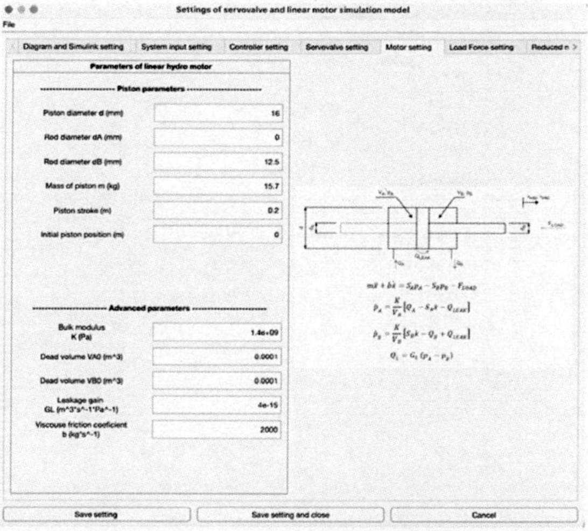

Fig. 2. Parameters setting of the hydraulic cylinder and used mathematical model.

The definition of the load force acting on the piston rod consists of two parts, the time depending force and the force depending on the piston position

$$F = F(t) + F(x). \tag{1}$$

The driven mass m can consists of three parts. The first one is the mass of the piston and piston rod of the hydraulic cylinder m_{HM}, the second part is the final reduced mass m_{RED} of the driven bodies in relation to the hydraulic cylinder axis.

$$m = m_{HM} + m_{RED}. \tag{2}$$

The reduced mass can be defined using two components. The first one can depend on the time $m_{RED}(t)$ and the second one on the piston rod position $m_{RED}(t)$

$$m_{RED} = m_{RED}(t) + m_{RED}(x). \tag{3}$$

The servo valve parameters involve the magnitude of the input voltage, eigen frequency of the servo valve, damping ratio and maximal spool velocity expressed in relative opening per second. There are three options available for the definition of the flow characteristic. The valves, mainly the servo valves can be described only using the nominal flow by the nominal pressure drop across the valve or using from these values calculated flow gain of the control edge. The flow characteristic of the control valve can be described also using the table defining the flow in dependence on the valve opening for the given pressure drop. The characteristic can be nonlinear and different for positive and negative direction of the valve opening. In all cases the spool overlap can be taken into the account. The window for control valve parameters setting is shown in Fig. 3. The corresponding flow characteristic is drawn in the right part. If the definition using the flow characteristic is chosen, the given points must be set using the table below.

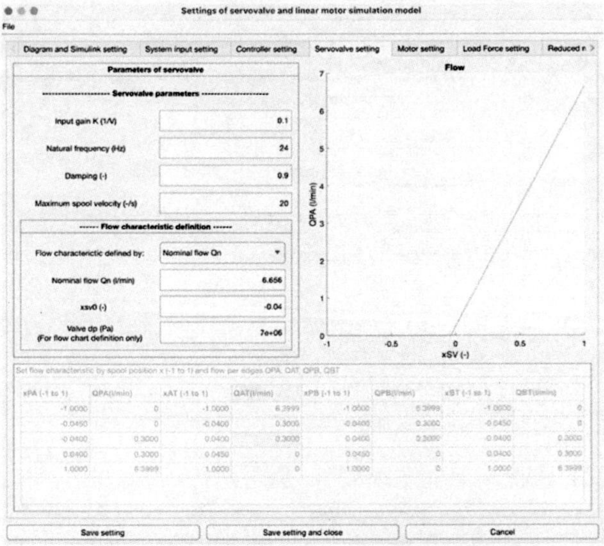

Fig. 3. Parameters setting of the used control valve (the option of setting using the flow characteristic is in the shown case not active).

The command value for the open loop simulation as well as the desired value for the closed loop simulation can be defined as a step input, ramp signal, sine wave, chirp signal or using the time table.

2.2 Model Parameterization Using the Experimentally Obtained Data

The experimentally obtained data for example the measured step response or outputs other identification experiments can be evaluated using the identification methods implemented in the System Identification toolbox of the MATLAB-Simulink program. The identified models in form of transfer function or state space model can be imported in the Hydraulic System Control application. If the model was prepared in advance and is active in workspace of MATLAB, it can be specified using the used model parameters or matrices in the Application using the sub window for parameters setting. The standard format of the model parameters is used for the transfer of the models from the MATLAB workspace into the Application.

2.3 Controller Tuning

The controller tuning is supported for the closed loop control system with PID-type controller or with state controller. For each option the block diagram is depicted in the sub window for parameter setting. In the first case tuning of the PID-type controller tuning is performed using the Control System Designer implemented in MATLAB. This efficient tool involves different tuning methods and enables plotting of the characteristics, which are important for the evaluation of the quality control, e.g., Bode Editor, PID Tuning and others. The state feedback controller design using the Ackerman's formula as well as the calculation of the observability matrix and controllability matrix are supported. The pole placement is supported using the eigen values position plot and tool for setting of their new desired position.

3 Practical Use of the Application for Hydraulic System Control and Experimentally Obtained Results

The practical use of the developed MATLAB-Application was tested on the design of the position controller for the hydraulic drive installed on the test rig in the laboratory, the structure of the test rig is shown in Fig. 4. The tested drive consists of the control valve and hydraulic cylinder equipped with the position transducer and can be controlled using the analogue or digital controller. The drive is using the plug-in data acquisition card (DAQ card MF624) connected to the personal computer (PC) which allows the data acquisition, their visualisation and digital controller realization.

The control valve properties are characterized by the data in Table 1 obtained from the data sheet. The hydraulic cylinder has the stroke 0.2 m and the piston diameter 0.016 m, the one side piston rod has the diameter 0.01 m. The moving mas of the cylinder and connected plate–the slide unit is 15.7 kg.

Automatic choice of the suitable controller is done after the calculation of the eigenfrequency of the hydraulic cylinder and setting of the ratio of the eigenfrequency of the control valve and hydraulic cylinder.

$$\kappa = \frac{f_{sv}}{f_{HM}}. \tag{4}$$

The proportional controller was chosen automatically based on the recommendations in Table 2 [7] for the evaluated value of $\kappa = 0.85$.

For the controller tuning the parameterized non-linear model was numerically linearized and the transfer function (5) form was obtained

$$G_{HM}(s) = \frac{2.395}{1.255 * 10^{-9} * s^5 + 2.94 * 10^{-7} * s^4 + 1.03 * 10^{-4} * s^3 + 8.552 * 10^{-3} * s^2 + s} \tag{5}$$

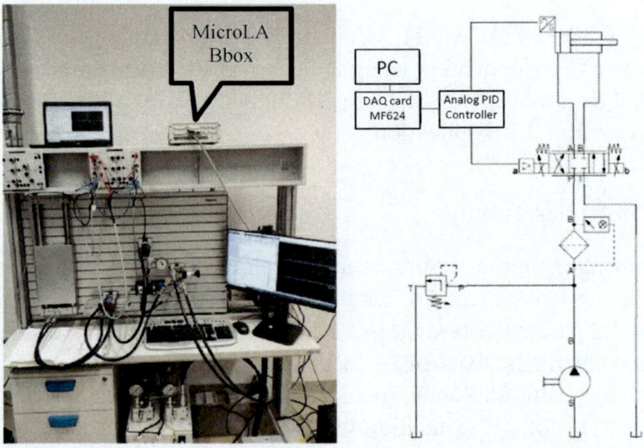

Fig. 4. Test rig for the closed loop control of the hydraulic drives.

Table 1. Control valve parameters

Gain	Eigenfrequency	Time constant	Damping ration	Max. Spool velocity	Nominal flow	Pressure drop
K_{sv} (V^{-1})	F_{sv} (Hz)	T_{sv} (ms)	ξ_{sv} (-)	v_{svmax} (-/s) (relative)	Q_n (l/min)	Δp (MPa)
0.1	24	41.6	0.9	20	6.4	12

Table 2. Controller type selection criterion

κ	from 0.2 to 0.5	from 0.5 to 1	from 1 to 3	$\kappa \geq 3$
controller	PDT1	P	PT1	State Control

and finally, the for the requested phase margin 70° the controller gain $K_p = 11.84$ was set. Three responses presented in Fig. 5 were obtained using the MIL simulation for linear mathematical model without saturation of the command variable, for the linear model with command variable saturation at ± 10V and for the complex nonlinear model. The results obtained for the decreased value of the controller gain to $K_p = 6.5$ are shown in Fig. 6.

The HIL simulation was done in the next step. The nonlinear complex simulation model of the hydraulic drive was implemented in the dSPACE MicroLABbox unit. The MicroLABbox was connected using the electrical signals corresponding to the control value and output value to the PC with the plug-in data acquisition card MF624. The non-linear model of the controlled hydraulic drive is running in real time and the digital

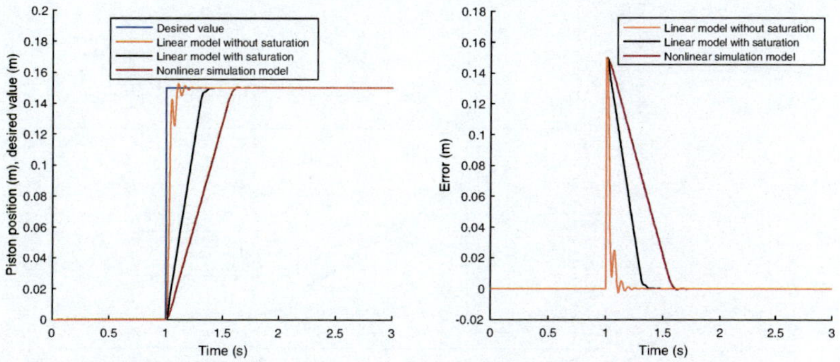

Fig. 5. MIL simulation using different models for $K_p = 11.84$.

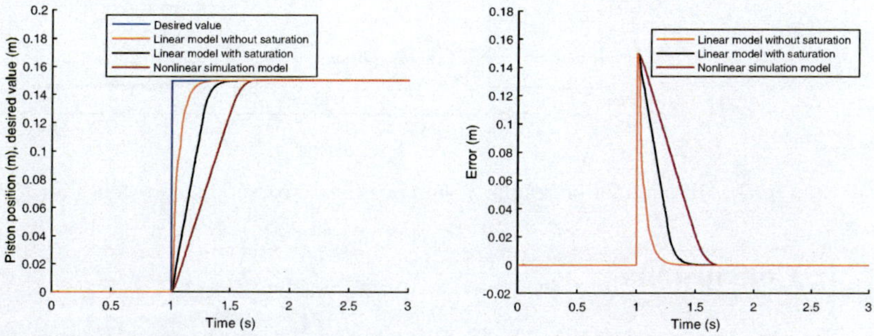

Fig. 6. MIL simulation results using different model for $K_p = 6.5$.

controller is running in Simulink Real-Time on PC. The obtained results for MIL and HIL simulation and real control are shown in Fig. 7 and confirm the expectations of the fast aperiodic step response of the drive.

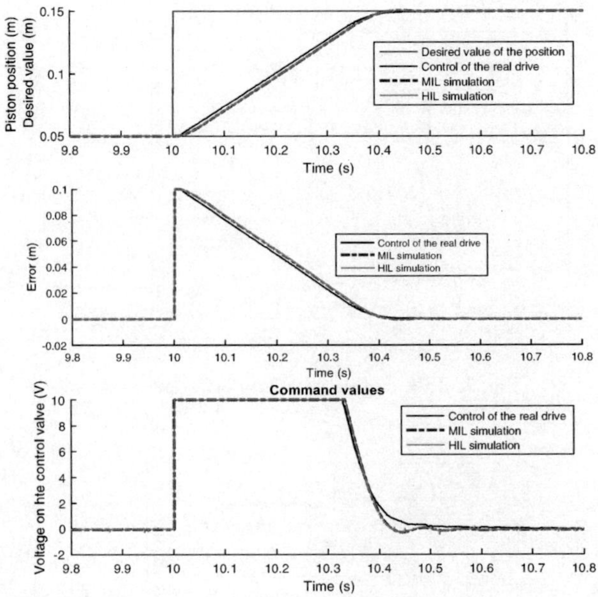

Fig. 7. HIL simulation results–piston position, error, command values for $K_p = 11.84$.

4 Conclusions

The new developed MATLAB-Simulink based user application Hydraulic System Control was presented in the paper. The application supports design and verification of the controllers for closed loop controlled hydraulic drives. The applied methodology is the Model-Based Design supported using the simulation models prepared in Simulink and powerful toolboxes for System Identification and Control Design. The practical use of the Application was demonstrated on the design of the position controller for the linear hydraulic drive installed in the laboratory. The simulated results were verified using the Hardware-in-the-Loop simulation and on the real drive. The real time simulation of the controlled drive was done using the dSpace MicroLABbox real time computer. The measured results on the real drive confirm the outputs of the simulations and efficiency of the Model-Based Design approach.

Acknowledgment. This work was supported by the European Regional Development Fund in the Research Centre of Advanced Mechatronic Systems project, CZ.02.1.01/0.0/0.0/16_019/0000867 within the Operational Programme Research, Development, and Education and the project SP2023/074 Application of Machine and Process Control Advanced Methods supported by the Ministry of Education, Youth and Sports, Czech Republic.

References

1. Watton, J.: Fluid power systems. Modeling, simulation, analog and microcomputer control. Prentice Hall International (UK) Ltd. (1989). ISBN 0–13–323197–6

2. Backé, W.: Servohydraulik. RWTH Aachen (1984)
3. Murrenhoff, H.: Servohydraulik - Geregelte hydraulische Antriebe. IFAS RWTH Aachen. Shaker Verlag GmbH (2008). ISBN 978-3-8322-7067-4
4. Jelali, M., Kroll, A.: Hydraulic Servo-systems: Modelling, Identification and Control. Springer, London (2004). ISBN 1852336927
5. Dorf, R.,C., Bishop, R.H. Modern Control Systems. 8 edn.Addison Wesley Longman, Inc. California 1998
6. Milata, J.: Model based Design of Control of Mechatronic System. Master thesis, Faculty of Mechanical Engineering, VŠB-Technical University of Ostrava 92022)
7. Noskievič, P.: Simulation, analysis and laboratory control of hydraulic servo drives using MATLAB-Simulink. In: Proceedings of the 2nd International Workshop on Computer Software for Design, Analysis and Control of Fluid Power Systems, pp. 124–132. Tanger s.r.o., Ostrava (2001)
8. Socci, V.: Implementing a model-based design and test workflow. In: 2015 IEEE International Symposium on Systems Engineering (ISSE), Rome, Italy, 2015, pp. 130–134 (2015).https://doi.org/10.1109/SysEng.2015.7302745
9. Preuße, H., Lapp, S.-C., Hanisch, H.-M.: Closed-loop system modeling, validation, and verification. In: Proceedings of 2012 IEEE 17th International Conference on Emerging Technologies & Factory Automation (ETFA 2012), Krakow, Poland, pp. 1–8 (2012). https://doi.org/10.1109/ETFA.2012.6489679
10. Dalton, S.W., Wagner, F., Bergmann, A., Bock, B.: Using MATLAB V&V-toolbox for target-specific model-based design. In: 2018 Third International Conference on Engineering Science and Innovative Technology (ESIT), North Bangkok, Thailand, pp. 1–5 (2018). https://doi.org/10.1109/ESIT.2018.8665319
11. Capponi, L.: Direct model predictive control using model based design. In: 2021 IEEE International Conference on Predictive Control of Electrical Drives and Power Electronics (PRECEDE), Jinan, China, pp. 680–685 (2021). 1109/PRECEDE51386.2021.9681020
12. Saglietti, F., Föhrweiser, D., Winzinger, S., Lill, R.: Model-based design and testing of decisional autonomy and cooperation in cyber-physical systems. In: 2015 41st Euromicro Conference on Software Engineering and Advanced Applications, Madeira, Portugal, pp. 479–483 (2015). https://doi.org/10.1109/SEAA.2015.68
13. Dalton, S.W., Wagner, F., Bergmann, A.: Embedded hardware design with model-based design and CONSENS. In: 2017 International Conference on Research and Education in Mechatronics (REM), Wolfenbuettel, Germany, pp. 1–5 (2017). https://doi.org/10.1109/REM.2017.8075235
14. Zhang, J., Liu-Henke, X.: Model-based design of the vehicle dynamics control for an omnidirectional automated guided vehicle (AGV). In: 2020 International Conference Mechatronic Systems and Materials (MSM), Bialystok, Poland, pp. 1–6 (2020). https://doi.org/10.1109/MSM49833.2020.9202248
15. Houhua, J., Haifeng, L., Yihang, G.: Model based design and experimental test of brake-by-wire controller. In: 2020 4th CAA International Conference on Vehicular Control and Intelligence (CVCI), Hangzhou, China, pp. 255–258 (2020). https://doi.org/10.1109/CVCI51460.2020.9338622

Comparative Analysis of the Hydrostatic Drive System for a Midi Wheel Loader

Jakub Takosoglu, Łukasz Chłopek(✉), Kamil Ziejewski, Ryszard Dindorf, and Piotr Woś

Kielce University of Technology, Kielce, Poland
{qba,lchlopek,kziejewski,dindorf,wos}@tu.kielce.pl

Abstract. The article presents a qualitative and quantitative comparative analysis of the currently used hydrostatic-mechanical wheel drive systems on the example of a wheel loader. In addition to the conventional systems in common use, the analysis also takes into account the future system of direct, individual wheel drive by means of a variable displacement hydrostatic pump-motor. The object generating the boundary conditions for comparison is a typical midi class wheel loader weighing 8 tonnes. Current hydrostatic drive systems for wheeled chassis have a number of limitations due to the configuration of the entire drive system as well as the design of the hydrostatic pumps and motors themselves. Beginning with the typical drive made up of axle and an intermediate gearbox driven by a bent axis variable displacement motor and ending with the aforementioned pump-motor solution, there are significant differences in the actual useful functionality of all drive systems. This functionality translates into the static and dynamic characteristics of the individual drive systems, significantly influencing wheel loader performance, operating economy, reliability and operator comfort. At present, individual direct wheel drive systems are built using slow-speed motors with stepwise or infinitely variable displacement, but within a given range, which imposes limitations on the system's use. One of the fundamental parameters is the maximum speed of the wheeled chassis equipped with such a system, limited to several km/h, while an innovative system based on a hydrostatic pump-motor with variable displacement does not have this limitation.

Keywords: variable displacement hydrostatic pump-motor · hydrostatic-mechanical wheel drive · hydrostatic slow-speed motor · direct individual wheel drive

Abbreviations

FT	Single wheel traction force
FN	Total chassis traction force
FM	$=m_c 9.81[m/s2] \rightarrow$ Running weight
iC	Overall gear ratio
iD	Dynamic ratio
iK	Kinematic ratio
mC	Total weight of machine, Including load

J. Stryczek and U. Warzyńska (Eds.): NSHP 2023, LNME, pp. 114–127, 2024.
https://doi.org/10.1007/978-3-031-43002-2_11

MS	Torque on the shaft of hydrostatic motors
Mg	Torque at the hydrostatic pump shaft
nS	Hydrostatic motor shaft speed
ng	Hydrostatic pump shaft speed
Δp	Pressure in the system
Q	Hydrostatic pump/motor capacity
qs	Motors displacement
qg	Pump displacement
R	Dynamic radius of the drive wheel
εS	Hydrostatic motor setting parameter
εg	Hydrostatic pump setting parameter
ηhmg	Pump hydraulic-mechanical efficiency
ηhms	Hydraulic-mechanical motor efficiency
ηvg	Volumetric efficiency of the pump
ηvz	Volumetric efficiency of the pump adjuster
ηvs	Volumetric efficiency of the motor

1 Introduction

Wheeled loaders are now a widespread group of machines which, thanks to their characteristic design (in classic terms), satisfy the needs of a wide range of customers, from the agricultural sector to open-cast mines. The subject of this article is to analyse the various hydrostatic drive system architectures that can be used on wheeled loader chassis manufactured around the world today, with reference to the classic transfer case solution that has been on the market for decades and to a future system based on individual direct wheel drive. When analysing this group of machines, it is necessary to define the object of consideration itself. According to ISO 7131:2009 [1], a wheel loader is a *„Self-propelled crawler or wheeled machine which has front-mounted equipment primarily designed for loading operation (bucket use) and which loads or excavates through forward motion of the machine". Wheel loader manufacturers such as Liebherr and Komatsu rank their products in most cases as tonnage classes* [2, 3] *which is in accordance with ISO 7131:2009* [1] *and ISO 6165:2012* [4] *where you can find the definition of a compact wheel loader in Subsect. 4.2.3.* Manufacturers themselves very often divide their products into different classes, which is intuitive for the customer and effective for marketing. According to a number of reports from Ritchie-Bros [5], for example, the most popular tonnage class among customers of, for example, construction companies, is the midi class, which Volvo, one of the world's leading manufacturers of this type of machine, defines as [6] a mid-sized loader with a maximum lifting capacity of ~ 5000 kg. The high prevalence among purchasers of machines in this class equates to a large number of manufacturers and therefore drive system solutions, which is a natural argument for carrying out an analysis of hydrostatic drive system architectures on this group of machines. One of the best-known brands on the Polish market, as well as one of the best perceived, is Atlas Weyhausen, which some time ago changed its logo to Weycor. The midi model in this manufacturer's product range is the AR530 with a lifting capacity of 5198 kg. This machine (Fig. 1) is a good starting point for this analysis as its drive system is a perfect example of a typical classic hydrostatic drive system with intermediate gearing.

Fig. 1. Weycor AR 530 wheel loader, formerly known as Atlas Weyhausen AR95.

2 Wheel Loader with Classic Drive System-The Benchmark

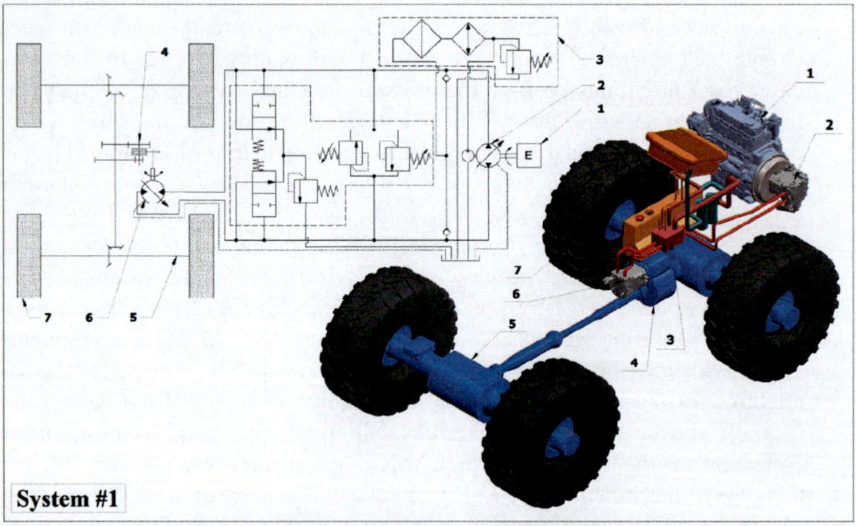

Fig. 2. Weycor AR 530 wheel loader, formerly known as Atlas Weyhausen AR95 (bottom left corner of manufacturer's material).

The reference point for comparing the properties of the drive systems is a real object in the form of a midi wheel loader (Fig. 1), i.e. with a maximum load capacity of ~ 5000 kg and a loader bucket volume of 1.6 m^3. The original machine (**solution #1**) (Fig. 2) is equipped with a drive system consisting of a diesel engine (Deutz TCD 3.6 L4) _1_ coupled to an axial piston swash plate pump (Linde HPV-02-A2-055R) _2_ controlled by a proportional signal from an operator-controlled foot controller. Pump _2_ is equipped with a constant power (master) capacity control system. The pressurised fluid stream is processed by a high-speed axial piston swash plate motor (Linde HMV02-A2-055N) _6_. The swash plate motor's angle of rotation depends on the current external load and is controlled by an electronic controller. Motor _6_ is bolted to transfer case (Dana model 367) _4_, which is equipped with two operating gears–fast/transport (ratio 1.09) and slow/reverse (ratio 4.53). The drive torque is then directed to the drive axles (Dana 123 26 R 56 with overall gear ratio 22) _5_ in which two gears are built in: a bevel gear and a planetary gear mounted in the wheel hub. The drive axle _5_ is also equipped with a differential that is locked 100% by an electric actuator. Wheels fitted with radial tyres _6_ are bolted to the axles. The hydrostatic system is equipped with an obligatory cooling and filtering system _3_. The technical data of the AR 530 wheel loader adopted for the comparative calculations is given in Table 1.

Table 1. Technical data of the wheel loader-parameters adopted for calculations.

Description	Parameter
Diesel engine output	80 kW at 2200rpm
Diesel engine max. Torque	430 Nm at 1600rpm
Diesel engine displacement	3621 cm^3
Tyres size	Nokian TRI 2 440/80 R 24
Tipping load, straight	5577 kg
Weight of machine with bucket 1.6m^3	7900 kg
Mass of excavated material at density 2700 kg/m^3	4320 kg → standard ISO 8313 i EN 474–3
Max. Lifting capacity	5198 kg
Machine weight with load in bucket	12220 kg

3 Overview of the Drive Systems Under Study

3.1 Drive System #2 with Integrated Motor

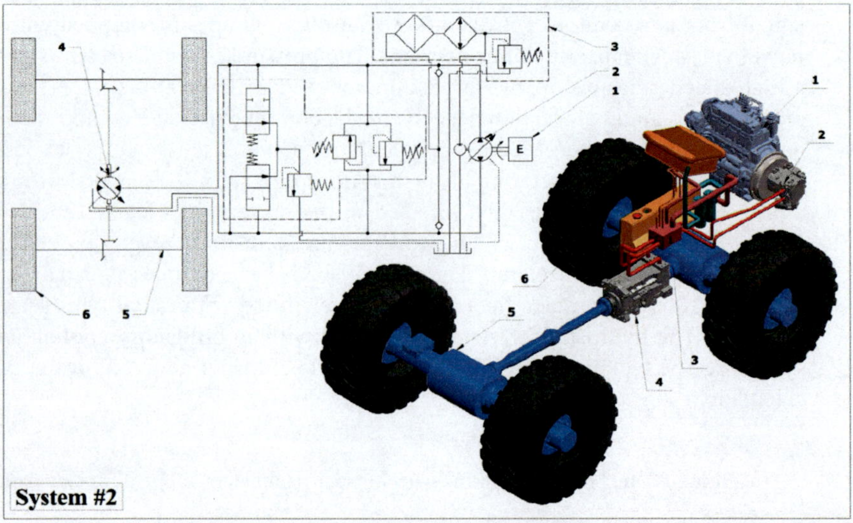

Fig. 3. Drive system #2 with integrated motor

The figure (Fig. 3) shows a drive system with a hydrostatic integrated motor. The twin-rotor axial piston swash plate motor 4 is a Linde proprietary design -Linde model HMV165D E6. The axial piston swash plate pump 2 (Linde HPV-02-A2-055R), which generates the oil flow directed to motor 4, is driven by diesel engine (Deutz TCD 3.6 L4) 1. The torque generated by hydrostatic motor 4 is directed to drive axles (Dana 123 26 R 56 with overall gear ratio 22) 5 with wheels 6. The system is completed by a filtering and cooling system for working fluid 3.

3.2 Drive System #3 with ICVD Gearbox

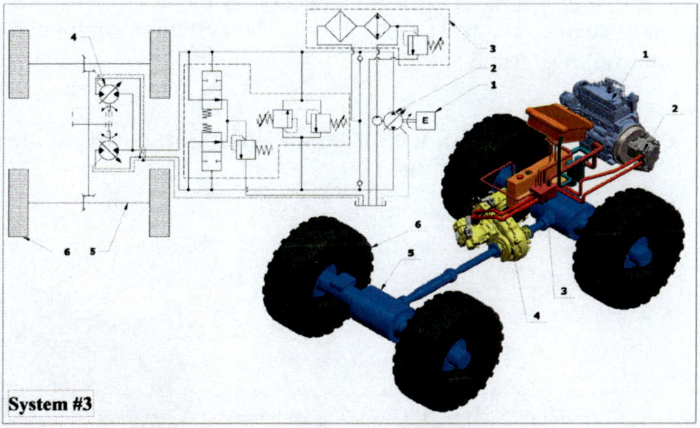

Fig. 4. Drive system #3 with ICVD gearbox

The drive system with the architecture as shown in the figure (Fig. 4) consists of a diesel engine (Deutz TCD 3.6 L4) 1 driving an axial piston swash plate pump 2 (Linde HPV-02-A2-055R), which pressurises oil to the GKN-Walterscheidnumber ICVD GT-S2 A 233V gearbox 4, from which power is transmitted to the drive axles 5 with wheels 6. The system is completed by a filtering and cooling system for the working fluid 3.

3.3 Drive System #4 with Bent Axis Motor Coupled to High Ratio (1:35.9) Planetary Gearbox

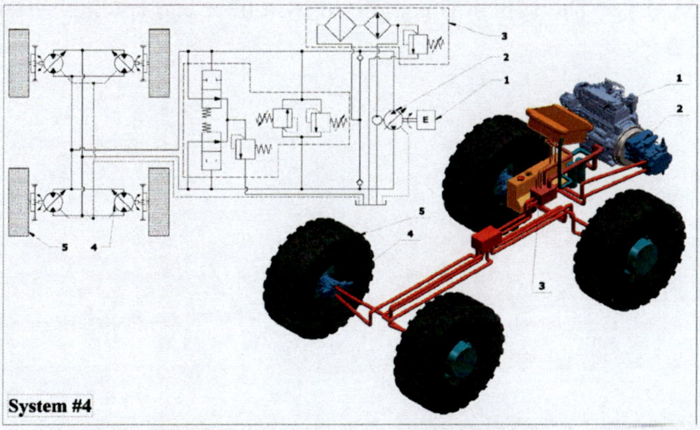

Fig. 5. Drive system #4 with bent axis motor coupled to high ratio (1:35.9) planetary gearbox

The figure below (Fig. 5) shows a drive system with individual wheel drive with a single pump. Planetary gear reducers (Rexroth GFT) with a large gear ratio are integrated with piston bent axis motors (Rexroth A6VE 107) _4_ directly driving the wheels _5_. The motors _4_ are powered by a single axial piston swash plate pump _2_ (Rexroth A4VG71EP4) driven by a diesel engine (Deutz TCD 3.6 L4) _1_. The system is completed by a working fluid filter and cooling system _3_.

3.4 Drive System #5 with with Radial Piston Motor with External Cam

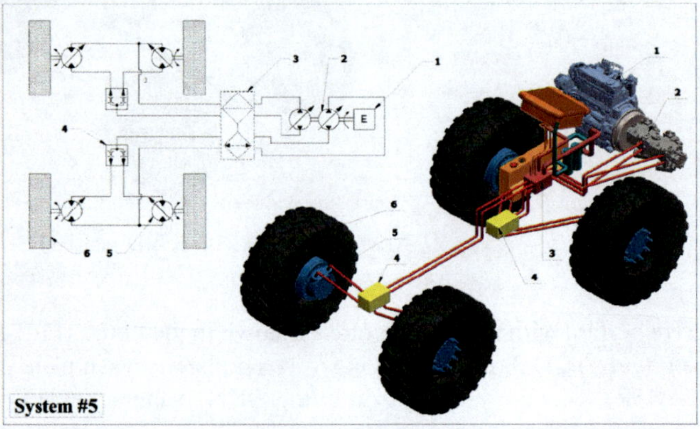

Fig. 6. Drive system #5 with with radial piston motor with external cam

The figure above (Fig. 6) shows an individual wheel drive system with a tandem axial piston swash plate pump (2x Linde HPV-02-A2-055R) _2_ coupled to a diesel engine (Deutz TCD 3.6 L4) _1_. Power is transmitted to the wheels _6_ via two flow dividers (Poclain FD-H2) _4_ to four two-speed radial piston motors with a external cam (Poclain MS35AE) _5_. The system is completed by a filter and cooling system for the working fluid _3_.

3.5 Drive System #6 with Axial Piston Swashplate Motos with Small Ratio Planetary Reducers (1:5)

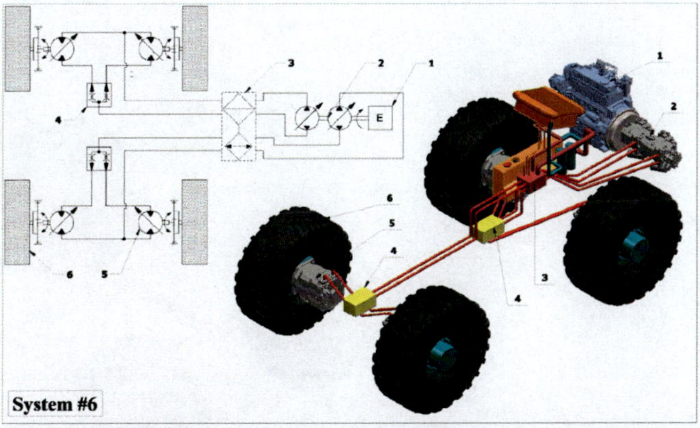

Fig. 7. Drive system #6 with axial piston swashplate motos with small ratio planetary reducers (1:5)

The figure above (Fig. 7) shows an individual wheel drive system with a tandem axial piston swash plate pump (2x Linde HPV-02-A2-055R) 2 coupled to a diesel engine (Deutz TCD 3.6 L4) 1. Power is transmitted to the wheels 6 via two flow dividers (Poclain FD-H2) 4 to four axial piston swash plate motors (Linde HMV02-A2-280N) interlocked with small ratio planetary gears 5 in the figure. The system is completed by a working fluid filtration and cooling system 3.

3.6 Drive System #7 with Hydrostatic Individual Direct Wheel Drive with Switchable Circuits

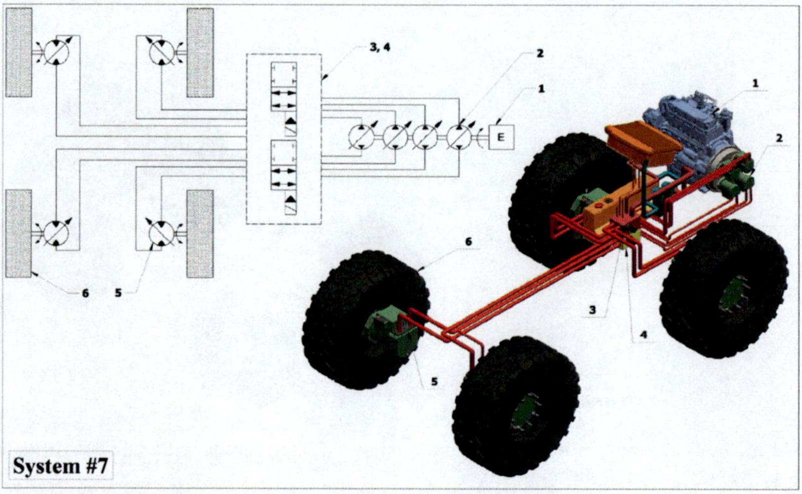

Fig. 8. Drive system #7 with individual direct wheel drive with switchable circuits

The figure above (Fig. 8) shows a hydrostatic drive system with individual independent direct wheel drive. Four pumps _2_ according to paper [7] are coupled to a diesel engine (Deutz TCD 3.6 L4) _1_ and supply oil to four hydrostatic motors _5_ according to paper [7]. Each pump _2_ feeds oil to the slow-speed motor _5_ individually. The system also allows the expenditure of the pumps _2_ to be combined to power the engines _5_ when the machine is moving in transport mode. In this mode, the front 'axle' is supplied by four pumps _2_, the rear 'axle' is disconnected and operates in freewheeling mode ($M_S = 0[Nm]$). The total expenditure of pumps _2_, oil filtering and cooling is the responsibility of modules _3_ and _4_.

4 Comparison of Drive Systems

4.1 Qualitative Method of Assessing the Capabilities of Propulsion Systems

The qualitative method of assessing the functionality of drive systems is very important in the comparison presented in this article. This assessment has a direct impact on the quantitative method by clearly defining the existence or absence of a given function/capability in the drive system. The existence or absence of a given function/capability in a given analysed system is due to:

- Operational constraints defined by the manufacturer of a given component included in the propulsion system - Technical parameters: hydrostatic pumps/motors; drive axles; transfer cases; planetary gears, etc.

- The architecture/configuration of the propulsion system that arbitrarily defines how it operates and therefore the performance of the chassis
- Functionalities proposed by wheel loader manufacturers
- Engineering principles, particularly concerning the selection of hydrostatic drive components

Table 2. Table for qualitative assessment of the drive systems analysed

Criterion	Function/capabilities	Evaluation
I → Hydrostatic motor type	Continuously variable displacement motor	1
	Stepwise control of hydrostatic motor displacement	2
II → Hydrostatic differential lock	Flow dividers are present	1
	No flow dividers	2
III → Reducers in the wheels	Reducers used	1
	No reducers	2
IV → Drive axles	Yes	1
	No	2
V → Drive type of axles	Direct drive	1
	By means of a transfer case	2
VI → Method of power transfer to drive axles	Driveshafts, Cardan couplings	1
	Absent	2
VII → Possibility of transferring the entire power of the diesel engine to a single wheel	Yes	1
	No	2

Table 2 shows the evaluation criteria adopted without taking into account the characteristics of the drive system and the evaluation adopted for it. The figure (Figure 9) shows the results of the qualitative analysis carried out.

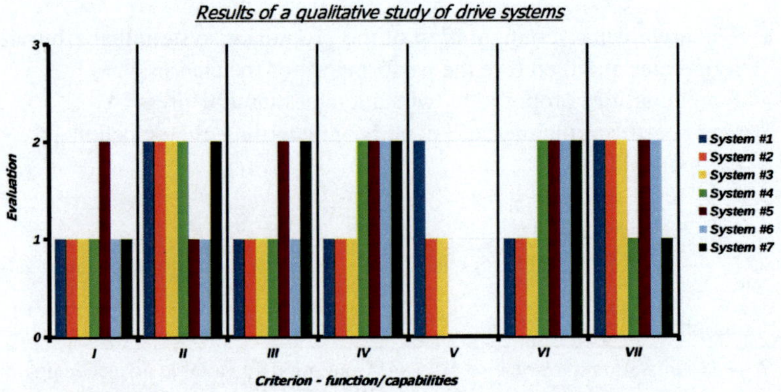

Fig. 9. Results of a qualitative study representing logical analysis of drive systems

4.2 Quantitative Method of Assessing the Capabilities of Propulsion Systems

The calculations made in modelling each of the propulsion systems being compared are the basis of the quantitative method. The key parameters under analysis included: speed v, single wheel traction force F_T, dynamic ratio i_D, kinematic ratio i_K and hill-climbing capacity β.

$$v = \left(\left(\frac{Q900}{q}\right)\frac{1}{i_c}\right)\frac{\pi \cdot R}{30} \cdot \left(\frac{36}{10}\right) \tag{1}$$

Equation (1) makes it possible to calculate the speed v of the machine.

$$F_T = \left(\frac{q_s \cdot \Delta p}{63}\right) \cdot i_c \cdot \left(\frac{1000}{R}\right) \cdot \left(\frac{1}{4}\right) \tag{2}$$

The F_T traction force of a single wheel is represented by Eq. (2), multiplying the calculated value by 4 leads to the total traction force of the chassis F_N. Equation (2) is valid for idealised conditions–no coefficient of rolling resistance, ideal efficiency of the hydrostatic unit, etc.

$$i_D = \frac{M_s}{M_g} = \frac{\varepsilon_S q_s \cdot \Delta p}{\varepsilon_S q_s \Delta p}\eta_{hmg} \cdot \eta_{hms} \tag{3}$$

Equation (3) represents the torque ratio (ratio of moments on hydrostatic unit shafts) of the actual gearbox, i.e. the dynamic ratio i_D. The hydraulic-mechanical efficiencies of the hydrostatic units reduce the gear ratio.

$$i_k = \frac{n_g}{n_s} = \frac{\varepsilon_s \cdot q_s}{\varepsilon_g \cdot q_g \cdot n_{vg} \cdot n_{vz} \cdot n_{vs}} \tag{4}$$

The speed ratio of the actual gearbox (speed ratio on shafts of hydrostatic units) is represented by Eq. (4), this ratio otherwise known as the kinematic ratio i_K increases as the volume loss of the hydrostatic units increases.

$$sin(\beta) = \frac{4 \cdot F_T}{m_c \cdot 9.81} \because sin(\beta) = \frac{F_N}{F_M} \supseteq F_N = 4 \cdot F_T \tag{5}$$

Algorithm (5) is a recipe for calculating the climbing ability parameter β. It is given in degrees or percent. With the traction force of the loader's chassis F_N (four times the traction force of a single wheel F_T) and the running weight of the vehicle m_C (F_M [N]), it is possible to calculate the gradient of the road β that it can theoretically climb at the lowest gear ratio, i.e. in the range of maximum driving torques, the algorithm is presented in paper [8]. Figures 10 to 12 (Figs. 10, 11, 12) below show the families of characteristics obtained from the calculations carried out.

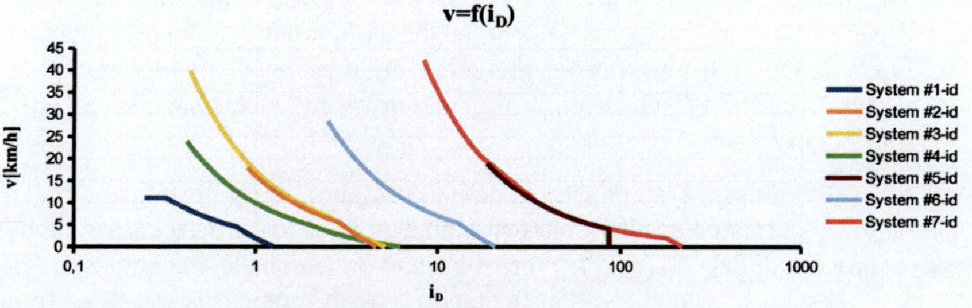

Fig. 10. Family of speed characteristics of wheel loader with analysed drive systems #1-#7 as a function of the dynamic transmission ratio

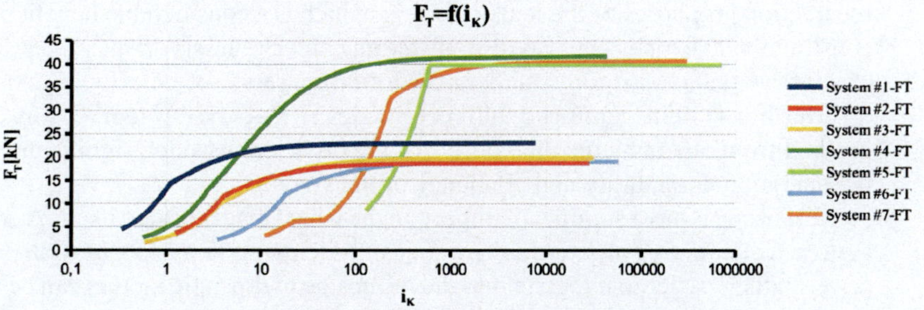

Fig. 11. Family of characteristics of the tractive force of a single wheel with the analysed drive systems #1-#7 as a function of the kinematic ratio of the drive system.

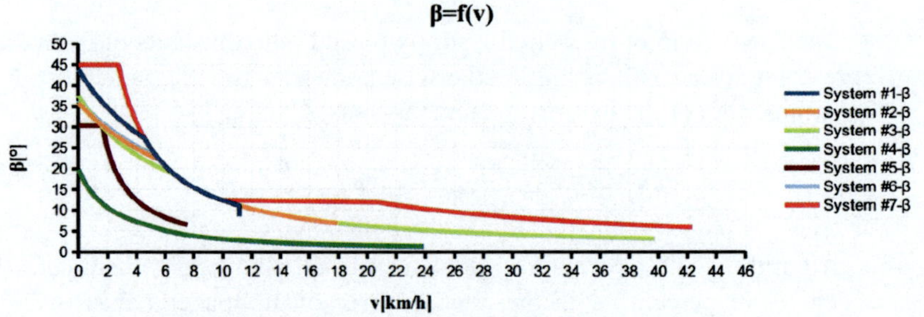

Fig. 12. Family of characteristics of the climbing ability of a wheel loader with the analysed drive systems #1-#7 as a function of the machine speed

5 Conclusions

The qualitative analysis presented in Sect. 4.1 clearly defines the characteristics of the drive architectures analysed. One of the most important criteria presented in Table 2 and Fig. 9 is the ability to transfer the entire drive power of the diesel engine to a single wheel - criterion *VII*. Only propulsion system *#4* and *#7* enable this but with two caveats for system *#4*:

- In a propulsion system without active control of the hydrostatic motors in the wheels Fig. 5 number 4's ability to transfer all the power to the wheel is a completely random process. It depends on the type of ground on which the machine is moving
- A system in which the displacement of each motor 4 is managed by a PLC with feedback from the hydrostatic motor shaft speed removes the randomness of the drive process and guarantees a good level of traction force control

The #7 system, thanks to its architecture, allows active continuous adjustment of the traction properties of the loader chassis, which is a considerable benefit not only for construction machines but also for forestry machinery chassis drives - the priority being to preserve the natural forest undergrowth (minimisation of wheel slip). An important criterion for systems equipped with drive axles is *V*. Drive system *#2* Fig. 3 is distinguished from others by the direct drive of the drive axles, which significantly increases the operational reliability and efficiency of the system.

Criterion *IV* has a significant impact on the wheel loader design itself. A construction system which does not require drive axles (systems up to *#4* to *#7*), transfer cases or drive shafts is easier to integrate into the frame, the hydrostatic motors can be attached to the frame in a variety of ways - the builder is only limited by his or her imagination. The weight of the drive system is considerably less than a classic system with drive axles.

Looking at the families of characteristics from Fig. 10. to Fig. 12 created during the implementation of quantitative analysis 4.2, one can significantly refine the performance achieved by individual drive systems *#1-#7*.

Looking at Fig. 10 it can be seen that systems *#3, #6* and *#7* have the best characteristics although they are offset from each other on the abscissa axis. This leads to the conclusion that it is appropriate to increase the displacement of the hydrostatic motors driving the wheels and at the same time reduce the ratio of the reduction gears until they are completely eliminated from the system. The set of characteristics shown in Fig. 11 illustrates the magnitude of the tractive force of a single wheel with the qualitative assessment from Sect. 4.1 in the background. This force translates directly into the working characteristics of the wheel loader - the force on the bucket teeth and the terrain characteristics of the chassis in the form of overcoming obstacles of various types or working on surfaces with high rolling resistance (wetlands). Systems #4, #5 and #7 are characterised by the highest single wheel traction force - these architectures can be considered worthy of consideration for heavy applications. The characteristics according to Fig. 12 are important from the point of view of wheel loader operation. This is especially true for transporting the machine over longer distances in hilly and mountainous terrain where it is necessary to maintain a high transport speed despite the slope of the terrain. Another characteristic of the chassis may be the operation in working mode on inclines. Systems *#1* and *#7* have the best properties.

References

1. ISO 7131:2009, page 6, subsection 3.1.1
2. https://www.liebherr.com/en/pol/products/construction-machines/earthmoving/wheel-loaders/overview/hrc/hrc.html. Accessed 24 Mar 2023
3. https://komatsupoland.pl/katalog-produktow/ladowarki. Accessed 24 Mar 2023
4. ISO 6165:2012, page 8, subsection 4.2.3
5. https://blog.rbauction.com/market-trends. Accessed 24 Mar 2023
6. https://www.volvoce.com/asia/en-as/about-us/blog/2021/everything-you-need-to-know-about-wheel-loaders/. Accessed 24 Mar 2023
7. Doctoral dissertation by Łukasz Chłopek, M.Sc., entitled -_Analysis and synthesis of fluid properties of a hydrostatic pump-motor with active control system_, under the supervision of Jakub Takosoglu, Ph.D., Prof. PŚk
8. Reimpell, J.: Grundlagen (Vogel-Fachbücher). Vogel Communications Group GmbH & Co. KG, Deutsch (2000), Chapter 6.4.2
9. Stryczek, S.: Napęd hydrostatyczny. tom II. WNT, Warszawa (1995)
10. Brach, I.: Gustaw Tyro: Maszyny ciągnikowe do robót ziemnych. WNT 1986
11. Pellegri*, M., Green, M., Macpherson, J., McKay, C., Caldwell, N.: Applying a multi-service digital displacement pump to an excavator to reduce valve losses. In: 12th International Fluid Power Conference, Dresden (2020)
12. Achten, P.A.:Power density of the floating cup axial piston principle, IMECE2004–59006. In: Proceedings of IMECE04 2004 ASME International Mechanical Engineering Congress and Exposition November 13–20. Anaheim, California USA (2004)
13. Linde. https://www.linde-hydraulics.com/product/hmv-02-d/. Accessed 24 Mar 2023
14. Rexroth A Bosch Copmpany, Axial piston variable pump A4VG Series 40 manual, RE 92004/2020–03–30
15. Rexroth. https://store.boschrexroth.com/Hydraulics/Motors?cclcl=en_GB. Accessed 24 Mar 2023
16. Walterscheid-group. https://www.walterscheid-group.com/icvd/#technical-data. . Accessed 24 Mar 2023
17. Linde. https://www.linde-hydraulics.com/product/hpv-02. Accessed 24 Mar 2023
18. Poclain . https://poclain-hydraulics.com. Accessed 24 Mar 2023
19. https://www.deutz.com/en/products/engines?tx_deutzengines_pi1%5Baction%5D=list&tx_deutzengines_pi1%5Bcontroller%5D=Engine&tx_deutzengines_pi1%5Bengine%5D=143&cHash=dcc3a16e53a121dc37f8459173e7bc6e. Accessed 24 Mar 2023

Innovative Modular Load Sensing B-type System for Industrial Standard ISO 4401 Subplate

Janusz Rajda[1](✉), Edward Lisowski[2], and Grzegorz Filo[2]

[1] Ponar Wadowice S.A., Wojska Polskiego 29, 34-100 Wadowice, Poland
janusz.rajda@wp.pl

[2] Faculty of Mechanical Engineering, Cracow University of Technology, Jana Pawla II 37, 31-864 Cracow, Poland
{lisowski,filo}@mech.pk.edu.pl

Abstract. Subject of the article is an innovative modular load sensing (*LS*) system, which allows for free creation of hydraulic control systems in terms of the structure of functions and control logic, as well as in terms of their design, enabling free valve placing and adjustment of the valve block dimensions for installation. So far known mobile-type *LS* systems have mainly form of block manifolds, most often multi-section ones. The innovative *LS* modular system has been designed for the commonly used industrial standard, often called CETOP, which meets ISO 4401 mounting subplate pattern. To transmit *LS* signal the system uses port (*X*), which is dedicated to control line. The *LS* system (B type) presented in the article consists of two basic modules: an element for obtaining and processing the *LS* signal and an element for controlling the delivery of a constant pump, which for the *LS* type variable pump is unnecessary. As part of the developed system, few modified and adapted valve types and accessories were proposed, including e.g. directional control valves as an option. All other elements for building the system, the 4-way directional valves (classical and proportional) and modular (sandwich) valves are available on the CETOP valve market. The proposed *LS* system allows for cheap and easy way to apply the *LS* technology in industrial standard, as well as to easily modify existing systems, which enables energy savings. The article presents operation characteristics and also comparative analyzes of the energy efficiency of control systems with the applied *LS-B* system vs similar standard ones for several selected scenarios of system operation. The presented solution is patent pending.

Keywords: hydraulic system design · reduction of energy consumption · load sensing system

1 Introduction

For mobile devices, especially working machines produced in large series, dedicated systems of integrated load-sensing hydraulic drives (*LS*) are developed. These systems are characterized by a compact structure and a strictly defined configuration. The operation of the *LS* system is carried out through specially designed, usually multi-section

J. Stryczek and U. Warzyńska (Eds.): NSHP 2023, LNME, pp. 128–139, 2024.
https://doi.org/10.1007/978-3-031-43002-2_12

directional valves as well as variable-capacity pump controllers or control blocks for constant-capacity pumps. *LS* systems are more complex structures than standard solutions, but they give measurable benefits in energy savings. They can also positively affect dynamic properties and precision of working movement. Subsystems and elements of the *LS* system are the subjects of analysis in multiple leading research centers. Casoli et al. in [1] indicated the effects of the practical application of *LS* in a heavy-duty excavator, while Shang et al. [2] developed a novel integrated load-sensing valve-controlled actuator characterized by high efficiency and low energy consumption for a decrease in both overflow and throttling losses. Proposals of similar solutions leading to a reduction in energy consumption in load-sensing systems were made by Lisowski et al. [3] for a hydraulic system with multiple receivers connected in parallel, Wang et al. [4] for an electro-hydraulic servo valve systems and Chao et al. [5] related to a load-sensing pump design in an electro-hydrostatic actuator system. Significant energy losses may occur in hydraulic throttling control systems without *LS*, which are converted into heat. *LS* systems reduce energy losses and thus decrease the heat supplied to the working fluid. Replacing the classic control system with *LS* often allows you to avoid extended cooling systems. A disadvantage of the currently used elements of the *LS* systems is the lack of interchangeability with the standard power hydraulics components of general application. In industrial applications, hydraulic element assembling system based on the ISO 4401 standard [6] has been widely developed. This standard defines the dimensions for connecting elements, such as pressure, flow control, relief, and directional valves, and grouping these elements into blocks performing complex functions.

The base element for this analysis is the connection block in the form of a thick-walled plate with properly made channels and ports. It allows the interchangeability of components from different manufacturers and the installation of additional valves in a sandwich system without needing external connections. So far, this system lacks the channels and connectors necessary to create the *LS* line. The article indicates the possibility of easily expanding the system following the ISO standard while maintaining all its properties. It has also been demonstrated that minor structural modifications to the standard hydraulic components enable their application in the *LS* system. The *LS* system designed in this way was called *LS-B*. The article presents proprietary studies on which outcomes can be used to create different *LS-B* systems using standard hydraulic elements according to the ISO 4401 standard.

2 An ISO Mounting Plate for the LS Signal

Mounting blocks, together with ISO standard internal channels, enable the assembly of various types of valves in a tandem arrangement. Considering that the hydraulic signal does not require the transmission of significant energy but only information about a load of a given branch of the system, it was proposed to make additional small-diameter channels in the mounting plates and appropriate outputs for this signal. Due to the small diameters of these channels, there is no need to increase the dimensions of the existing ISO standard significantly. Figure 1 shows a schematic diagram of the modified mounting block for enabling the installation of four directional valves and, optionally, other components. A 3D model of such a solution is shown in Fig. 2. The existing

X connecting port was used to transmit the *LS* signal since it was not used for other purposes.

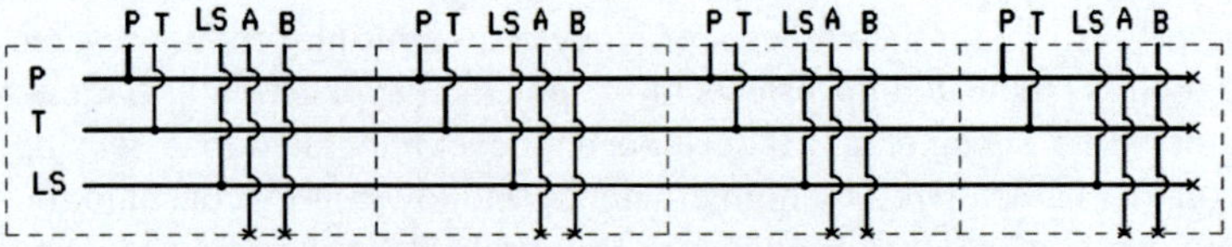

Fig. 1. Connection diagram inside the mounting block for four-way valves with LS signal

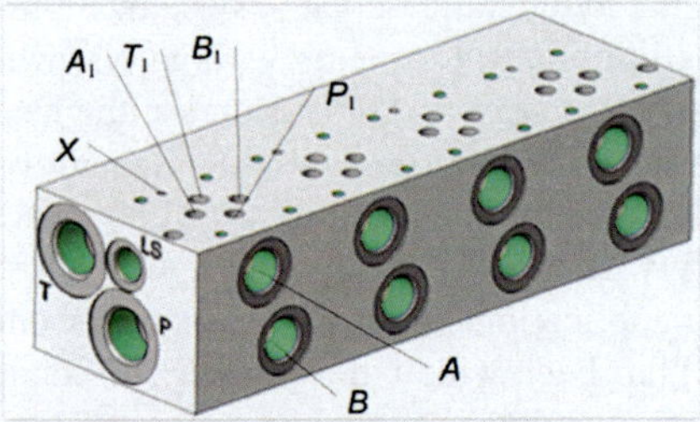

Fig. 2. 3D model of the mounting block with *LS* signal: *P* – supply port, *T* – return line, *LS* – load signal collection line, *A*, *B* – connecting ports, *X* – control signal, P_1 – pressure line port (1-st section), T_1 – return line port (1-st section), A_1, B_1 – actuator ports (1-st section)

3 LS Signal Acquisition

In *LS* systems, there is a need to acquire information about the load on particular receivers, regardless of the direction of the working movement. With multiple receivers, information is sought on which receiver is most heavily loaded at a given moment. Therefore, the signals from individual receivers should be compared, and the maximum value should be determined. Then the selected signal should then be transmitted to the variable displacement pump controller or fixed displacement pump release valve. For the method of operation adopted in this way, the following assumptions were adopted for the construction of the connection block:

a. selection of the highest pressure among all outputs of receivers,
b. resetting the *LS* signal value (discharging the pressure in the *LS* line) when all directional valves are in neutral positions (closed centre),
c. the use of standard, 4-way directional valves in the connection system of the closed centre type, according to ISO 4401,
d. cooperation with a variable-capacity pump with an LS controller or a fixed-capacity pump with an appropriate release valve.

For such assumptions, small-sized logic valves for placing in the stack of the connection block were developed. The connection dimensions were made following the ISO standard, and the channels were shaped in such a way that they could receive a receiver load signal. Then, the signal is compared with signals from other receivers, and its highest value is sent to the *LS* line. Figure 3 shows the construction of a logic valve intended for mounting under the directional valve body.

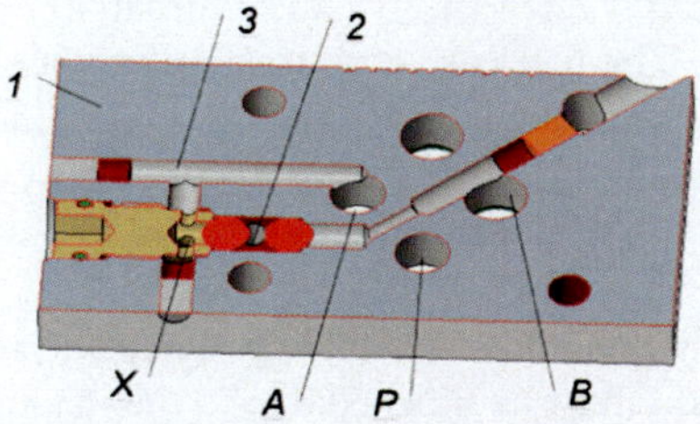

Fig. 3. Cross-section of a logic valve for the *LS* signal collection: 1 – body, 2 – movable elements, 3 – *LS* channels, *P* – supply port, *A*, *B* – connecting ports

3.1 Signal Transmission Paths

The presented system of collecting pressure signals from the receiver outputs A and B by directing the shut-off sides of the check valves towards each other. This enables a small fluid flow from the port with higher pressure and, thus, the transmission of the signal. At the same time, it blocks the connection with the lower pressure at the second valve port. As a result, we obtain the selection of a higher pressure value on a given operating section. In the further process of signal formation, the highest pressure value from all the sections should be chosen. As you can easily see, the selection of the highest pressure value will be guaranteed after connecting all the shut-off sides of the check valves to one common line. Moreover, the remaining valves will be blocked, not allowing the highest pressure to be discharged from all other outlets. To avoid the usage of external hoses for load signal transmission, a series of internal channels were made to supply signals to the main LS line in the connection block. Hence, all connected valves must have inner channels for the LS signal transmission. For this purpose, the X-signal channel defined in the ISO standard was used.

3.2 Method of *LS-B* Signal Discharge in the Directional Valve Neutral Position

When designing the system using the described method, it is necessary to ensure the *LS* line's zeroing, i.e. the pressure discharge in the neutral (middle) position of the directional control valve. Solving this problem for 4-way directional valves, assuming no interference in their construction and its implementation only by means of hydraulic control, requires additional assumptions and analysis. Therefore, it should be assumed that the directional valve's connection scheme in the neutral position ensures the pressure discharge from both channels (*A*, *B*) to the return line (*T*). In this aspect, two primary cases may be identified. The first case is a situation in which the receiver remains unloaded, and its position does not need to be locked, e.g. a hydraulic motor driving a device which performs a "kinematic" task (the device is loaded only during movement) or is equipped with its own brake. In this case, there is no need to use any additional receiver-blocking valves.

The second case includes a situation when the receiver remains loaded while at rest, and its position must be blocked due to the inadmissibility of its movement, e.g. a hydraulic cylinder lifting the load. Such cases will constitute a significant number and perhaps even the majority of practical applications. In such cases, controlled check valves (hydraulic locks) are usually used. To ensure the correct closing operation, the check valves should be supplied by directional valves, which relieve the operating channels (*A*, *B*) to the return line (*T*) in the neutral position. Hence, the connection scheme in the neutral position should be exactly the same as the one mentioned in the first case. The connection diagram of a directional control valve that meets the requirements of the *LS-B* system is shown in Fig. 4.

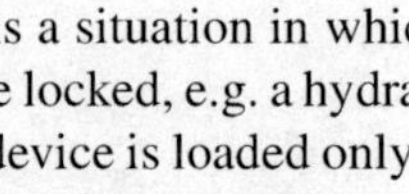

Fig. 4. Scheme of the directional control valve meeting the requirements of the *LS-B* system.

There is still the issue of the design solution of the LS-B system component, which is responsible for blocking the position of the loaded receiver (hydraulic lock, usually a controllable check valve or a brake valve). Each hydraulic lock should be installed in the supply line of the individual receiver after the directional valve, and its installation method is assumed to be sandwich-type according to ISO 4401. Hence, the method of signal transmission from the signal-drawing valve to the LS collector in the connecting plate must be determined. Because in the standard sandwich-operated check valves of the smallest nominal sizes (NS6 and NS10), there is usually no X-channel available, creating this additional flow path is necessary. Based on the analysis of such valves from different manufacturers, and taking into account that the signal can be sent through a small diameter channel, it is possible to drill additional passages in the bodies of these valves. Figure 5 shows a standard double-controlled check valve modification with the created channel for the LS signal transmission.

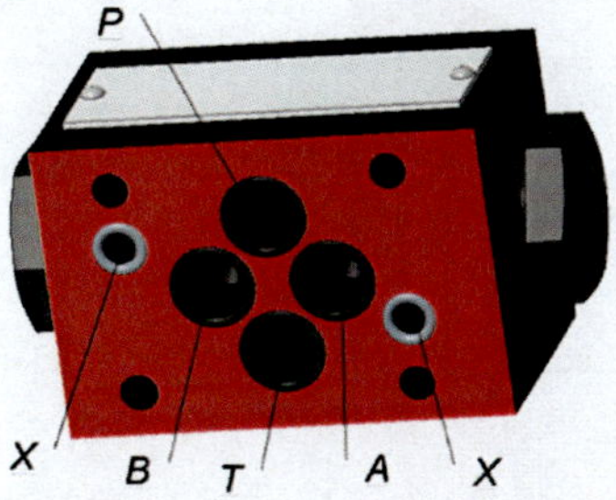

Fig. 5. Double pilot-operated check valve Z2S6 with the X channel [7].

4 Elements and Configuration of LS-B Systems

Theoretically, it is possible to install different types of valves in the ISO system. When using a connecting plate adapted to multiple valve sections, there are many combinations of the hydraulic system configuration. The sandwich-type components most often installed under the main directional control valve include a wide assortment of available relief valves, check valves, throttling and throttling-non-return valves, reducing and reducing-overflow valves, flow regulators, switching valves or pressure transmitters. All these elements can be used directly in the designed *LS-B* system without any modifications.

4.1 Directional Control Valves for *LS-B* Systems

Various kinds and types of directional valves from different manufacturers, with a plate connection by ISO 4401, may be used to control the receivers in the LS-B systems. Among them, there are commonly used electromagnetically-controlled WE spool valves [8] as well as electrically-controlled USAB proportional valves without feedback from the spool position [9] or similar USEB solutions with feedback [10]. The feedback provided by the lvdt sensor allows the spool of proportional valve to maintain a stable position, regardless of changing hydrodynamic forces acting on it. In this way, a constant value of the set throttling is ensured.

An important issue for *LS* systems is matching the pump capacity to the system demand. In the case of throttling control, proportionally-controlled valves are best suited for this function. In turn, in the case of the classic directional valve, this effect can be achieved by adjusting the chokes or reducing the opening of the flow area by limiting the spool stroke. Here, the useful accessories for standard control valves to achieve the assumed purpose are worth mentioning. Among such accessories, the UNE spool stroke adjuster deserves attention [11]. Using this element, it is possible to cheaply obtain manual or electromagnetically-switched throttling on the selected supply path of the receiver. It should be noted that by introducing the spool stroke limiter, the set throttling of the flow will be implemented both on the inflow to the receiver and the outflow. Figure 6 shows easy-to-apply spool stroke adjusters that are mounted (screwed) on a sleeve of the electromagnet.

Fig. 6. Directional control valve spool stroke adjuster, UNE6 type [11].

Another variant of a classic valve that can be used in the proposed *LS-B* system is the directional control valve with hydraulically adjustable speed (soft-shift valve) shown in Fig. 7. By adjusting the throttling in the channel which connects the electromagnet chambers with the use of two miniature throttling-return valves, it is possible to influence the spool speed independently in both directions and enable soft start-up and braking of receivers without introducing significantly more expensive proportional directional valves.

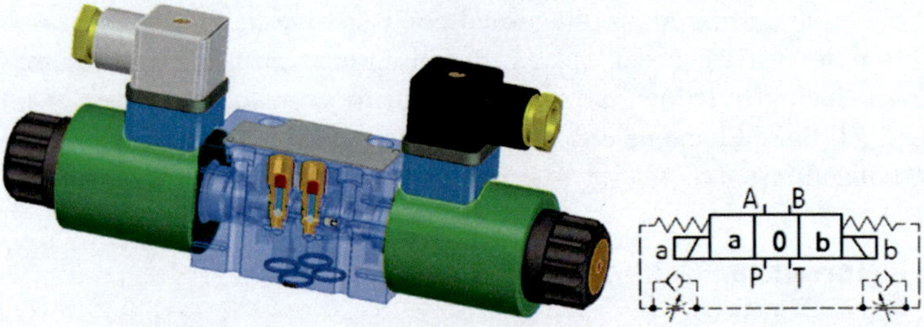

Fig. 7. Soft-shift type directional control valve and its scheme.

4.2 Example Configuration of a Hydraulic System Based on the *LS-B* Solution

Figure 8 shows the *LS-B* system with three directional valves controlling the operation of three hydraulic cylinders (6). The system comprises three proportional directional valves (1), *LS* logic valves (2), controlled check valves (3) (blocking the movement of actuators in the absence of power), *LS-B* release valve (4) (for the fixed-capacity pump) and the connection block (5).

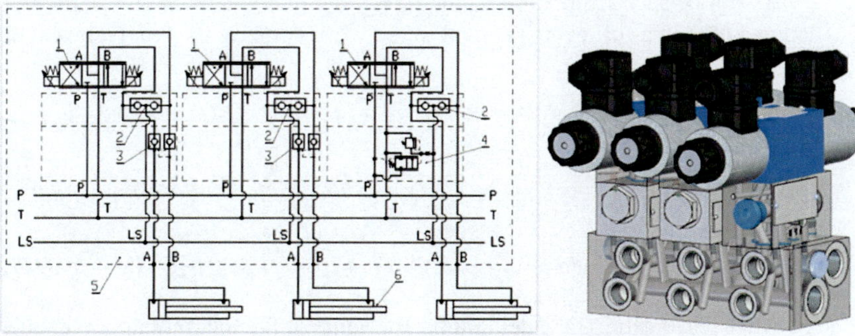

Fig. 8. An exemplary *LS-B* system for controlling the operation of three actuators powered by one fixed-capacity pump; hydraulic diagram and control block.

5 Energetic Analysis of Selected Applications of the *LS-B* System

The carried-out analyses compared the proposed *LS-B* system with a standard design, including a fixed-capacity pump and a variable-capacity pump with the load-sensing controller. These analyses aimed to determine the suitability of the *LS-B* system for various parameters and operating modes. The reference point was the theoretical energy demand of a standard hydraulic system without the load-sensing feature.

5.1 Theoretical Energy Demand of a Standard System

A system consisting of four receivers with different flow rates, loads and working cycle times was adopted for the analysis. To simplify the calculations, constant values of the speed and load of receivers were assumed in individual steps of each working cycle. Due to the expected primary application scenario of the proposed system, which is a system with small or medium pump capacity and medium or high operating pressure, the following parameter values were adopted for the analysis: constant capacity pump $Q = 20$ dm^3/min and nominal working pressure of the most loaded receiver $p = 20$ MPa. The parameters of individual receivers are listed in Table 1.

Table 1. Parameters of receivers for calculating energy demand.

No	Flow rate Q_i (dm^3/min)	Load pressure p_i (MPa)	Cycle time T_i (s)
Receiver 1	5	20	5
Receiver 2	10	15	10
Receiver 3	15	10	5
Receiver 4	20	5	10
Standstill	0	0	5

As can be seen from Table 1, the full cycle time is $T = 35$ s. The theoretical energy demand (E_t) in a full cycle is equal to the work done by the receivers and amounts to:

$$W = \sum_{1}^{4} W_i, \text{ where } W_i = \int_0^{T_i} p_i(t)\, Q_i(t)\, dt = p_i Q_i T_i \tag{1}$$

$$p_i(t) = const = p_i, Q_i(t) = const = Q_i \; therefore W_i = p_i Q_i T_i \tag{2}$$

$$E_t = W, \; where \; W = p_1 Q_1 T_1 + p_2 Q_2 T_2 + p_3 Q_3 T_3 + p_4 Q_4 T_4 \tag{3}$$

After performing the calculations, the value of the theoretical energy demand, equal to the effective work, is obtained and amounts to $E_t = 62.5$ kJ. The result in the graphic form is shown in Fig. 9.

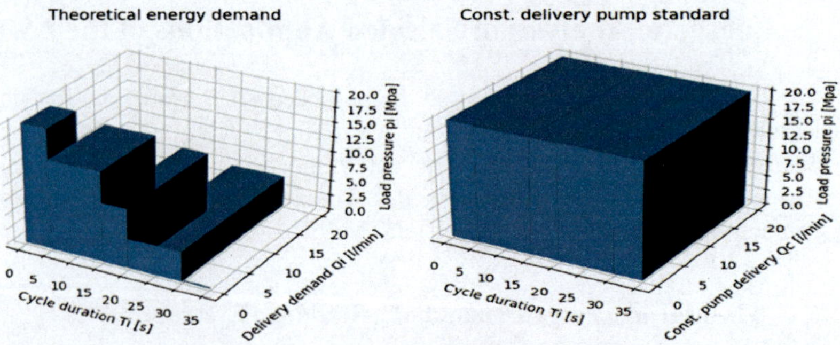

Fig. 9. Theoretical energy demand vs actual energy consumption in the working cycle

The theoretical required hydraulic power of the drive with a fixed-capacity pump (N_t) was determined based on the maximum flow demand (receiver number 4) and the maximum load (receiver number 1):

$$N_t = p_1 Q_4 \tag{4}$$

After performing the calculations, the theoretical requirement for the power of the motor drive is obtained: $N_t = 6.67$ kW. Obviously, after taking into account the efficiency of the drive, the motor will be selected with a correspondingly higher power, e.g. $N = 7.5$ kW. Assuming for simplicity that there is no pump discharge during a short standstill, which would have to be implemented by an additional valve and a suitable control system, the actual energy consumption E_r for the entire working cycle T of the standard system is:

$$E_r = N_t T \tag{5}$$

Which results in $E_r = 233.33$ kJ. Thus, the energy efficiency of the cycle is calculated as $\eta_1 = E_t/E_r \cdot 100 = 26.8\%$. This implies that 73.2% of the energy (170.83 kJ) is converted into heat, which must be removed from the system for the working fluid temperature to remain stable. This involves a significant investment in the cooling system. In the case of using a more expensive variable-capacity pump with a pressure regulator, the energy required to complete the entire working cycle (E_p) is:

$$E_p = pQ_1T_1 + pQ_2T_2 + pQ_3T_3 + pQ_4T_4, \tag{4}$$

which results in $E_r = 133.33$ kJ, and the theoretical energetic efficiency of the cycle is $\eta_2 = 46.8\%$.

5.2 Calculations for LS-B System with the Fixed-Capacity Pump

Calculations for the *LS-B* system powered by a fixed-capacity pump are carried out in a similar way. In the event of an excess of working fluid capacity over the system demand, the relief valve discharges the excessive flow rate to the tank at the load pressure of the receiver increased by the relief valve activation pressure, which was assumed for the calculations as $\Delta p = 1.0$ MPa:

$$\begin{aligned} E_c = {} & (p_1 + \Delta p)Q_cT_1 + (p_2 + \Delta p)Q_cT_2 + (p_3 + \Delta p)Q_cT_3 + \\ & (p_4 + \Delta p)Q_cT_4 + \Delta pT_0 \end{aligned} \tag{7}$$

In this case, the energy requirement is $E_c = 128.33$ kJ. Thus, the efficiency equals $\eta_3 = 48.7\%$ and is almost twice as high as for a standard fixed-capacity pump system without load sensing. Compared to the variable displacement pump, the result is similar (slightly better). Figure 10b shows a graphic interpretation of the E_c energy demand for the whole working cycle of the LS-B system with the fixed capacity pump.

5.3 Calculations for LS-B System with the Variable-Capacity Pump

In the *LS-B* system powered by a variable-capacity pump, in the event of an excess flow rate over the system demand, the load-sensing controller reduces the pump output at the load pressure of the receiver increased by the activation pressure of the controller, which for calculations was assumed $\Delta p = 1.0$ MPa:

$$E_v = (p_1 + \Delta p)\,Q_1T_1 + (p_2 + \Delta p)\,Q_2T_2 + (p_3 + \Delta p)\,Q_3T_3 + (p_4 + \Delta p)\,Q_4T_4 \tag{8}$$

In addition, it was assumed that internal leakage is negligible in the state of zero pump capacity. The resultant energy demand is $E_v = 69.17$ kJ. Therefore, the theoretical energy efficiency of the working cycle is equal to $\eta_4 = 90.4\%$, which is more than three times higher than for the base system. The graphs of the theoretical demand and estimated energy consumption for all considered variants are presented in Fig. 10.

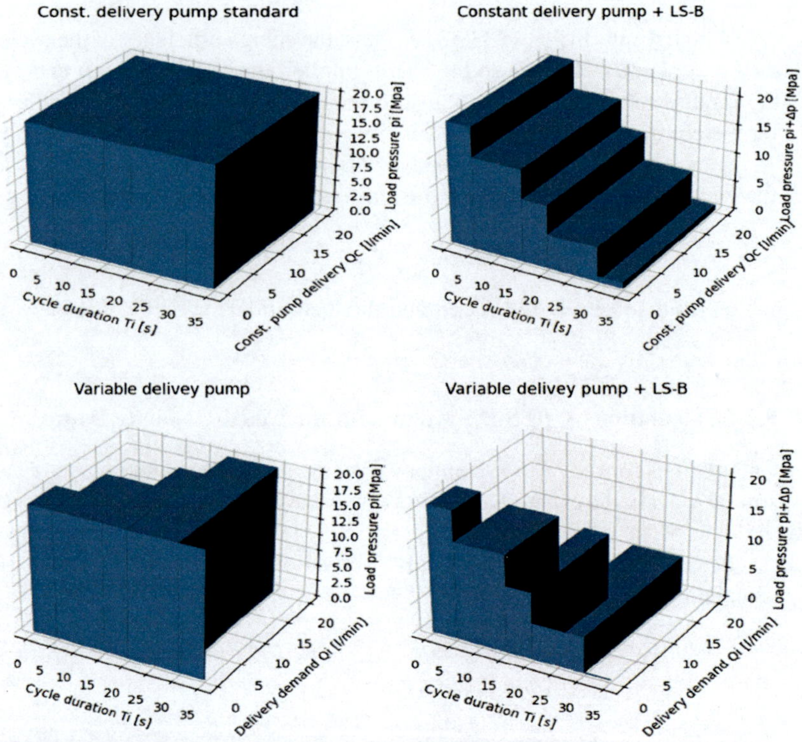

Fig. 10. Estimated energy requirement for the considered hydraulic system types while performing the assumed working cycle: a) standard system with a fixed-capacity pump, b) *LS-B* system with a fixed-capacity pump, c) variable-capacity pump with the pressure controller, d) *LS-B* system with a variable-capacity pump.

6 Summary and Conclusions

The article presents a method of building load sensing systems using typical hydraulic elements compliant with the standard ISO 4401. The system built in this way is briefly called LS-B. A geometrical analysis of widely produced hydraulic elements, such as directional control valves, throttle and throttle-check valves, flow regulators, pressure control valves (relief, reduction, sequence), check and pilot operated check valves, etc. was carried out. It was shown that after introducing an additional path X in some elements or using the existing one, it is possible to use them in the proposed load sensing system (LS-B). The structural and functional analysis of the innovative LS-B load sensing system, presented in the article and dedicated mainly for use in industry, showed its ease of application and the use of available elements compliant with the ISO standard. So far, there are no solutions of industrial systems implementing load sensing solutions to such extent. Energy analysis of a simulated case showed that savings after a simple modification of existing standard systems can be as high as 48% for variable displacement pump systems with an LS controller and up to 45% for fixed displacement pump systems.

The proposed LS-B system can be used in a simple and low-cost way to modify the existing systems. An additional benefit of lower energy losses in load sensing (LS-B) type drives is the ability to eliminate cooling systems of working fluid that are expensive in investment and operation.

References

1. Casoli, P., Scolari, F., Vescovini, C.M., Rundo, M.: Energy comparison between a load sensing system and electro-hydraulic solutions applied to a 9-ton excavator. Energies **15**, 2583 (2022). https://doi.org/10.3390/en15072583
2. Shang, Y., Liu, X., Jiao, Z., Wu, S.: An integrated load sensing valve-controlled actuator based on power-by-wire for aircraft structural test. Aerosp. Sci. Technol. **77**, 117–128 (2018). https://doi.org/10.1016/j.ast.2018.02.030
3. Lisowski, E., Filo, G., Rajda, J.: Analysis of the energy efficiency improvement in a load-sensing hydraulic system built on the ISO plate. Energies **14**(20), 6735 (2021). https://doi.org/10.3390/en14206735
4. Wang, W., Wang, B.: An energy-saving control strategy with load sensing for electro-hydraulic servo systems. Strojniški vestnik – J. Mech. Eng. **62**, 709–716 (2016). https://doi.org/10.5545/sv-jme.2016.3685
5. Chao, Q., Zhang, J., Xu, B., Shang, Y., Jiao, Z., Li, Z.: Load-sensing pump design to reduce heat generation of electro-hydrostatic actuator systems. Energies **11**(9), 2266 (2018). https://doi.org/10.3390/en11092266
6. ISO 4401:2005 Hydraulic Fluid Power—Four-Port Directional Control Valves—Mounting Surfaces (2005). www.iso.org. Accessed 07 Apr 2023
7. Double pilot-operated check valve Z2S, Ponar Wadowice catalogue.https://www.ponar-wadowice.pl/en/type/273-cut-offcheck-valves-check-valves-pilot-operated-sandwich-twin-z2s. Accessed 11 Apr 2023
8. Directional electrically operated spool valve WE, Ponar Wadowice catalogue.https://www.ponar-wadowice.pl/en/type/14-directional-control-valves-directional-control-valves-subplate-cetop-on-off-electrically-controlled-we. Accessed 11 Apr 2023
9. Proportional directional valve USAB, Ponar Wadowice catalogue. https://www.ponar-wadowice.pl/en/type/399-directional-control-valves-directional-control-valves-subplate-cetop-proportional-proportionally-controlled-usab. Accessed 11 Apr 2023
10. Proportional spool valve USEB, Ponar Wadowice catalogue. https://www.ponar-wadowice.pl/en/type/401-directional-control-valves-directional-control-valves-subplate-cetop-proportional-proportionally-controlled-useb. Accessed 11 Apr 2023
11. Adjustment of the electromagnet stroke, UNE type, Ponar Wadowice catalogue. https://motorimpex.ua/files/downloads/UNE6.pdf. Accessed 11 Apr 2023

Rehabilitation Tricycle with Pneumatic Drive System

Jakub Takosoglu[1](✉), Ryszard Dindorf[1], Piotr Woś[1], Jacek Jegier[2], Andrzej Sternik[2], Henryk Woliński[2], Jan Marciniak[2], Jacek Pusz[2], and Jacek Krolski[2]

[1] Kielce University of Technology, Aleja Tysiąclecia Państwa Polskiego 7, 25-314 Kielce, Poland
{qba,dindorf,wos}@tu.kielce.pl

[2] Beta Cosmos Sp. z o.o., ul. Sąsiedzka 11, 03-168 Warszawa, Poland
jjegier@interia.pl, hwolinski@wp.pl

Abstract. Rehabilitation of people with lower limb dysfunctions concerns people after orthopedic injuries, strokes and at present mainly elderly people. As a result of the extension of human life, civilization progress and improvement in the quality of life in the world, the percentage of people in post-working age is systematically growing. This trend is particularly visible in highly developed countries. The most common of musculoskeletal disorders in the elderly is osteoarthritis of the large joints of the limbs. Currently more than 55 million people live with dementia worldwide, and there are nearly 10 million new cases every year. Degenerative changes are mostly caused by obesity, excessive overload in earlier periods of life (hard physical work) and metabolic disorders. Rehabilitation of people is aimed at preventing the dangerous consequences of long-term immobilization, maintaining the physiological range of motion in the joint, preventing the formation of contractures and preventing degeneration of the articular cartilage. An important challenge is the activation of the elderly through devices supporting movement and rehabilitation devices. One of the possibilities are three-wheeled rehabilitation bicycles. The article presents the design of a three-wheeled rehabilitation bicycle with an innovative pneumatic drive. Thanks to the solutions used, the tricycle is environmentally friendly. Heavy and problematic to dispose of gel or lithium-ion batteries were eliminated and replaced with composite cylinders for high-pressure compressed air. The tricycle is characterized by many positive features, such as: quiet operation, low weight, the ability to adjust the load on the rehabilitated limbs, the ability to move freely in the absence of a power source and travel at high speed.

Keywords: rehabilitation · rehabilitation tricycle · pneumatic motor · pneumatic servo-drive

1 Introduction

The gradual increase in human longevity, the increasing progress of civilisation, the development of new high technologies and the improvement in the quality of life in the world mean that the number of persons at the post-working age continues to increase. This

J. Stryczek and U. Warzyńska (Eds.): NSHP 2023, LNME, pp. 140–151, 2024.
https://doi.org/10.1007/978-3-031-43002-2_13

is particularly evident in highly developed countries. For instance, in Japan, estimates show that in 2030 as many as 30% [1] of citizens will be over 65 years old [3]. In the European Union (EU), almost one in five persons (19.4%) or nearly 100 million people, are aged 65 years and older [1]. The effect of an ageing population is a high old-age dependency ratio, which was as high as 29.2% in 2017 in EU countries [1], resulting in approximately three persons at an working age less for each person aged at least 65. The issue of an ageing population its consequences also affect Poland. In accordance with the demographic projections made by the Central Statistical Office (CSO), in Poland in 2050, the proportion of people at working age in the population is going to equal 57% [1]; the share of persons over 65 years of age will be as high as 32.7% [1], while the share of persons at pre-working age is going to be lower than 11% [1].

The World Health Organization (WHO) states that at a global scale, 55 million people suffer from dementia, with 10 million new patients added each year. The total count is expected to triple in 30 years. Canadian researchers state that the sedentary lifestyle, lack of exercise and physical activity are factors of importance similar to genetic predispositions. A study covering a sample of over 1,600 adults over the age of 65 has shown that a sedentary lifestyle is associated with the same risk of dementia as having a mutation in apolipoprotein E (APOE) which increases the risk of dementia.

Issues related to rehabilitation affect not only persons with senile dementia, but also persons with dysfunctions of lower limbs, those who suffered with injuries, traffic accidents, as well as the elderly. One of the possibilities for rehabilitation and increased physical activity involves the use of a bicycle [7, 9, 10] – a common solution since it is a type of equipment which is the least demanding on the human body. Older persons are using bicycles increasingly more often. “KIELCE BIKE – EXPO 1/2016” Polish Bicycle Market states that Polish citizens regularly use bicycles. Approximately 1 million people report that they commute to work and school every day. Currently, rehabilitation therapists use stationary equipment [2, 8] for lower and upper limb rehabilitation and electric assisted bicycles, albeit to a much lesser extent. The tricycle can be used for rehabilitation and physical activity due to its many positive features, such as quiet operation, low weight, the ability to adjust the load on the rehabilitated limb, the ability to continue moving after exhausting pneumatic pressure supply, and the ability to move at high speed.

2 The Design of a Rehabilitation Tricycle

The key issue to be solved involves selecting the drive unit and its system design in a way which ensures the following characteristics:

- power of 250W (peak power 500W),
- torque of 35–50 Nm,
- high efficiency (low air consumption),
- the low weight of the drive train,
- a relatively simple and failure-free design,
- an easily reducible speed range at a cadence of approximately 60 rpm,
- main gearbox speed 200–300 rpm,
- maximum speed of 15 km/h.

The following types of pneumatic actuators were selected based on the analysis and initial design assumptions:

- vane motor GAST 1AM-NRX-64A,
- pneumatic cylinder SMC CD85N25–250-B,
- piston motor Globe Benelux RM012-SXXCW,
- piston motor PMO 1800.

For each of the selected drive solutions, idealised pneumatic diagrams were developed and a selection of components was made to meet the project requirements. Figure 1 shows the pneumatic control system schematic [4, 5] for the drive in the form of a vane motor 1AM-NRX-64A manufactured by GAST. Table 1 presents a list of the pneumatic system components.

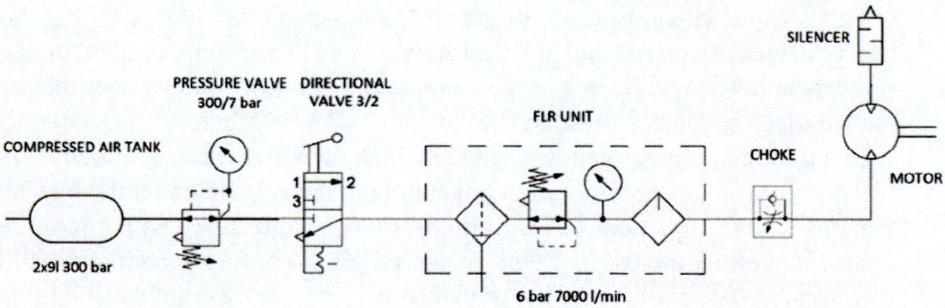

Fig. 1. Pneumatic diagram of the drive system with 1AM-NRX-64A motor.

Table 1. List of components of the pneumatic system for the 1AM-NRX-64A motor.

No	Description	Producer	Type
1	Compressed air tank	SAFER	Cylinder 9.0 L
2	Pressure valve	PNEUMAT	Futura FU 7210
3	Directional valve 3/2	PNEUMAT	322 LL90
4	FLR unit	PNEUMAT	FR+L G1/4 0-12bar 20 μm
5	Choke	SMC	AS2002FS-06
6	Motor	GAST	1AM-NRV-63A
7	Lubricator	GAST	AH102L
8	Silencer	GAST	AF350

Figure 2 shows the schematics of the pneumatic control [4, 5] for a drive in the form of a pneumatic double-acting cylinder with a single-sided piston rod CD85N25–250-B manufactured by SMC. Table 2 presents a list of components of the pneumatic system shown in Fig. 2.

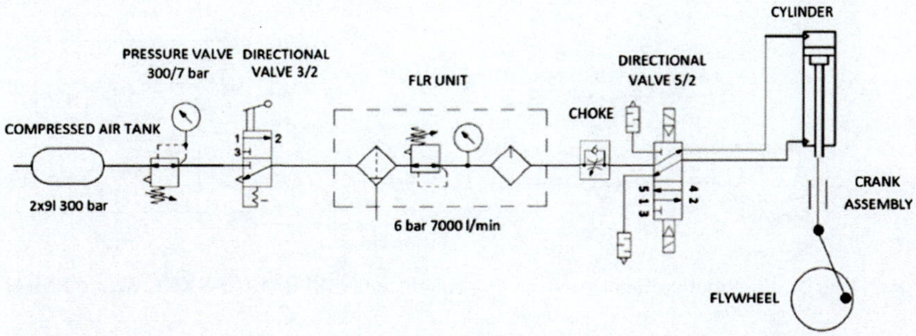

Fig. 2. Pneumatic diagram of the control system with the CD85N25–250-B actuator.

Table 2. List of components of the pneumatic system of the CD85N25–250-B actuator.

No	Description	Producer	Type
1	Compressed air tank	SAFER	Cylinder 9.0 L
2	Pressure valve	PNEUMAT	Futura FU 7210
3	Directional valve 3/2	PNEUMAT	322 LL90
4	FLR unit	PNEUMAT	FR+L G1/4 0-12bar 20 μm
5	Choke	SMC	AS2002FS-06
6	Directional valve 5/2	SMC	SY3220-6LOU-M5-Q
7	Silencer	SMC	AN05-M5
8	Pneumatic cylinder	SMC	CD85N25–250-B

Figure 3 shows the pneumatic control system schematic [6] for the RM012-SXXCW multi-piston motor drive manufactured by GLOBE BENELUX and the PMO1800 manufactured by PTM Mechatronics. Table 3 presents a list of components of the pneumatic system shown in Fig. 3.

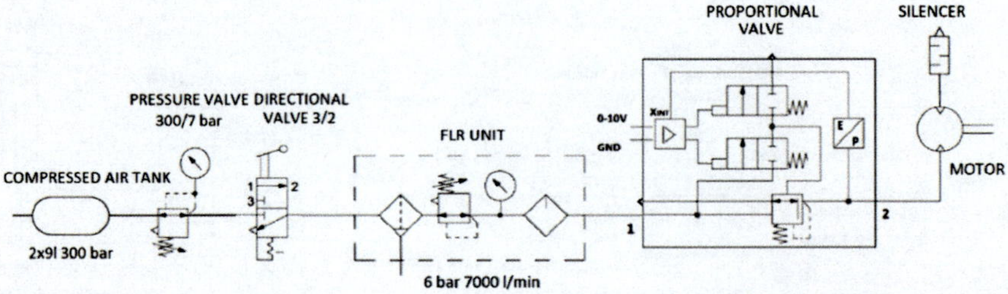

Fig. 3. Pneumatic diagram of the control system with RM012-SXXCW and PMO1800 drives.

Table 3. List of pneumatic components for RM012-SXXCW and PMO1800 drives.

No	Description	Producer	Type
1	Compressed air tank	SAFER	Cylinder 9.0 L
2	Pressure valve	PNEUMAT	Futura FU 7210
3	Directional valve 3/2	PNEUMAT	322 LL90
4	FLR unit	PNEUMAT	FR+L G1/4 0-12bar 20 μm
5	Proportional valve	Festo	VPPE-3-1_8–6-010
6	Motor RM012-SXXCW/PMO 1800		
7	Silencer		

For the control systems shown above, it was necessary to design a system that would reduce the rated speed to values suitable for the designed bike. The necessary reduction mechanism calculations were performed. The summary of the data is summarised in Table 4.

A direct transmission was assumed in the case of the control system with the CD85N25–250-B actuator.

Table 4. Calculation of reduction mechanisms for RM012-SXXCW and PMO1800 motors

Wheel no	G1	G2	G3	G4	G5	Speed
Number of teeth	20	29	20	18	18	km/h
RPM	100	100	31	34,1	34,1	3,87
			53	47,4	47,4	5,38
			64	71,6	71,6	8,11
			75	67,3	67,3	7,63
			100	111,1	111,1	12,60
			122	110,1	110,1	12,48
			142	157,7	157,7	17,88
			162	145,4	145,4	16,48
Type	motor axis	hub	hub	intermediate axis	freewheel	

3 Prototyping

In accordance with the design brief, the pneumatic systems were built as shown in Figs. 1, 2 and 3, mechanical designs for the various drive types were made, solid models of the tricycle drives were made and electrical schematic designs were made to control the drives. No electrical control is required for the drive system with 1AM-NRX-64A motor manufactured by GAST. Based on the laboratory tests carried out and the analysis of the results obtained, the following section presents drive systems using RM012-SXXCW and PMO1800 motors, which are the only ones that meet the design parameters. Figure 4 shows a solid model of the drive system for the RM012-SXXCW motors from GLOBE BENELUX and PMO1800 from PTM Mechatronics, while Fig. 5 shows a diagram of the control system.

Fig. 4. Solid model of the drive train for the RM012-SXXCW and PMO1800 motors.

The system includes the following components: proportional pressure valve [6], brake with sensor, LCD display, Arduino Nano microprocessor, support with pressure sensor, potentiometer, voltage converter, PWM converter, voltage stabiliser, switch, battery cage.

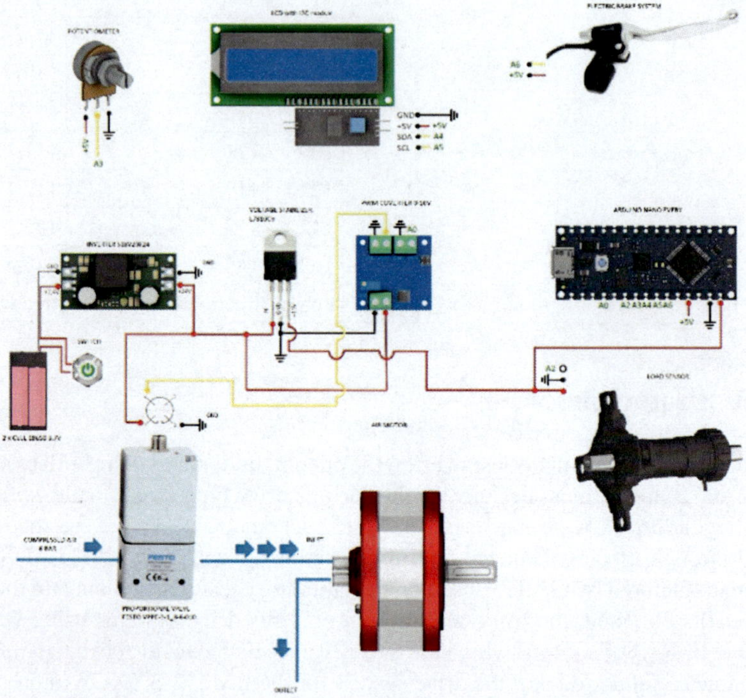

Fig. 5. Control circuit diagram for RM012-SXXCW and PMO1800.

3.1 Operation Mode Description

The system is powered by two 18650 cells with 3.7V 3000mAh. The control system uses three voltage levels of 5V, 10V, 24V DC. Appropriate inverters were used to generate the voltages. The key component of the entire system is the contact force sensor. It generates a voltage value ranging from 1.5V (no pressure) to 3V (maximum pressure). This signal is then sent to the Arduino module, where it is converted into a digital value. As the voltage rises above a certain fixed value, the PWM signal is increased. This signal is then passed to the LCT-167 converter. The converting circuit converts the PWM signal to a voltage ranging from 0-10V. The pressure proportional valve that supplies compressed air to the motor is controlled by the 0-10V voltage. A mechanical brake with sensor is also fitted to the system. The control algorithm manages the PWM signal in such a way that there is no pneumatic assistance when the brake is engaged. The braking function has priority over the other functions. The other component is the potentiometer, which acts as a voltage divider in the range of 0–5 V. The signal from the potentiometer is sent to the microcontroller. The microcontroller interprets the voltages in the following manner:

- 0–1.25 V as level 0 (no support),
- 1.26 V–2.5 V as level 1 support,
- 2.6–3.75 V as a level 2 support,
- **3.76-5V as level 3 support (the PWM signal has a fill of 100%, the proportional valve** is fully open with the exception for when the brake is engaged).

The proposed control system of an air-assisted rehabilitation tricycle is shown schematically in Fig. 6.

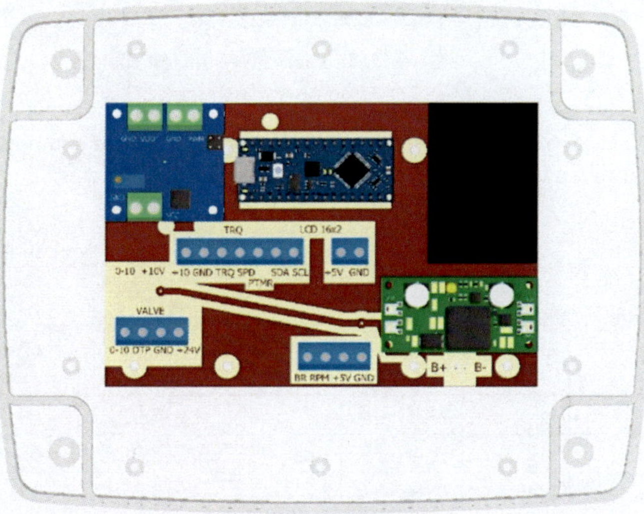

Fig. 6. Diagram of the complete tricycle control system.

4 Laboratory Test

A permanent magnet generator producing 24V DC and rated at 250W at 2,700 rpm was **used for laboratory testing. Highly efficient gear units have been used for the tests. Five** 60W 24VDC incandescent lamps connected in parallel provided the power. The voltage drop across the bulb terminals was measured as a function of the power output from the bike's drive system. The measurement was carried out over a period of 2 min, during which the drop in pressure of the compressed air in the cylinder was read, from which the instantaneous air consumption was determined. The optimum load parameters for **the air motor and a prototype for field testing were determined from the results.** Figure 7 shows a functional diagram of the test stand.

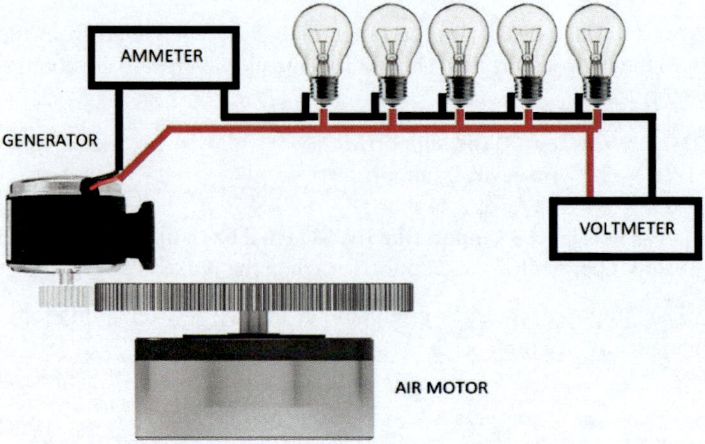

Fig. 7. Diagram of the test stand.

Tables 5, 6 and 7 show the results of laboratory tests of compressed air consumption depending on the generated power for the used drive systems.

Table 5. Laboratory test results for RM012-SXXCW motor

Pressure (bar)	Motor speed (rpm)	Voltage (V)	Air consumption (l/s)	Current (A)	Power (W)	Coefficient (W/l)
3	94	20,1	1,7	3,0	60	35,29
3,5	112	21,4	1,88	3,0	65	34,57
4	124	22,8	2,1	3,6	83	39,52
4,5	130	23,1	2,3	4,1	95	41,30
5	147	24,2	2,45	4,8	115	46,94
5,5	185	24	2,8	5,0	120	42,86
6	201	24,1	3,1	5,2	126	40,65

Table 6. Laboratory test results for PMO1800 motor

Pressure (bar)	Motor speed (rpm)	Voltage (V)	Air consumption (l/s)	Current (A)	Power (W)	Coefficient (W/l)
3	47	18	1,8	4,6	83	46,11
3,5	57	18,6	1,9	4,9	92	48,42
4	75	19,7	1,97	5,1	101	51,27
4,5	83	20	2,05	5,5	110	53,66
5	98	22,7	2,2	6,2	140	63,64
5,5	127	23	2,8	6,5	150	53,57
6	153	24,1	3,4	6,8	165	48,53

Table 7. Laboratory test results for 1AM-NRX-64A motor

Pressure (bar)	Motor speed (rpm)	Voltage (V)	Air consumption (l/s)	Current (A)	Power (W)	Coefficient (W/l)
3	1800	16	3,65	6,3	100	27,40
3,5	2700	18,1	4,45	8,3	150	33,71
4	3600	22,1	5	7,7	170	34,00
4,5	4250	24,2	5,5	8,7	210	38,18
5	5050	25,1	6,1	8,8	220	36,07
5,5	6200	24,6	6,9	10,0	245	35,51

5 Field Tests of the Prototype

Tests, including laboratory tests, were carried out on a prototype built in accordance with the design specification. A number of rides were made with different bike settings, i.e. different levels of power steering and different gear settings. Results were recorded using a GPS-based application. The surveys were carried out on a paved tarmac surface with hills of up to 50 m in height and gentle bends. Figure 8 shows a general view of a complete air-assisted rehabilitation tricycle that has been prepared for testing in the field. Tables 8 show the results of field tests.

Table 8. Field test results of tricycle

Amount of used air (dm^3)	Hub (gear no)	Derailleur (gear no)	Support (level)	Distance (km)	Ride time (min)	Avr speed (km/h)
5310	8	7	3	20,8	91	13,7
5220	8	7	2	7,93	35	13,6
2520	7	7	2	2,46	14	10,2
2610	8	7	3	10,8	50	12,9
5310	8	n/d	4	5,56	19	17,5
5175	6	6	3	15,1	89	10,1
5310	6	5	3	12,2	79	9,2
4770	5	n/d	4	7,01	28	15,1
5310	8	7	1	4,62	21	13,5
5220	1	7	3	3,18	16	12,2

Fig. 8. Air-assisted rehabilitation tricycle prototype.

6 Conclusions

The laboratory and field tests carried out confirmed the achievement of the assumed operational and performance parameters of the prototype of the air-assisted rehabilitation tricycle. Riding an assisted bike makes moving around much easier and makes cycling enjoyable and not physically demanding. The bike has a wide range of settings that can have an impact on the length of the ride, the speed of the ride and the distance that the rider can cover. The control system makes it possible to drive on difficult terrain (e.g. to navigate on grassy paths, to get out of a dip or to drive up a kerb). It makes it significantly easier to start and stop riding. Testing has confirmed that the bike can be used for rehabilitation purposes.

References

1. Starzenie się społeczeństwa – wyzwanie dla rynku pracy, aktywizacja pracowników 50+. Raport tematyczny. Instytut Analiz Rynku Pracy Sp. z o.o., PARP (2020)
2. Maculewicz, J., Serafin, S., Busk Kofoed, L.: A Stationary bike in virtual reality - rhythmic exercise and rehabilitation. In: Conference: DCBIOSTEC 2015 – Doctoral Consortium on Biomedical Engineering Systems and Technologies (2015)
3. Belforte, G., Eula, G., Ivanov, A., Sirolli, S.: Soft pneumatic actuators for rehabilitation. Actuators **3**(2), 84–106 (2014). https://doi.org/10.3390/act3020084
4. Dindorf, R.: Napędy Płynowe. Wydawnictwo Politechniki Świętokrzyskiej, Kielce (2009)
5. Dindorf, R., Takosoglu, J.E., Wos, P.: Development of Pneumatic Control Systems, Kielce (2018)
6. Blasiak, S., Takosoglu, J.E., Laski, P.A.: Optimizing the flow rate in a pneumatic directional control valve. In: Fuis, V. (ed.) Engineering Mechanics 2014. 20th International Conference on Engineering Mechanics, pp. 96–99. Praque (2014)
7. Kotler, D., Babu, A.N., Robidoux, G.: Prevention, evaluation, and rehabilitation of cycling-related injury. Curr. Sports Med. Rep. **15**(3), 199–206 (2016)
8. Yum, H., Kim, H., Lee, T., Park, M.S., Lee, S.Y.: Cycling kinematics in healthy adults for musculoskeletal rehabilitation guidance. BMC Musculoskelet. Disord. **22**, 1044 (2021). https://doi.org/10.1186/s12891-021-04905-2
9. McLeod, D., Blackburn, T.A.: Biomechanics of knee rehabilitation with cycling. Am. J. Sports Med. **8**(3), 175–180 (1980). https://doi.org/10.1177/036354658000800
10. Fidelis, O.P., Falola, A.O., Odebiyi, D.O., Nwoye, E.O.: Development of bi-modal exercise bicycle for physical fitness and rehabilitation. Niger. J. Technol. **37**(3), 846–852 (2018)

Design of the Stand for Experimental Tests of Pneumatic Bellows Actuators

Jakub Takosoglu, Kamil Ziejewski(✉), Ryszard Dindorf, Piotr Woś, and Łukasz Chłopek

Department of Mechatronics and Armaments, Faculty of Mechatronics and Mechanical Engineering, Kielce University of Technology, al. Tysiąclecia Państwa Polskiego 7, 25-314 Kielce, Poland

kziejewski@tu.kielce.pl

Abstract. The article contains an analysis of the possibility of precise position control of a bellows actuator. For this purpose, a special research stand was built. It allows an accurate determination of the static and dynamic characteristics of air springs under variable loads generated by a pneumatic cylinder. The bellows is placed on a prototype base, enabling, apart from precise height measurement, determination of changes in the volume of the bellows immersed in liquid. Based on the experimental data, a formula for the effective area of the bellows was derived, depending on its height and the applied load. Subsequently, a mathematical model of the stand was made which was used to carry out further simulation tests in the MATLAB /SIMULINK environment.

Keywords: Pneumatic · Elastic · Bellows Actuator

1 Introduction

The bellows actuators belong to the group of flexible pneumatic actuators. They are used in applications where a linear mass shift is necessary and where it is necessary to eliminate vibrations. The bellows are used mainly for flexible mounting of vibration and impact-generating equipment, such as presses, power hammers, textile equipment, hoists, conveyors, and heavy equipment. The bellows are also used in passenger cars and trucks in pneumatic suspension control systems or for flexible mounting of vehicles and semi-trailers, which allows for the isolation of shocks and vibrations of vehicle components. An important application of pneumatic bellows is the elimination of vibrations in technological processes. They are widely used to isolate laboratory equipment from vibration. Pneumatic pillows have many specific properties that allow them to be used in conditions that do not allow the use of classic actuators. Their installation height is much lower than that of a typical pneumatic actuator. The natural compensation of assembly inaccuracies and the adaptation to tilt angles of up to 30° (depending on the type) mean that complicated connecting elements or hinges are usually not required. The bellows are suitable for compressed air or other gaseous media such as nitrogen. They

J. Stryczek and U. Warzyńska (Eds.): NSHP 2023, LNME, pp. 152–161, 2024.
https://doi.org/10.1007/978-3-031-43002-2_14

can also be used in systems with oil-free compressed air in the food industry. The bellows actuator is also able to operate reliably in low-pressure ranges with hydraulic media such as water or glycol. It is practically maintenance-free even in the presence of dirt, dust, granules, or deposits and does not require lubrication to ensure proper operation. The bellows are characterized by high resistance to dynamic loads and long service life, even in extreme operating conditions. They are resistant to weather conditions and many other negative environmental factors, including aggressive chemicals. Figure 1 shows a typical construction of a single-fold bellows. The durability of the bellows depends on their use. The guarantee of trouble-free operation is perfect workmanship and no material defects. To ensure safe use and long-term operation, its technical condition and properties must be constantly checked. The standard service life of a bellows actuator for a typical application is several years. Unlike pneumatic piston cylinders, they do not have sliding seals, which are the weak point of cylinders, because the penetration of abrasive particles into the sealing nodes causes wear of the surface of the cylinder elements and the seals themselves [1].

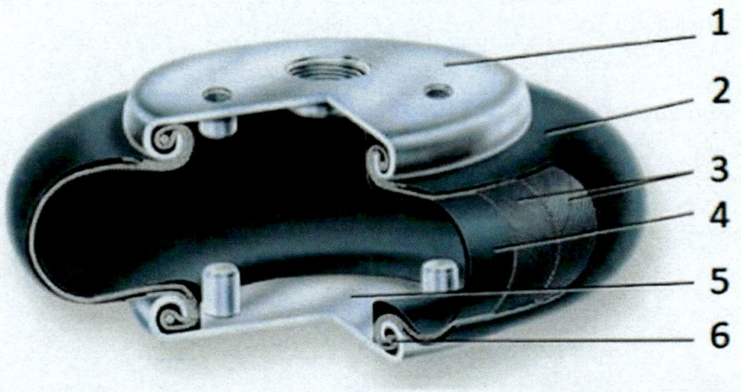

Fig. 1. Cross section of a typical single-fold bellows actuator: 1 - upper steel rolled fixing plate, 2 - outer elastomer layer, 3 - reinforcement, 4 - inner elastomer layer, 5 - lower fixing plate, 6 - steel rod. Source: machinedesign.com [2].

To experimentally determine the static and dynamic properties of the bellows actuator, a dedicated experimental stand was built. The main structural element of the system is the reaction frame, in which a double-acting cylinder with a single-sided piston rod is mounted, which is an element that loads the tested bellows actuator. The bellows is placed on a base, the construction of which enables the testing of liquids. In preliminary studies, water was used, but oil is planned to be used (due to its higher viscosity and thus the reduction of surface undulations). The use of liquids enables the analysis of changes in the volume of the bellows.

2 Experimental Stand Elements

The subject of the described tests is a single-fold Aventics air spring of the BCP series (Fig. 2). Its basic parameters and dimensions are presented in Table 1.

Table 1. Parameters of the tested bellows [3].

Version:	Bellow actuator with cover
Functional principle:	Single-acting, retracted without pressure
Working pressure min./max.:	0–8 bar
Ambient temperature min./max.:	−40–70 °C
Working medium:	Compressed air
The permissible angle of inclination max.:	15°
Recommended working pressure:	6 bar
Maximum effective stroke:	54 mm
Maximum force.:	4500–7500 N
Weight:	1.2 kg
H_{min}:	51 mm
H_{max}: _	105 mm
Max. Rubber diameter:	165 mm
Cover diameter:	108 mm
Compressed air connection:	G ¼
Min. Radial installation space:	180 mm

Fig. 2. Aventics air spring 0822419002. Source: Manufacturer's catalogue [3].

Fig. 3. Three-dimensional model of a bellows base with polycarbonate tube F180 mm

Figure 4 shows the diagram according to which the experimental stand was built.

The Asco Joucomatic 60590100 pressure servo valve was selected to control the pressure inside the bellows. Changes in pressure within the bellows are recorded by the Peltron PXW-6 pressure transducer. The force loading the bellows was controlled by the Rexroth R414002009 pressure servo valve, which was connected to the piston chamber actuator (Rexroth 523–407-020–0) with a piston diameter of 80 mm.

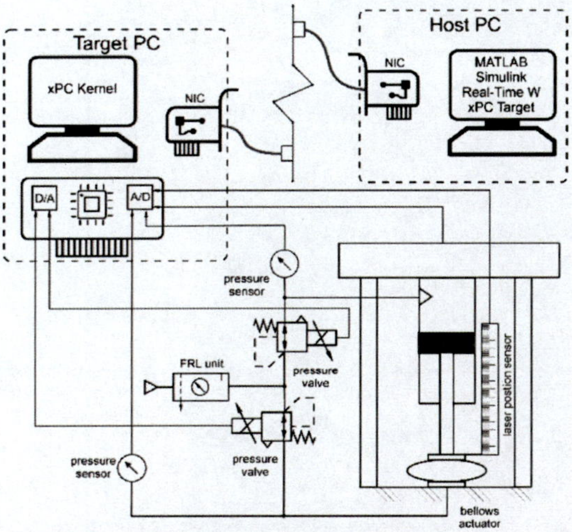

Fig. 4. Schematic of the experimental stand.

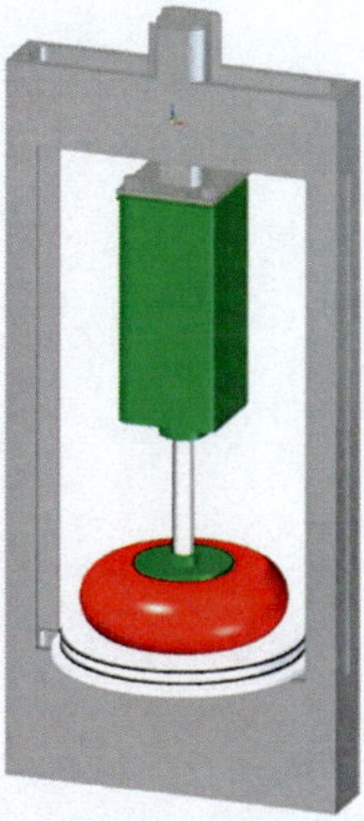

Fig. 5. Three-dimensional model of the reaction frame with actuators.

The data acquisition and control system consists of two host-target computers. The Matlab Simulink package is installed on the host computer on which the control was prepared and the measurement data were saved. A SpeedGoat card with A/D and D/A converters was installed in the target computer, due to which the converters are physically connected to the measuring and control elements of the experimental stand. In the course of design work, a three-dimensional model of the reaction frame connecting the executive elements was developed (Fig. 5), **which was then welded from steel profiles. Then a** bellows base was designed and made on a CNC milling machine, enabling (together with a matching polycarbonate pipe) to measure the volume of the bellows. Its design is shown in Fig. 3. Due to the favourable properties, polyacetal was used as the material. Height was measured using a Baumer CH-8501 laser distance sensor attached to the reaction frame. In addition, the following were used to build the stand:

- 24 V stabilized power supply,
- air preparation unit.

All elements were assembled as shown in Fig. 6.

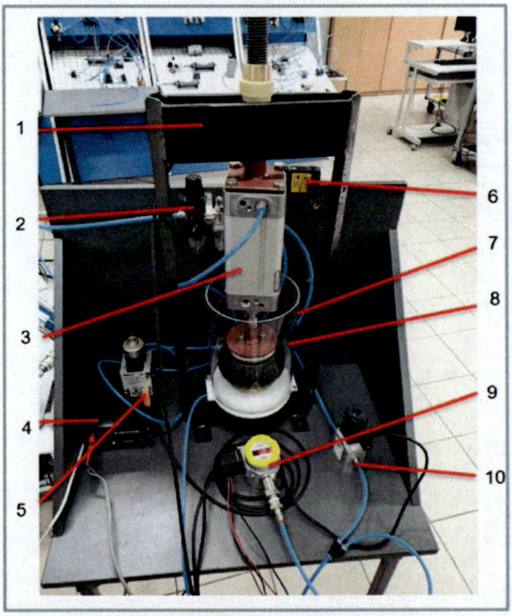

Fig. 6. Experimental stand: 1 - reaction frame, 2 - FLR, 3 - loading cylinder, 4 - power supply, 5 - Rexroth servo valve, 6 - distance sensor, 7 - polycarbonate tube, 8 - bellows, 9 - pressure transducer, 10 - ASCO Jumatic servo valve.

3 Mathematical Model

Since the tested bellows serves as an actuator and not as an element of vibration isolation, spring forces and damping were omitted at this stage, which at the speed of movement used during the tests of 0.1 m/min have very low values, unnecessarily complicating the model. However, the friction force of the piston seals and the cylinder rod was taken into account, which is important both under static and dynamic conditions. The starting point for the mathematical description of the forces acting in the system is Newton's second law of motion:

$$m \cdot \dot{v} = A \cdot p + F_{\text{ext}} \tag{1}$$

As the reduced mass m_c, the mass of the piston, piston rod, and nut were taken into account. The friction coefficient for cylinders with a piston diameter of 80 mm is between 80 and 110 Ns/m. Its effective area is constant and amounts to 50.3 cm^2. The actuator is connected to the power supply only on one side, so the side of the force generated from the piston rod is omitted [4].

An analogous equation was adopted for the bellows actuator. However, in this case, the effective area is variable depending on the volume and height of the bellows, which was adopted as a generalized coordinate. Friction, which in the case of the bellows occurs only in the shell, has also been omitted. Therefore, Eq. 1 will take the form:

$$m \cdot \ddot{h} = A_e \cdot p_3 + F_{ext} \tag{2}$$

The effective area of the bellows was calculated for stationary conditions, where. $\ddot{h} = 0$, therefore:

$$A_e(h) = \frac{F_m}{p_3} \tag{3}$$

where F_m is the force generated by the bellows as a function of time and p_3 is the supply pressure. **Due to the nonlinear nature of changes in the effective area, coefficients were** used to describe it, which were then estimated experimentally. Therefore, the equation will take the form (Fig. 7):

$$m \cdot \ddot{h} = (b_2 \cdot h^2 + b_1 \cdot h + b_0) \cdot p_3 + F_{ext} \tag{4}$$

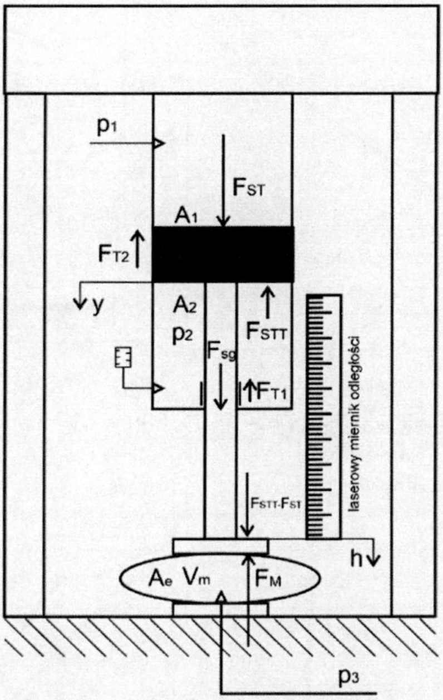

Fig. 7. Test bench forces diagram: h – bellows height, generalized coordinate, y – stroke of the loading cylinder, m_c – reduced piston and rod mass, p_1 – supply pressure in the working chamber of the loading cylinder, p_2 – pressure in the piston rod chamber connected to the atmosphere, p_3 – supply pressure in the bellows chamber, A_1 – effective area of the piston in the working chamber, A_2 – effective area of the piston in the piston rod chamber, A_e – effective area of the air bellow, V_m – working volume of the air bellow, F_m – force generated by the pressure in the bellows, F_{sg} – gravitational force of the reduced mass mc, F_{T1}, F_{T2} –friction forces of the piston and rod seals, F_{ST} – force acting on the piston of the loading cylinder in the working chamber, F_{STT} – force acting on the piston of the loading cylinder in the piston rod chamber

4 Results

After the experimental stand was prepared, several tests were performed to determine the static and dynamic characteristics of the bellows. Figure 8 shows the change in air pressure in the bellows actuator during its inflation for various loading forces. The initial aberrations seen in the graph are due to the operation of the pressure servo valve. Figures 9 and 10 show, respectively, the bellows height vs. time graphs for various internal pressures and the graphs of the force generated during the next test vs. height – this time for constant internal bellows pressures. Due to the limited ability of the bellows load by the pneumatic cylinder, pressure values of 1, 2 and 3 bars were selected to determine the model.

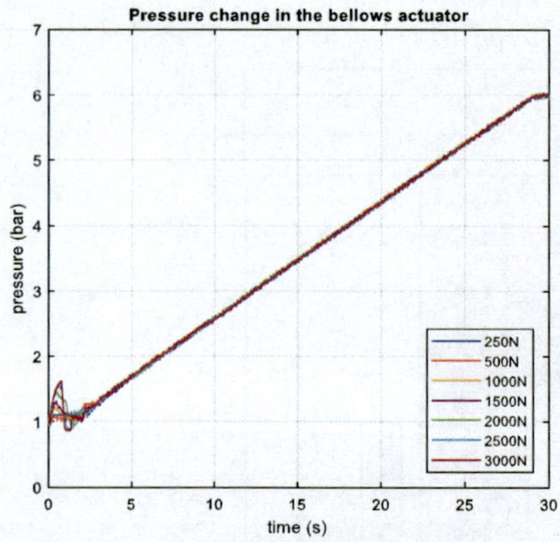

Fig. 8. Dynamic characteristics of pressure changes in the bellows for different forces loading the bellows.

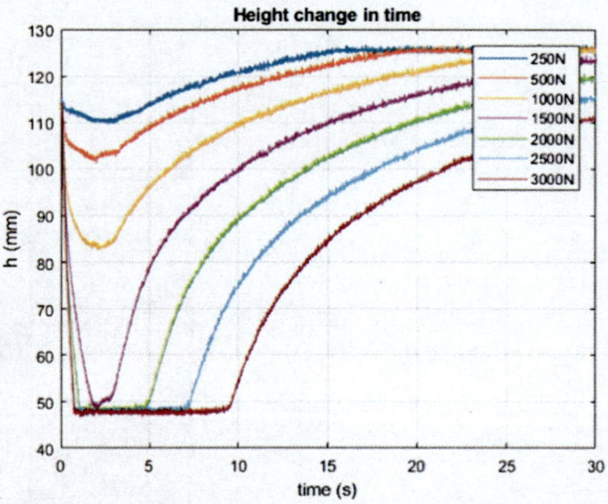

Fig. 9. Dynamic characteristics of the height of the bellows as a function of time for various loading forces.

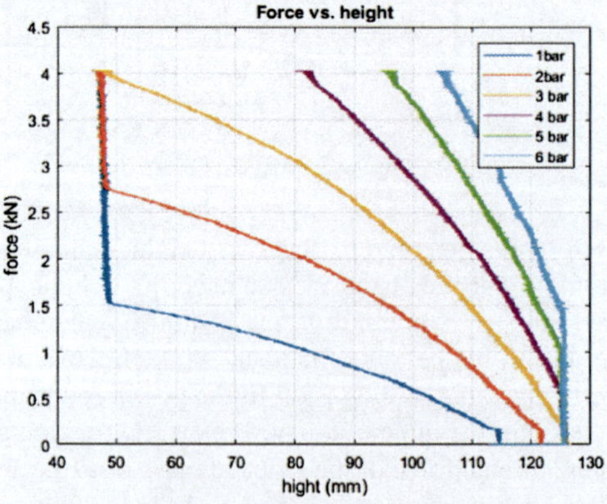

Fig. 10. Dynamic characteristics of the force as a function of height. The axes in the graph were not limited to showing the achievable minimum and maximum stroke values.

A polynomial function was adopted as a mathematical model to describe the force generated by the bellows (5):

$$y(x_1, x_2) = b_0 + b_1x_1 + b_2x_2 + b_{12}x_1x_2 + b_{11}x_1^2 + b_{22}x_2^2 \tag{5}$$

Regression analysis was used to determine the coefficients of Eq. 5. After substituting the coefficients into formula 5, Eq. 6 was obtained that describes the force generated by the bellows as a function of pressure and height (stroke):

$$F(p, h) = 0.016 \cdot p^2 - 0.0003 \cdot h^2 + 1.727 \cdot p + 0.04 \cdot h - 0.01 \cdot p \cdot h - 1.04 \tag{6}$$

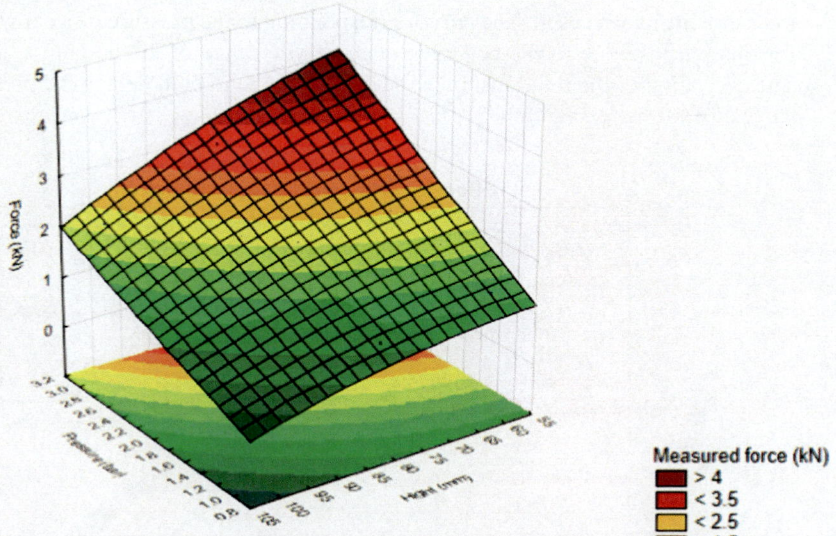

Fig. 11. Bellows force as a function of pressure and height.

The PARETO chart (Fig. 12) was used to analyse the significance of five coefficients that affect the force generated by the bellows. The most significant turned out to be the coefficient responsible for the linear effect of the supply pressure. Subsequently, the inverse effect of the height of the bellows on the force generated by it was found. In addition, the inversely proportional effect on the bellows' force was shown by the coefficients corresponding to the product of the pressure and height and the squared height of the bellows (Fig. 13). At the selected confidence level, the coefficient accompanying the squared pressure value turned out to be statistically insignificant. However, it was not eliminated from the equation to preserve the maximum coefficient of determination R^2. To visualize the model, Fig. 11 shows a 3D graph with the response surface outlined. Figure 14 shows the experimentally determined dependence of the effective area on the

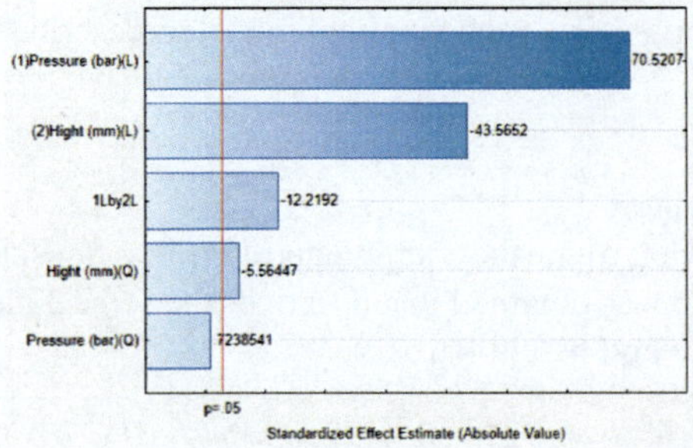

Fig. 12. PARETO chart.

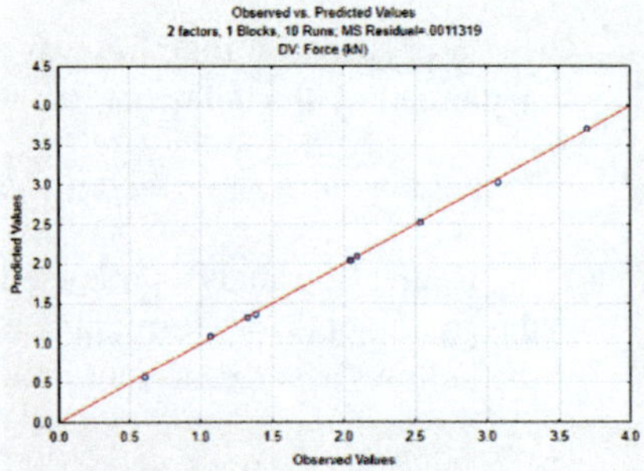

Fig. 13. Fitting the model.

height of the bellows actuator and the pressure. To determine the coefficients of Eq. 7 the curve fitting was used. The curve corresponding to the pressure of 3 bar was selected for this purpose due to the coverage of the entire range of the height of the bellows. The derivation of the relationship of the effective area to both height and pressure is the subject of further research.

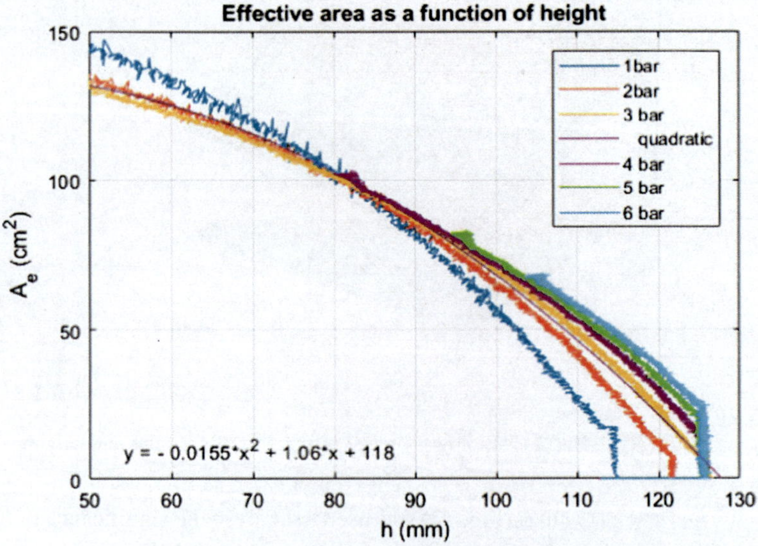

Fig. 14. Bellows effective area as a function of height. The curve was fitted at a significance level of 0.95. Determination factor $R^2 = 0.9884$

$$A_e = -0.0155 \cdot h^2 + 1.06 \cdot h + 118 \tag{7}$$

5 Conclusions

The article presented the construction of a stand that is used to test the bellows actuators. Position and force control systems may be used in the experimental stand. Several experiments were carried out on it, obtaining the characteristics of the tested bellows. The results were then subjected to statistical analysis. Using regression analysis, the formula for the force generated by the bellows was derived as a function of the supply pressure and the actual height of the bellows. The tests showed a strong linear pressure effect and an inversely proportional linear effect of the height of the bellows on the generated force. At the given level of significance, the inverse effect of the square of height and the effect of height and pressure on each other was also demonstrated. A simplified formula for the effective area of the tested bellows was also derived. The stand is prepared to measure changes in the volume of the bellows. Thanks to this, it will be possible to verify the mathematical models of the bellows with the change of the volume and effective area, which will be described in subsequent publications.

References

1. Dindorf, R.,: Elastyczne aktuatory elastyczne . Wydawnictwo Politechniki Świętokrzyskiej, Kielce (2013)
2. Machine design homepage. https://www.machinedesign.com/mechanical-motion-systems/pneumatics/article/21833374/primer-on-flexible-actuators-and-their-uses. Accessed 03 Apr 2023
3. Aventics BCP series catalogue. https://www.emerson.com/documents/automation/catalog-series-bcp-aventics-en-6912372.pdf. Accessed 03 Apr 2023
4. Beater, P.: Pneumatic Drives. Springer, Heidelberg (2007). https://doi.org/10.1007/978-3-540-69471-7

Check for updates

Effect of Simultaneous Valve Closures in Hydraulic Piping Systems

Kamil Urbanowicz[1(✉)], Igor Haluch[1], Anton Bergant[2,3], Adam Deptuła[4], and Paweł Śliwiński[5]

[1] Faculty of Mechanical Engineering and Mechatronics, West Pomeranian University of Technology, Szczecin, Szczecin, Poland
{kamil.urbanowicz,hi46746}@zut.edu.pl
[2] Litostroj Power d.o.o., Ljubljana, Slovenia
anton.bergant@litostrojpower.eu
[3] Faculty of Mechanical Engineering, University of Ljubljana, Ljubljana, Slovenia
[4] Faculty of Production Engineering and Logistics, Opole University of Technology, Opole, Poland
a.deptula@po.edu.pl
[5] Faculty of Mechanical Engineering and Ship Technology, Gdansk University of Technology, Gdańsk, Poland
pawel.sliwinski@pg.edu.pl

Abstract. The paper investigates wave interference (between pressure waves) occurring in simple hydraulic systems. Water hammer was induced by simultaneous closure of three valves located at the reservoirs of a "Y" type hydraulic system. Numerical simulations were carried out with the help of the freeware computer package Allievi enabling the reader to replicate results in a direct manner. The influence of the following quantities has been tested: Reynolds number *Re* (laminar and turbulent flow), the length of the pipe arms (L_1, L_2, L_3), the constant pipe internal diameter (D_i = const.) and variable internal pipe diameters D_i ($D_1 \neq D_2 \neq D_3$). The results of the research showed how unwanted interferences between pressure waves may occur and thus increase the possibility of piping system damage. Further numerical investigations are sought in consideration of unsteady skin friction losses and viscoelastic pipe wall effects.

Keywords: Water hammer · Transient flow · Three reservoir problem

1 Introduction

In hydraulic piping systems, transient events are commonly observed. In most cases, they occur during the start-up and shutdown of the system. The phenomena accompanying these flows are increasingly being studied because the pressure surges that can occur are large enough to damage the system components. Complex systems consist of various hydraulic devices that can contribute to the occurrence of water hammer, which is usually associated with damped dynamic pressure and bulk velocity waveforms. Such

J. Stryczek and U. Warzyńska (Eds.): NSHP 2023, LNME, pp. 162–173, 2024.
https://doi.org/10.1007/978-3-031-43002-2_15

waveforms have been extensively studied in simple systems (e.g., reservoir-pipeline-valve systems (RPV systems) with a single-pipe). However, as indicated by selected research [1, 2], in more complex systems (similar to real-world ones), they can be the cause of catastrophic damages. The sources of pressure waves and bulk velocity pulsation are controlled and uncontrolled valve closures (resulting from power failure). From the literature review [3–6], it is apparent that the closure of a single valve has been thoroughly investigated. In these systems, as long as there is no separation of the liquid column (transient cavitating flow), the maximum pressure should not exceed the pressure calculated using the Joukowsky formula. Significantly less attention has been given to situations inspired by actual events recorded in the Deltares laboratory (Delft, Netherlands), which are described in the work by Bergant et al. [7]. In that case, the simultaneous closure of two valves occurred due to a fault in an electronic conversion box caused by power failure, resulting in leakage of water at pipe joints and damage of the system components (metal supports). Research on the simultaneous closure of two valves since that time is still ongoing [8, 9]. Real systems consist of numerous valves, manifolds, pumps, and other hydraulic devices, which, during water hammer events, act as sources of primary and secondary waves. The goal of this study is the initial analysis that focuses on the impact of the simultaneous closure of three valves in a system containing a "Y" pipe junction. To the best knowledge of the authors, such research has not been conducted before. For modelling purpose in this paper, a freeware water hammer Allievi computer code [10] was used. The reader of this paper can then directly replicate the numerical results. The research will be conducted for two sub-variants, with the following parameters kept constant: a) the diameters of all pipes, b) the Reynolds numbers in all pipes during steady-state flow preceding the transient event.

2 Analyzed Scenarios

The paper presents cases of the influence of a number of closed valves on pressure increments that take place in simple water supply and hydraulic systems.

The tested scenarios were as follows:

A) Flow from the high-pressure reservoir to the low-pressure reservoir through the pipe connecting the two reservoirs, which is abruptly interrupted by the closing of a single valve located at the downstream end low-pressure reservoir (node N3 - Fig. 1a). The idea of this test was to critically evaluate the capabilities of the used Allievi code.

B) Flow as in Scenario A, except that the valves are installed at both pressure reservoirs (nodes N1 and N3 - Fig. 1b). The two valves are closed at the same time generating a water hammer with two initial pressure waves in the straight section of the conduit. The idea of this test was to initially study the possibility of wave interactions along the pipeline.

C) Flow from the high-pressure reservoir to the two low-pressure reservoirs. Here all three pipes are connected in a "Y" junction. The water hammer is forced by the simultaneous closing of three valves located just nearby the reservoirs (nodes N3, N7 and N11 - Fig. 1c). The proposed test stand is sufficiently simple to be built in a laboratory and used for future validation tests.

The first Scenario A is well known and frequently analyzed, the role to repeat this simulation in this paper is to investigate the performance of Allievi code. Scenario B was checked to analyze the interaction of two waves in a system in which two valves were closed simultaneously. However, Scenario C is the main object of our research as the effects of simultaneous instantaneous closure of three valves in a "Y" junction system have not been studied before.

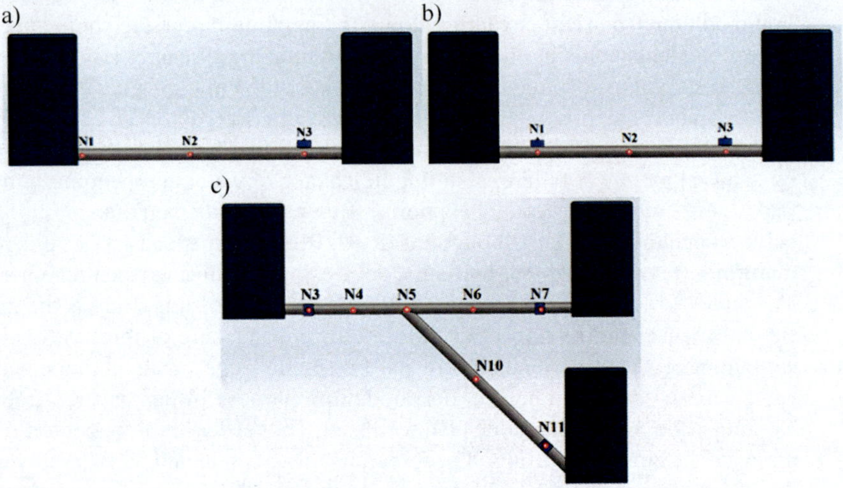

Fig. 1. Schematic drawing of tested systems.

The results obtained for the first two Scenarios (i.e. A and B) will be presented at the conference, while an extended investigation and discussion will be presented for selected results from Scenario C. Among a number of freeware computer programs available to the users (Allievi - Spain; WHAMO - USA; TSNet - USA; Hytran - New Zealand), the Allievi code (made at Polytechnic University of Valencia [10]) was selected. This choice was due to the fact that this program has been used in research by others [11, 12]. The program has a modular structure. It can include a number of water supply and hydraulic devices, as well as analyze both steady and transient states. Its operation is based on the method of characteristics, with which the following system of equations is solved, which describes both steady and unsteady liquid flow:

$$\begin{cases} \frac{\partial p}{\partial t} + \rho c^2 \frac{\partial v}{\partial x} = 0 \\ \frac{\partial p}{\partial x} + \rho \frac{\partial v}{\partial t} + \frac{4}{D}\tau = 0 \end{cases} \tag{1}$$

where: p – pressure [Pa], ρ – fluid density [kg/m^3], c – pressure wave speed [m/s], t – time [s], x – axial coordinate [m], v – bulk velocity [m/s], D – pipe inner diameter [m], τ - wall shear stress [Pa].

The wall shear stress in this program is calculated in a quasi-steady manner:

$$\tau = f\frac{\rho v|v|}{8} \tag{2}$$

where: f – Darcy-Weisbach friction factor.

For a single pipe, the system of Eqs. (1) can be numerically solved for given boundary conditions. The numerical solution at internal nodes is based on the values from previous time steps:

$$\begin{cases} p_{t+\Delta t,i} = \frac{1}{2}\left[\left(p_{t,i-1} + p_{t,i+1}\right) + c\rho\left(v_{t,i-1} - v_{t,i+1}\right) + \frac{4c\Delta t}{D}\left(\tau_{t,i+1} - \tau_{t,i-1}\right)\right] \\ v_{t+\Delta t,i} = \frac{1}{2}\left[\left(v_{t,i-1} + v_{t,i+1}\right) + \frac{1}{c\rho}\left(p_{t,i-1} - p_{t,i+1}\right) - \frac{4\Delta t}{\rho D}\left(\tau_{t,i-1} + \tau_{t,i+1}\right)\right] \end{cases} \quad (3)$$

where: Δt – time step.

3 Discussion of simulation results of Scenario C

The issue of liquid flow between three reservoirs is in itself a problem that has been analyzed many times over the years in the literature [13, 14]. The difficulty lies in determining from which reservoir (or reservoirs) the flow takes place, i.e. de facto determining the directions of the liquid flow in conduits in steady motion. In this work, the problem has been simplified, as it was assumed that the details of the flow will be known. Only water flows taking place in steel pipes are analyzed. It was decided to examine the flows in two sub-variants while maintaining the constant: C.1 for Reynolds number $Re = \text{const}$; C.2 for internal pipe diameter $D = \text{const}$. Assuming that the flow will take place from reservoir 1 to reservoir 2 and 3, it was possible to write linear pressure losses equations for steady motion:

$$\begin{cases} p_{R2} = p_{R1} - \Delta A \\ p_{R3} = p_{R1} - \Delta B \end{cases} \quad (4)$$

where ΔA and ΔB are the respective pressure losses along the pipe lengths, calculated as follows:

$$\begin{cases} \Delta A = \frac{\rho}{2}\left[\frac{f_1 L_1 v_1^2}{D_1} + \frac{f_2 L_2 v_2^2}{D_2}\right] \\ \Delta B = \frac{\rho}{2}\left[\frac{f_1 L_1 v_1^2}{D_1} + \frac{f_3 L_3 v_3^2}{D_3}\right] \end{cases} \quad (5)$$

where L_1, L_2 and L_3 are respective pipe lengths.

The minimum pressure in the analyzed system in steady motion was assumed to be calculated in a similar way as in Scenarios A and B, i.e. when $\Delta A > \Delta B$ then:

$$\begin{cases} p_{R2} = p_a + \rho c_{max} v_{max} \\ p_{R1} = p_{R2} + \Delta A \end{cases} \quad (6)$$

otherwise $\Delta B > \Delta A$:

$$\begin{cases} p_{R3} = p_a + \rho c_{max} v_{max} \\ p_{R1} = p_{R3} + \Delta B \end{cases} \quad (7)$$

After determining the above pressures with the help of Eq. (6) or Eq. (7) the pressure in the last reservoir need to be calculated using proper equation from a set of Eqs. (4).

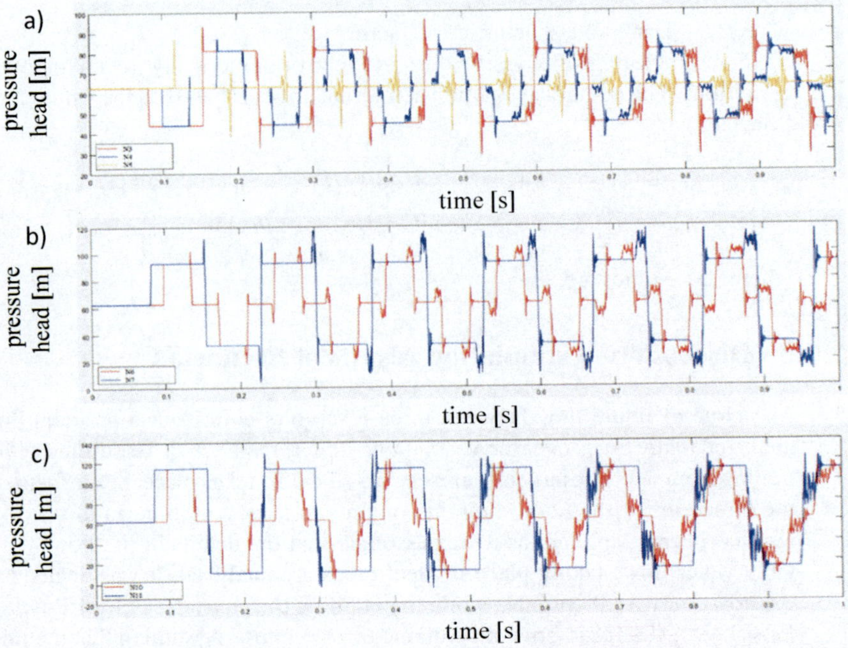

Fig. 2. Results for turbulent flow case in three-reservoir system – Sub-variant C.1.9 (L = const = 50 m)

The data necessary for implementation are listed in Tables 1 and 2 in Appendix A. The pipe lengths in Scenario C.1 were calculated to avoid the use of interpolation in the method of characteristics.

From all simulations, three runs were selected for qualitative analysis. Two runs concerned the analyzed sub-variant C.1 (C.1.9 and C.1.10 - Figs. 2 and 3) and one subvariant C.2 (C.2.8 - Fig. 4). Simulation results obtained for sub-variant C.1 showed that the influence of the type of flow preceding the water hammer (laminar, transient, turbulent) is only responsible for the increases in the values of the observed pressure and initial velocity. After rescaling and comparing waveforms in dimensionless forms, the obtained results were very similar (in systems with the same lengths and diameters of pipes). Thus, limiting the discussion to only selected waveforms is an acceptable assumption.

As can be seen from Fig. 2 in Subvariant C.1.9, which is characterized by the same length of all pipes, there are some local pressure pulsations that exceed the pressures calculated from the Joukowsky formula $\mathbb{Z} = \rho c|v_0|$. The observed pulsations are the result of the appearance of secondary waves resulting from the change in the diameter of the pipes at the "Y" junction. Similar pressure pulsations occur as a result of the mutual interaction of the liquid with the material of the pipes, commonly referred to as fluid-structure interaction FSI [15, 16]. This pulsation, which is initially characterized by a temporary increase in pressure at the N5 node ("Y" junction, see Fig. 2a - $t \approx 0.12$ s), is transferred in the system and is responsible for the pressure peaks appearing in the

L_1, L_2 and L_3 conduits, which can be seen on the beginnings and ends of the amplitudes (Figs. 2a, b and c). The duration of these pulsations begins to increase during the duration of the waveforms, which is related to the successive generation of secondary waves at the "Y" node. As a consequence, the time and number of these pulsations generated at the N5 node ("Y" junction) increase (see Fig. 2a - $t \approx 0.19$ s; $t \approx 0.25$ s; $t \approx 0.34$ s; $t \approx 0.41$ s…). These pulsations are responsible for the duration of reflected pulsations clearly visible at the beginnings and ends of tops and valleys of the analyzed pressure amplitudes.

Interestingly, in simulation C.2.5 (not shown graphically in the conference paper) in Subvariant C.2 (constant pipe diameters) this type of peaks does not occur. The situation in node N5 is similar to that observed in the middle node of the conduit from Scenario B, i.e. the pressure in node N5 remains at a constant level after closing the valves, identical to the value observed in steady motion. This situation is the result of the interference of the low-pressure wave generated in the L_1 pipe with the high-pressure waves generated at valves 2 and 3 in the L_2 and L_3 pipes, respectively.

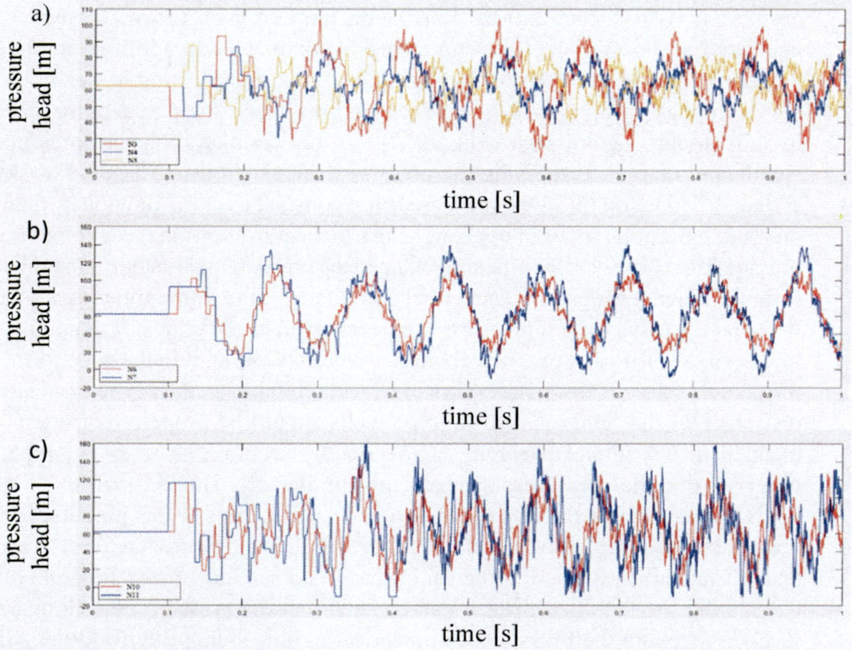

Fig. 3. Results for turbulent flow case in three-reservoir system – Sub-variant C.1.10 ($L_1 = 50$ m, $L_2 = 34.7$ m, $L_3 = 25$ m)

As it turns out from other studies, there is a much greater threat in systems with different lengths of pipes - let's focus on the example results of Subvariant C.1.10 (turbulent flow) shown in Fig. 3. In pipe L_1, at "Y" junction N5, the pulsation occurs in antiphase to the one present at junctions N3 and N4 (Fig. 3a). The highest pressures are observed at the junction with the third valve, i.e., at junction N11, where the pulsation intensifies during the first amplitude, thereby posing a greater threat to the pipes (Fig. 3c). The pressure increments at junctions N3 (Fig. 3a), N7 (Fig. 3b), and N11 (Fig. 3c) have values significantly higher than those estimated by the Joukowsky formula. In pipe L_2 (Fig. 3b), the pulsation waveforms in both analyzed sections are in the same phase. The waveforms observed in pipe L_3 (Fig. 3c) are highly chaotic. In the initial period of the impact, typical pressure amplitudes are not observed at junctions N10 and N11, which are caused by the intense interference of primary and secondary waves generated in this shortest pipe. At junction N11, near the valve 3, cavitation areas also appear (see Fig. 3c at $t \approx 0.325$ s and $t \approx 0.4$ s). Slight differences in pipe lengths (decimal places) are the result of rejecting the possibility of interpolation of results calculated using the method of characteristics. A constant time step was assumed, however the variable pressure wave speeds resulting from different pipe diameters forced the necessary division (and calculation of spatial step) into numerical elements, which was responsible for the delicate differences in pipe lengths noticeable in particular in Table 4.

Results obtained for sub-variants C.2.1 (laminar flow) and C.2.5 (turbulent flow) graphically are very similar to the situation observed in Fig. 2, although with no pressure peaks at the N5 node and on peaks on tops and valleys of pressure waveform in other cross-sections (which was associated with the lack of a source of secondary waves generated in the "Y" junction - no change of diameter). In addition, in node N5 ("Y" junction), the pressure remained equal to the pressure occurring in steady motion.

Exemplary results from the tests carried out when the pipe lengths were different in sub-variant C.2 are shown in Fig. 4. In this configuration, the pressure rises are significantly larger, surpassing the pressures calculated using the Joukowsky formula. At the "Y" junction (N5), the pressure is not constant as in the previously analyzed case; pressure pulsations occur in opposite phase to the pulsations in the L_1 pipe (Fig. 4a). The pressure rises observed significantly exceed the pressures estimated using the Joukowsky formula. Cavitation phenomena occur in this pipe after some time (at $t \approx 0.61$ s and $t \approx 0.71$ s). The maximum pressure rise is observed at $t \approx 0.96$ s. It would be interesting to investigate if further pressure increases would occur later in the waveform, indicating the phenomenon of resonance. However, for a reliable analysis, the unsteady frictional resistance of the fluid against the pipe walls should be properly considered, which is planned in our future research. At the middle section N4 of the L_1 pipe, it can be observed that the pressure pulsations are significantly smaller than those observed at the N3 junction near the first valve. No cavitation occurs at this junction (i.e., pressure does not drop to the vapor pressure). Figure 4b shows the pressure waveforms at the N6 and N7 junctions, located at the middle and end sections of the L_2 pipe, respectively. Comparing the recorded data, it can be seen that the pressure pulsations have similar levels, exceeding the pressures estimated using the Joukowsky formula. Additionally, it can be observed that at this section, the initial amplitudes of these pulsations do not exhibit significant damping. In the shortest pipe, L_3 (Fig. 4c), the pressure waveforms at the middle section and the valve almost overlap, with only slight differences observed in the peak and minimum values, which are slightly higher at the N11 section than those observed at the N10 section. In this pipe, significant pressure pulsations occur, and the peak values are similar to those observed in the L_1 and L_2 pipes. This indicates that the pressures in this pipe significantly exceed those predicted by the Joukowsky formula.

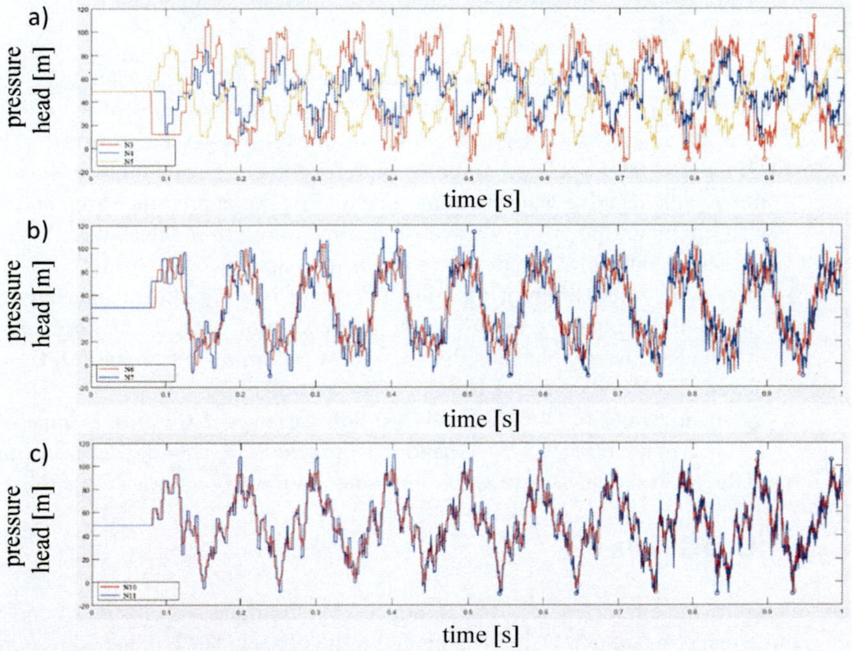

Fig. 4. Results for turbulent flow case in three-reservoir system – Sub-variant C.2.8 ($L_1 = 50$ m, $L_2 = 15$ m and $L_3 = 5$ m)

For qualitative analysis the special coefficient named Over Pressure Coefficient (OPC) was defined and calculated for all cases in Scenario C. This OPC parameter is the ratio of the actual pressure increase to the theoretical pressure increase predicted by the Joukowsky formula:

$$OPC = \frac{p_{max} - p_{initial}}{p_{max.J} - p_{initial}} = \frac{p_{max} - p_{initial}}{p_{initial} + \rho c|v_0| - p_{initial}} = \frac{p_{max} - p_{initial}}{\rho c|v_0|} \quad (8)$$

The obtained results of this parameter for Subcases C.1 and C.2 are collected in Tables 3 and 4 in Appendix. The main conclusions from quantitative analysis of transient flows caused by sudden closing of valves for constant Reynolds sub-variant C.1 (Table 3) are:

- in systems with different pipe diameters the overpressure above the Joukowsky pressure rise $\mathbb{Z}$ always occurs. It is a consequence of wave reflection taking place at pipe junctions;
- the highest and lowest value of OPC were obtained in pipe L_1 for tested run C.1.3 and in pipe L_3 for test C.1.9. No increases exceeding three times the Joukovsky pressure rise $\mathbb{Z}$ were recorded;
- the mean values of OPC for pipes were: in L_1 pipe $OPC_{mean} = 2.3$; in L_2 pipe $OPC_{mean} = 2.2$ and in L_3 pipe $OPC_{mean} = 2.0$;

The main conclusions from quantitative analysis for constant diameter sub-variant C.2 (Table 4) are:

- interaction of primary waves is minimized in systems with identical pipe lengths – see tested runs: C.2.1 and C.2.5. Here, the minimum values of the OPC parameter were obtained, which are at the level of the Joukowsky pressure rise. The value of $OPC_{min} = 0.99$ obtained in the L_1 pipe for the C.2.5 test is the consequence that in this pipe a negative water hammer occurs first, i.e. a pressure drop, and for positive water hammer the primary wave needs time to return in the analyzed system, which takes time during which the wave fronts are damped;
- in systems with different pipe lengths, there is a significant interference between the waves. Among the tested cases, the maximum value of the $OPC_{max} = 5.7$ was obtained in the L_3 pipe for the C.2.4 test. A large OPC value (OPC = 4.4) was obtained in the L_3 pipe for the C.2.8 test as well;
- the mean values of OPC for analyzed Sub-variant C.2 were: in L_1 pipe $OPC_{mean} = 1.6$; in L_2 pipe $OPC_{mean} = 2.3$ and in L_3 pipe $OPC_{mean} = 3.3$. The maximum values in the analyzed tests were always obtained in the L_3 pipe.

4 Conclusions

Pressure wave interference effects induced by simultaneous closure of end-valves in a three-reservoir system were investigated in this paper. The conducted research leads to the following conclusions:

- in a system consisting of three-reservoirs, where the flow occurs through three pipes connected in a "Y" junction, the pressures can significantly exceed the values estimated by a simple analytical Joukowsky formula commonly used in engineering practice;
- when the lengths and diameters of the pipes are kept identical, and all the valves are closed simultaneously, there will be no interference of waves beyond the Joukowsky increments. This is a result of the interference of the negative pressure wave generated at the valve 1 (upstream end valve) with the positive pressure waves generated at the valves 2 and 3 (downstream end valves);
- when the pipe lengths and diameters vary, the pressure increments in the pipes can significantly exceed the pressures estimated by the Joukowsky formula;
- freeware water hammer Allevi code includes quasi-steady friction losses and elastic-strain model only. Further research is needed to consider unsteady friction losses (unsteady friction dominated pipes) and viscoelastic effects of pipe walls (plastic pipes);
- in the near future, experimental studies are planned to confirm the occurrence of dangerous pressure surges resulting from the interference of pressure waves in systems with three reservoirs and a "Y" junction.

Appendix A

Table 1. Data for Sub-variant C.1 (Re = const).

Test	Flow type	L_1 [m]	L_2 [m]	L_3 [m]	v_1 [m/s]	v_2 [m/s]	v_3 [m/s]	H_1 [m]	H_2 [m]	H_3 [m]
C.1.1	laminar	50	50	50	0.01375	0.022	0.0366	15.5338	15.5324	15.5307
C.1.2	laminar	50	34.7	25	0.01375	0.022	0.0366	15.5324	15.5313	15.5307
C.1.3	laminar	50.3	30.1	15	0.01375	0.022	0.0366	15.5319	15.5309	15.5307
C.1.4	laminar	50	14.9	5	0.01375	0.022	0.0366	15.4315	15.4308	15.4308
C.1.5	transition	50	50	50	0.04125	0.066	0.11	25.9533	25.9450	25.9345
C.1.6	transition	50	34.7	25	0.04125	0.066	0.11	25.9450	25.9386	25.9345
C.1.7	transition	50.3	30.1	15	0.04125	0.066	0.11	25.9417	25.9358	25.9345
C.1.8	transition	50	14.9	5	0.04125	0.066	0.11	25.6390	25.6349	25.6350
C.1.9	**turbulent**	**50**	**50**	**50**	**0.1375**	**0.22**	**0.366**	**62.5024**	**62.4345**	**62.3478**
C.1.10	**turbulent**	**50**	**34.7**	**25**	**0.1375**	**0.22**	**0.366**	**62.4346**	**62.3817**	**62.3478**
C.1.11	turbulent	50.3	30.1	15	0.1375	0.22	0.366	62.4076	62.3591	62.3478
C.1.12	turbulent	50	14.9	5	0.1375	0.22	0.366	61.3829	61.3493	61.3502

c_1 = 1328 m/s; c_2 = 1365 m/s; c_3 = 1392 m/s; D_1 = 8 cm; D_2 = 5 cm; D_3 = 3 cm
laminar → Re = 1000; transition → Re = 3000; turbulent → Re = 10000

Table 2. Data for Sub-variant C.2 (D = const = 4 cm).

Test	Flow type	L_1 [m]	L_2 [m]	L_3 [m]	v_1 [m/s]	v_2 [m/s]	v_3 [m/s]	H_1 [m]	H_2 [m]	H_3 [m]
C.2.1	laminar	50	50	50	0.0605	0.04125	0.01925	18.8397	18.8283	18.8308
C.2.2	laminar	50	35	25	0.0605	0.04125	0.01925	18.8383	18.8283	18.8305
C.2.3	laminar	50	30	15	0.0605	0.04125	0.01925	18.8379	18.8283	18.8304
C.2.4	laminar	50	15	5	0.0605	0.04125	0.01925	18.8365	18.8283	18.8295
C.2.5	turbulent	50	50	50	0.275	0.165	0.11	49.1779	48.9631	48.9948
C.2.6	turbulent	50	35	25	0.275	0.165	0.11	49.1592	48.9631	48.9914
C.2.7	turbulent	50	30	15	0.275	0.165	0.11	49.1530	48.9631	48.9913
C.2.8	**turbulent**	**50**	**15**	**5**	**0.275**	**0.165**	**0.11**	**49.1343**	**48.9631**	**48.9788**

laminar: Re_1 = 2200 in pipe L_1; Re_2 = 1500 in pipe L_2; and Re_3 = 700
turbulent: Re_1 = 10000 in pipe L_1; Re_2 = 6000 in pipe L_2; and Re_3 = 4000
c = 1378 m/s

Table 3. Quantitative results in Sub-variant C.1 (Re = const).

Test	Flow type	*OPC* [-]			$\mathbb{Z}$ [Pa]		
		L_1 pipe	L_2 pipe	L_3 pipe	L_1 pipe	L_2 pipe	L_3 pipe
C.1.1	laminar	1.91	1.69	1.53	18260	30030	50947.2
C.1.2	laminar	2.22	2.41	2.12	18260	30030	50947.2
C.1.3	laminar	2.73	2.34	2.11	18260	30030	50947.2
C.1.4	laminar	2.42	2.41	2.46	18260	30030	50947.2
C.1.5	transitional	1.88	1.72	1.44	54780	90090	153120
C.1.6	transitional	2.19	2.44	2.06	54780	90090	153120
C.1.7	transitional	2.72	2.33	2.08	54780	90090	153120
C.1.8	transitional	2.42	2.41	2.46	54780	90090	153120
C.1.9	**turbulent**	**1.81**	**1.80**	**1.32**	**182600**	**300300**	**509472**
C.1.10	**turbulent**	**2.18**	**2.57**	**1.81**	**182600**	**300300**	**509472**
C.1.11	turbulent	2.42	2.34	2.23	182600	300300	509472
C.1.12	turbulent	2.60	2.25	2.32	182600	300300	509472

Table 4. Quantitative results in Sub-variant C.2 (D = const = 4 cm).

Test	Flow type	*OPC* [-]			$\mathbb{Z}$ [Pa]		
		L_1 pipe	L_2 pipe	L_3 pipe	L_1 pipe	L_2 pipe	L_3 pipe
C.2.1	laminar	1.00	1.00	1.00	83381	56851	26530
C.2.2	laminar	2.00	2.80	3.46	83381	56851	26530
C.2.3	laminar	1.80	2.57	3.64	83381	56851	26530
C.2.4	laminar	1.68	2.72	5.73	83381	56851	26530
C.2.5	turbulent	0.99	1.00	1.00	379005	227403	151602
C.2.6	turbulent	1.89	3.07	3.30	379005	227403	151602
C.2.7	turbulent	1.80	2.59	3.62	379005	227403	151602
C.2.8	**turbulent**	**1.70**	**2.87**	**4.35**	**379005**	**227403**	**151602**

References

1. Adamkowski, A.: Case study: lapino powerplant penstock failure. J. Hydraul. Eng. **127**(7), 547–555 (2001)
2. Ridha, W.K.M., Kashyzadeh, K.R., Ghorbani, S.: Common failures in hydraulic Kaplan turbine blades and practical solutions. Materials **16**(9), 3303 (2023)
3. Bergant, A., Simpson, A.R., Vítkovský, J.: Developments in unsteady pipe flow friction modelling. J. Hydraul. Res. **39**(3), 249–257 (2001)
4. Adamkowski, A., Lewandowski, M.: Experimental examination of unsteady friction models for transient pipe flow simulation. J. Fluids Eng. **128**, 1351–1363 (2006)
5. Urbanowicz, K., Stosiak, M., Towarnicki, K., Bergant, A.: Theoretical and experimental investigations of transient flow in oil-hydraulic small-diameter pipe system. Eng. Fail. Anal. **128**, 105607 (2021)
6. Karadžić, U., Bulatović, V., Bergant, A.: Valve-induced water hammer and column separation in a pipeline apparatus. Strojniški vestnik-J. Mech. Eng. **60**(11), 742–754 (2014)
7. Bergant, A., et al.: Water hammer and column separation due to accidental simultaneous closure of control valves in a large-scale two-phase flow experimental test rig. In: Proceedings of

the ASME 2010 Pressure Vessels & Piping Division/K-PVP Conference PVP2010, Bellevue, Washington, USA, 18–22 July, PVP2010–26131, pp. 923–932 (2010)
8. Bergant, A., Karadžic, U.: Developments in valve-induced water-hammer experimentation in a small-scale pipeline apparatus. In: BHR Group - 12th International Conference on Pressure Surges, pp. 639–652 (2016)
9. Karadžić, U., Janković, M., Strunjaš, F., Bergant, A.: Water hammer and column separation induced by simultaneous and delayed closure of two valves. Strojniški vestnik-J. Mech. Eng. **64**(9), 525–535 (2018)
10. Allevi. https://www.allievi.net/allievi-es.php. Accessed 20 May 2023
11. Lončar, G., Sreng, Ž., Kulić, T., Milicević, H., Ostojić, S.: Numerical models for analysis of hydraulic transients. In: 16th International symposium on Water Management and Hydraulic Engineering, WMHE, Skopje, North Macedonia, 05–07 September, pp. 133–145 (2019)
12. Lupa, S.I., Gagnon, M., Muntean, S., Abdul-Nour, G.: Investigation of water hammer overpressure in the hydraulic passages of hydropower plants equipped with Francis turbines. In: IOP Conference Series: Earth and Environmental Science, vol. 1079, p. 012003 (2022)
13. Rangaraju, K.G., Sathuraman, S.: Direct solution of three-reservoir problem. Civ. Eng. J. **11**(3–4), Paper no CE 143, 15–23 (1969)
14. Loganathan, G.V., Kuo, C.Y.: A direct solution for the classical three reservoir problem. Civ. Eng. Pract. Des. Eng. **4**, 1–11 (1985)
15. Andrade, D.M., de Freitas Rachid, F.B., Tijsseling, A.S.: A new model for fluid transients in piping systems taking into account the fluid–structure interaction. J. Fluids Struct. **114**, 103720 (2022)
16. Bayle, A., Plouraboué, F.: Spectral properties of Fluid Structure Interaction pressure/stress waves in liquid filled pipes. Wave Motion **116**, 103081 (2023)

Hydraulic Modules of Mobile Robotic Bricklaying System

Ryszard Dindorf(✉), Jakub Takosoglu, Piotr Woś, and Łukasz Chłopek

Kielce University of Technology, Kielce, Poland
dindorf@tu.kielce.pl

Abstract. This article deals with the evaluation of the hydraulic modules used in the ZSM (Zrobotyzowany System Murarski) mobile robotic bricklaying system. This robotics bricklaying system was developed as a research project in cooperation between a scientific leader, the research and development partner, and the industry partner. Mobile ZSM is the first robotic brickwork system in Poland, designed primarily for the construction of facades and partitions in office and residential buildings, as well as in industrial buildings. A mobile ZSM has been designed and developed as an innovative demonstration solution for the construction industry to automate heavy, labour-intensive and burdensome masonry work traditionally done manually by masons. The ZSM mobile set was made up of an ABB industrial robot with six degrees of freedom (6DoF), a Hinowa tracked undercarriage, a robot support frame, a hydraulic lifting-levelling unit, a brick warehouse, a brick feeder, a mortar applicator, a control cabinet, an operator panel, and a hydraulic gripper. The evaluation involved the specified hydraulic modules of the mobile ZSM, such as the hydraulic power and control unit, the hydraulic drive of the tracked undercarriage, hydraulic control of the lifting-levelling unit, and the hydraulic robot gripper. The human-machine interface (HMI) operator touch panel was presented with the start screen and control screens for the individual hydraulic modules of the mobile ZSM. The HMI touch panel was visually programmed to manage the mobile ZSM.

Keywords: Hydraulic power and control module · Hydraulic driven track undercarriage · Hydraulic lifting-levelling module · Hydraulic robot gripper

1 Introduction

The main problem in the construction industry is not only improving productivity, but also the lack of an adequate number of skilled construction workers. The implementation of advanced automated and robotic construction technologies significantly accelerates bricklaying work and improves the quality of the work performed. At present, the construction industry is approaching a revolutionary period of robotics of technological processes and therefore of automation of traditionally manual works, such as bricklaying and plastering. Automation and robotization are considered the best solution to improve the efficiency of the construction industry. There have been many reports in

J. Stryczek and U. Warzyńska (Eds.): NSHP 2023, LNME, pp. 174–183, 2024.
https://doi.org/10.1007/978-3-031-43002-2_16

the literature on the use of robots for various masonry and plastering tasks. The study [1] presents research on the possibility of automation of bricklaying. A robotic bricklaying system with the ability to build walls from slag blocks was designed. Pritschow et al. demonstrated the automated process of building walls on a construction site using a mobile robot [2]. Madsen analyzed the advantages and disadvantages of semi-automated bricklaying (SAM100) in collaboration with a mason to smooth excess mortar [3]. In the article [4], the author introduced a comprehensive methodical process to implement automation and robotics in residential buildings. An alternative to automation of bricklaying on a construction site is pre-fabrication of brick panels produced in manufacturing plants [5].

As part of the research project, the authors have developed an innovative mobile ZSM [6–10]. Mobile ZSM was created in cooperation between Kielce University of Technology as the leading researcher, the CBRTP research and development partner, and the STRABAG industry partner. Mobile ZSM is the first Polish robotic bricklaying system. It is designed to build facades and walls in office, residential, and industrial buildings. Figure 1 shows the 3D CAD design of the mobile ZSM.

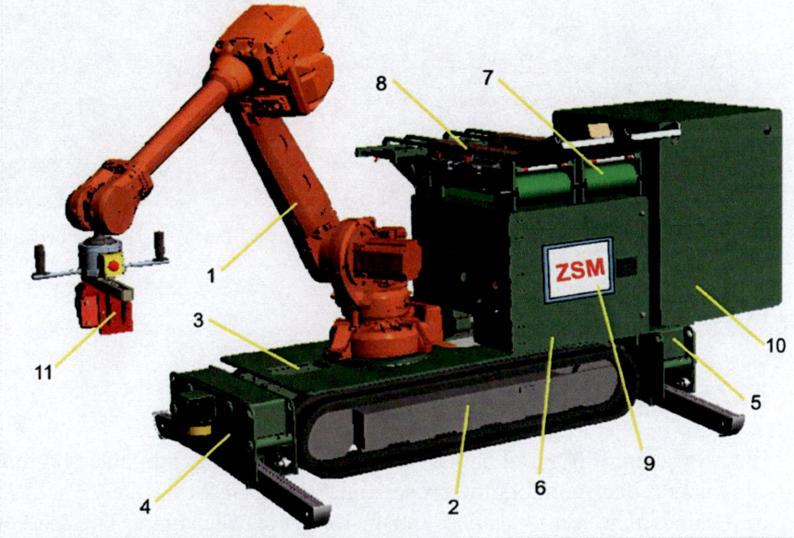

Fig. 1. 3D CAD design of mobile ZSM: 1 – ABB industrial robot, 2 – Hinowa tracked undercarriage, 3 – robot support frame, 4 – front hydraulic lifting-levelling module, 4 – rear hydraulic lifting-levelling module, 6 – hydraulic power and control module, 7 – brick warehouse, 8 – brick feeder, 9 – control panel, 10 – control cabinet, 11 – hydraulic robot gripper [11, 12].

The mobile ZSM consists of an ABB industrial robot with six degrees of freedom (6DoF), replaceable hydraulic gripper, Hinowa tracked undercarriage, robot support frame, front and rear hydraulic lifting-levelling module, hydraulic power and control module, brick warehouse, brick feeder, control panel, and control cabinet. The mobile ZSM enables the bricklaying of walls in large workspaces, limited by the robot's working range. The ABB industrial robot of type IRB4600 was used, with weights of 450 kg, a maximum vertical range of 3.055 m, a maximum horizontal range of 2.55 m and a repeatability of 0.06 mm.

Bricklaying begins with the drive, positioning, lifting, levelling, and anchoring of the ZSM on the bricklaying site, the determination of the bricklaying area using a laser sensor, and the programming of the robot for the automatic bricklaying. Robot programming takes into account how bricks are placed on the walls, the dimensions of the wall, the dimensions of the bricks, and the number of bricks in the wall. During the brickwork process, the bricks are picked up from the feeder, the mortar is applied to the brick by the applicator, and then the bricks are placed in the right place on the wall. Figure 2 shows the robotic bricklaying process under laboratory conditions.

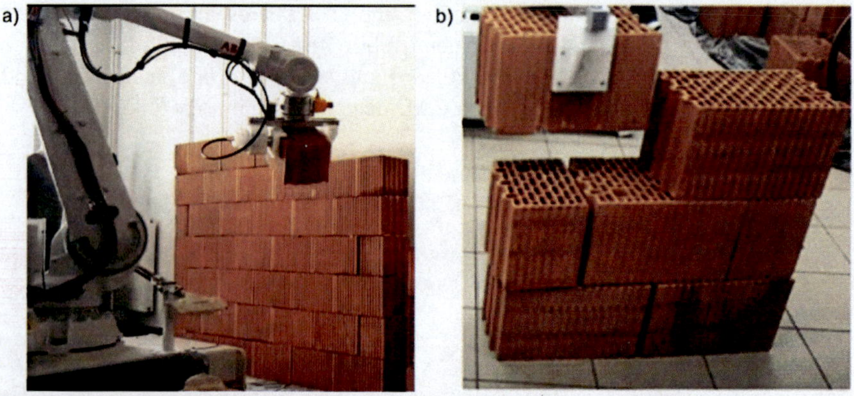

Fig. 2. View of the bricklaying process on a partition wall: a) robot during bricklaying, b) laying the brick on the wall [12].

2 ZSM Control Panel

The ZSM control is based on the Simatic S7-1500 programmable controller in the fail-safe version that monitors the proper functioning of all devices. The ZSM controllers are located in the control cabinet and the robot control cabinet. The ZSM controller is in constant communication with the robot control system. The control system operates on the Profinet communication network. The operating elements of the electric, hydraulic, and mechanical systems are used.

The Simens Simatic HMI touch panel allows the operator to communicate visually with the ZSM management controller. When power-up is activated, the control panel with the application developed from WinCC Advanced V16 starts. The Simatic HMI touch panel has several screens that can be called by the corresponding buttons. Figure 3 shows the Simatic HMI touch panel with a function menu.

The ABB FlexPendant is a portable (hand-held) operator unit with a touch panel that is used to perform many of the tasks while operating a robot system, such as running

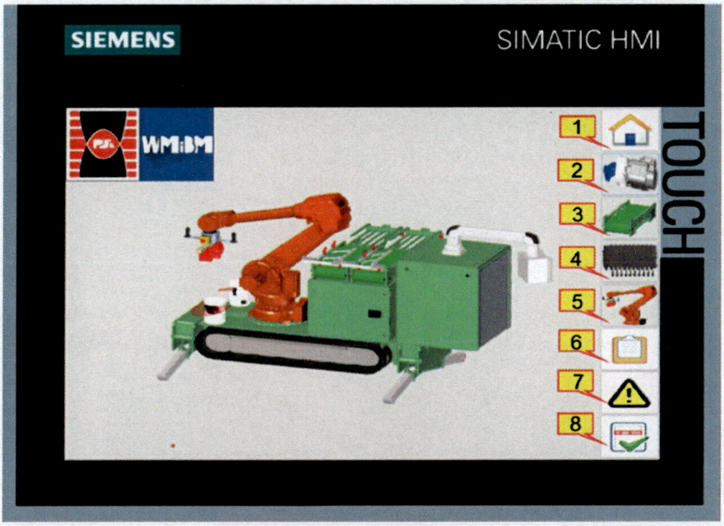

Fig. 3. Simatic HMI touch panel with a function menu: 1 – Start Home, 2 – settings of the hydraulic power and control module, 3 – control of the building material feeder, 4 – control of the tracked undercarriage drive and lifting-levelling module, 5 – robot control, 6 – programming of technological procedures, 7 – emergency events, 8 – reset settings [12].

programmes, jogging the robot, modifying robot programmes, etc. Figure 4 shows the ABB FlexPendat hand-held operator.

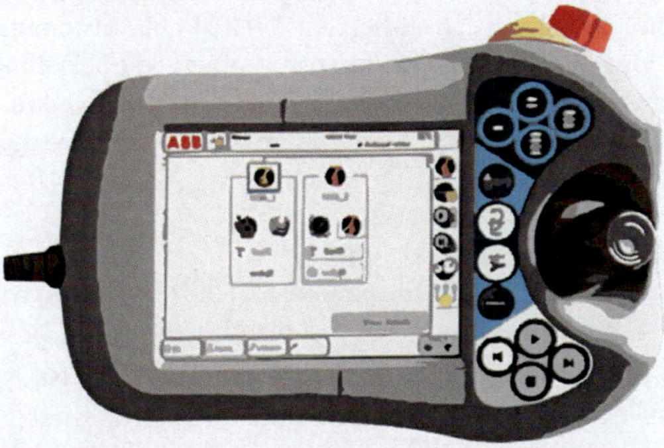

Fig. 4. ABB FlexPendant hand-held operator.

The FlexPendant consists of both hardware and software and is an independent computer. The FlexPendant is an integral part of the ABB robot, connected to the controller via a cable and connector. The hot-plug option allows the FlexPendant to be connected and disconnected at any time, even during automatic control of the robot.

3 ZSM Hydraulic Power and Control Module

The main devices of the ZSM hydraulic power and control module are a servo-pump consisting of a Simens Simotics electric servo-motor and a Hydro-Leduc fixed displacement hydraulic pump. The servo-pump power system is supplemented by hydro-pneumatic accumulators. Figure 5 shows the 3D CAD design of the ZSM hydraulic power and control module.

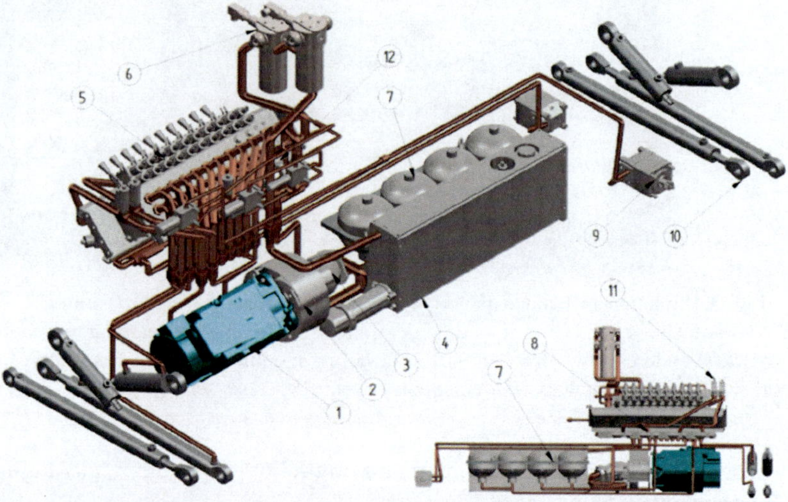

Fig. 5. 3D CAD design of the ZSM hydraulic power and control module: 1 – Simens servo-motor, 2 – clutch, 3 – Hydro-Leduc fixed displacement pump, 4 – oil tank unit, 5 – multi-section proportional directional valve, 6 – filter unit. 7 – hydraulic accumulator, 8 – hydraulic accumulator manifold, 9 – low-speed hydrostatic motors to drive the tracked undercarriage, 10 – cylinders of lifting-levelling modules, 11 – pressure switches in the hydraulic accumulator manifold, 12 – pressure transducer for pressure monitoring in hydraulic actuators [11].

The Siemens Simotics M type 1PH8103-10002-0GA1 compact electric asynchronous servo-motor with nominal speed 1500 rpm, maximum speed 9000 rpm, nominal torque 34 Nm, maximum torque 60 Nm, nominal power 6.3 kW was used. The Simotics M-1PH8 servo-motor is freely configurable for specific applications thanks to its modular design. The Simotion control system is based on the drive control chart (DCC), which allows variable speed operation of the hydraulic pump. The Simotics servo-motor provides an innovative and energy-efficient solution concept for ZSM hydraulic drive and control. The operating principle of the hydraulic system is to control the pressure and volumetric flow rate not by valve technology but directly by the speed and torque of the Simotics servo-motor. As a result, hydraulic power is supplied only by the servo-pump when it is necessary to operate the hydraulic actuators (motors and cylinders). In the open-circuit control system, a single servo-pump provides sequential control for multiple hydraulic actuators. The advantages of the Simotion control system are high energy efficiency and economy thanks to the use of speed-controlled hydraulics, as well as the availability of the solution based on the servo-pump for both single- and multi-axis systems, high dynamics thanks to the use of Simotics servo-motors with low rotor inertia, and high precision of synchronization of hydraulic multi-axis drives. The servo pump unit uses a Hydro-Leduc type XPi 12 0523820 fixed displacement piston hydraulic pump with bent axis with geometric displacement 12 cm^3/rev, flow rate of 24 L/min, maximum pressure 380 bar, maximum speed 3150 rpm and maximum torque 76 Nm. Hydro Leduc XPi bidirectional bent-axis pumps are specifically designed to meet the needs of mobile devices. This pump model has a 7-piston design to ensure optimal flow regularity and can withstand continuous working pressures up to 380 bar and 420 bar peaks. The pump automatically sets to required direction of rotation. Four Hydac diaphragm accumulators of type SBO210 - 2E1/112U 330 AK 030 with a capacity of 2 L were used. Figure 6 shows the control circuits of the ZSM hydraulic modules.

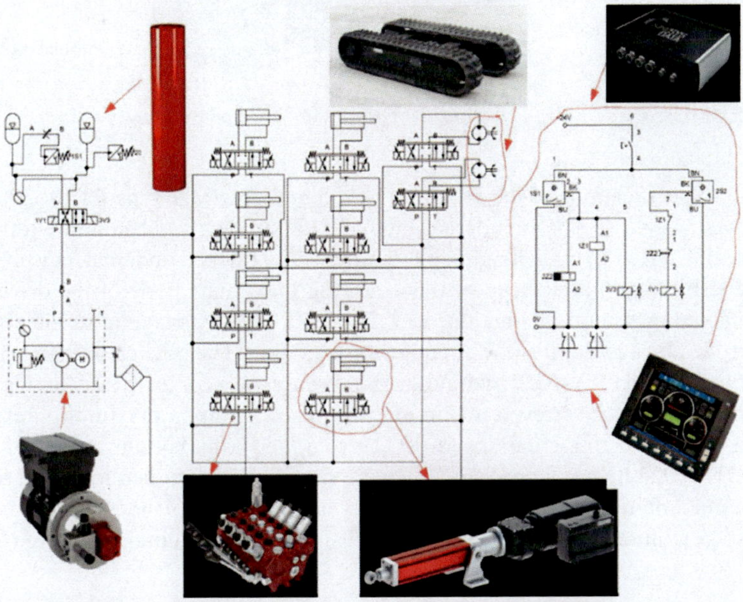

Fig. 6. Control circuits of the ZSM hydraulic modules [11].

The Hydac LX-610 multi-section directional proportional valve is the main hydraulic power distribution element in the ZSM hydraulic control system. This valve plays an important role in the stability and reliability of hydraulic system control even under higher loads. The Hydac valve LX-610/B0 with a nominal pressure of 350 bar and a nominal flow rate of 160 L/min can independently control the operating functions of ten ZSM hydraulic actuators. The pressure in the hydraulic actuators is monitored by a pressure transducer located behind the multi-section proportional directional valve.

4 Tracked Undercarriage Hydraulic Drive Module

A conventional Hinowa rubber track undercarriage type PT20GL was chosen for mobile ZSM (Fig. 7). The Hinowa tracked undercarriage with rubber tracks, due to its highly maneuverable design, can move efficiently over longer distances between construction sites. It can move for short distances and is placed precisely on the bricklaying site. The built-up tracked undercarriage has been patented as utility models [13, 14].

Fig. 7. Hinowa tracked undercarriage.

Specifications of the Hinowa tracked undercarriage type PT20: 2000 kg maximum total weight (robot + undercarriage), 1935 mm length, 1100 mm width, 250 mm track width, 370 mm track height, 5 + 5 rollers. The tracked undercarriage can move forward and backward at a speed of 0.56 m/s, the maximum slope of the driveway ±15°, and maximum lateral inclination ±8°. The track undercarriage motor is powered and controlled by a hydraulic power and control module. The tracked undercarriage is driven by two Bravini CTM1022 hydraulic orbital gear motors of the BRZV series with a displacement of 80.4 cm^3/rev, a maximum flow of 65 L/min, a maximum pressure of 210 bar, a maximum rotational speed of 119 rpm, and a maximum torque of 1280 Nm. The CTM1022 hydraulic gear motors were specifically designed for small track drives with a maximum weight of 2500 kg. These units feature a planetary gearbox (one reduction stage), fail-safe brakes (optional), built-in motor, and a braking valve (optional).

5 Hydraulic Lifting-Levelling Module

The robot support frame is built on the track undercarriage, on which hydraulic lifting-levelling modules are mounted at the front and rear (Fig. 8). These two hydraulic lifting-levelling modules were used to stabilize the position of the robot during automatic bricklaying. A single hydraulic lifting-levelling module is constructed using two double-acting differential cylinders Ponar Luban UCJ2F-16-40/25/500z+U for the extension of support legs and two double-acting differential cylinders Ponar Luban UCJ2F-16-50/28/100z for the lifting of the robot support frame. Figure 9 shows the working principle of the lifting-levelling module.

Lifting and levelling the robot support frame takes place in two steps. In the first step, the hydraulic cylinders extend the support legs without external load until they come

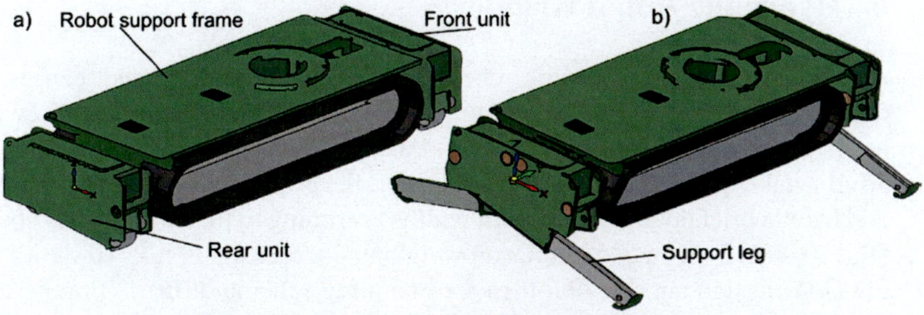

Fig. 8. Tracked undercarriage build-up: a) folded support legs, b) extended support legs [11].

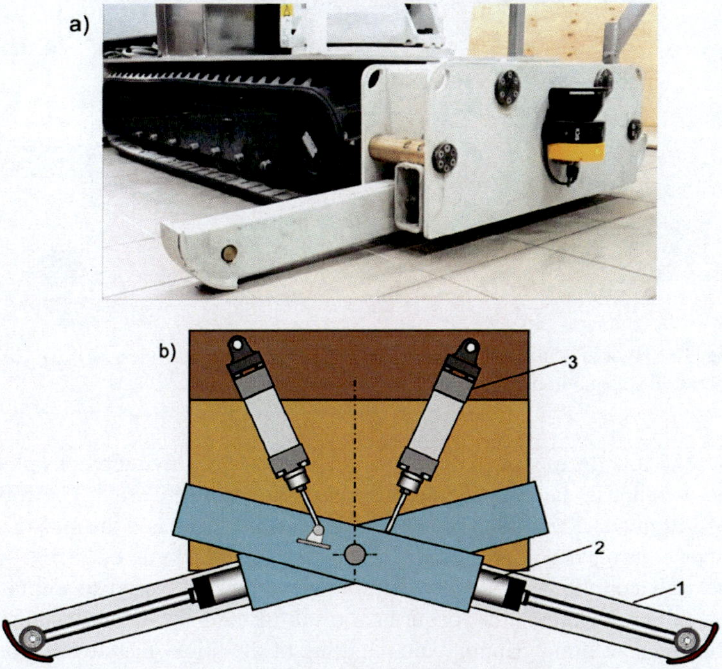

Fig. 9. Hydraulic lifting-levelling module: a) view of the extended support leg, b) design model of the hydraulic lifting-levelling module: 1 – support leg, 2 – double-acting cylinders for the extension of support legs, 3 – double-acting cylinders for the lifting of robot support frame.

into contact with the ground. In the second step, the hydraulic cylinder was pressing on the support legs until the robot's frame was raised to the required height. A robot support frame levelling control system was developed using feedback from the support leg position errors. [15]. When the robot support frame is level, the support legs are locked mechanically.

6 Hydraulic Robot Gripper

A special adjustable hydraulic robot gripper was constructed that can be adopted for different sizes of materials for building construction (brick, hollow brick, ceramic hollow brick, gas concrete blocks, cellular concrete, etc.). This is a two-jaw parallel gripper with an adjustable jaw and a moving jaw. The parallel gripper has five replaceable jaws to grip and handle bricklaying blocks with widths according to the dimensional specification of 80, 120, 180, 240, and 250 mm, and weighing up to 25 kg. Figure 10 shows the view, the 3D CAD design and the on/off mode operating scheme of the hydraulic robot gripper for the bricklaying blocks.

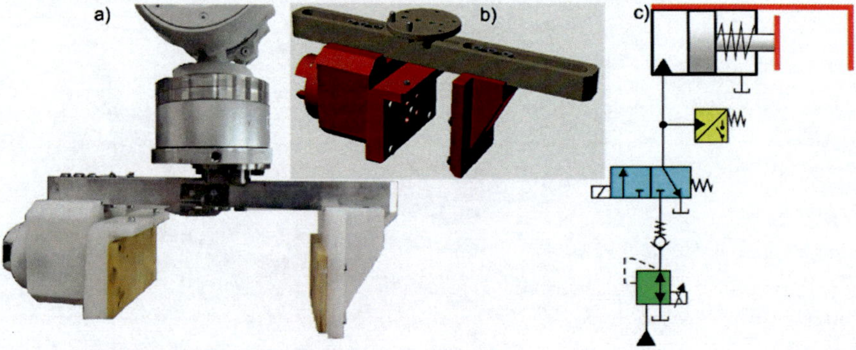

Fig. 10. Hydraulic robot gripper for bricklaying blocks: a) view, b) 3D CAD design [11], c) operation scheme in on/off mode.

The robot gripper adapted to the size of the bricklaying blocks picks them up from the warehouse, handles them, and places them on the wall. The robotic bricklaying algorithm considers the size of the bricklaying blocks and the size of the wall. During robot control, the gripper with the bricklaying blocks is correctly orientated against the wall coordinates. A laser calibration system was used to eliminate inaccuracies and minor deviations that occur as a result of the brick being misaligned by the robot gripper. Inaccurate gripping and handling of the bricks negatively affects the accuracy of the bricklaying, and therefore the flatness and uniformity of the masonry wall.

7 Conclusions

The ZSM mobile robotic bricklaying system was created as part of the POIR.04.01.02-00-0045/18-00 research project, funded by a grant from the National Centre for Research and Development in Poland within the framework of the Smart Growth Operational Programme 2014–2020. The mobile ZSM is designed to perform bricklaying tasks on construction sites. The hydraulic modules of the mobile ZSM perform their tasks effectively and energy efficiently. The hydraulically driven tracked undercarriage can efficiently travel the ZSM around construction sites. Hydraulic control of the tracked undercarriage allows for accurate positioning and displacement of the ZSM in place of the bricklaying work. Hydraulic lifting-levelling modules stabilize the position of the ZSM subjected to impact loads during robotic bricklaying. The stable position of the ZSM is of great importance for the quality of the brickwork made by the robot. The accuracy of the trajectory of the robot's hydraulic gripper depends not only on the mass of the bricklaying blocks being handled, but also on the precision of their gripping.

References

1. Dakhli, Z., Lafhaj, Z.: Robotic mechanical design for bricklaying automation. Cogent Eng. **4**, 1–22 (2017)
2. Pritschow, G., Dalacker, M., Kurz, J., Gaenssle, M.: Technological aspects in the development of a mobile bricklaying robot. In: 12th ISARC International Symposium on Automation and Robotics in Construction, Warsaw, Poland, 30 May–06 July 1995, pp. 1–8 (1995)
3. Madsen, A.J.: The SAM100: Analyzing labor productivity. Construction Management Department, California Polytechnic State University, San Luis Obispo, USA (2019)
4. Pan, W.: Methodological development for exploring the potential to implement on-site robotics and automation in the context of public housing construction in Hong Kong. Ph.D. Thes. Lehrstuhl für Baurealisierung und Baurobotik, Technische Universität München, München, Germany (2020)
5. Rihani, R.A., Bernold, L.E.: Methods of control for robotic brick masonry. Autom. Constr. **4**, 281–292 (1996)
6. Wos, P., Dindorf, R., Takosoglu, J.: The electro-hydraulic lifting and leveling system for the bricklaying robot. In: Stryczek, J., Warzyńska, U. (eds.) NSHP 2020. LNME, pp. 216–227. Springer, Cham (2021). https://doi.org/10.1007/978-3-030-59509-8_19
7. Wos, P., Dindorf, R.: Hydraulic leveling control system technology of bricklaying robot. In: 16th International Conference on Dynamic Systems, Theory and Application, Lodz, Poland, 6–9 December 2021, pp. 582–583 (2021)
8. Dindorf, R., Wos, P.: Vibration damping performance of the electrohydraulic self-levelling unit of the tracked platform of robotic bricklaying combine. In: IOP Conference Series: Materials Science and Engineering, vol. 1247, pp. 12–20 (2022)
9. Dindorf, R., Wos, P.: Energy efficiency of pressure shock damper in the hydraulic lifting and levelling module. Energies **15**(17), 1–28 (2022)
10. Dindorf, R., Wos, P.: Innovative solution of mobile robotic unit for bricklaying automation. J. Civil Eng. Transp. **4**(4), 21–32 (2022)
11. Dindorf, R., Takosoglu, J., Wos, P., Chłopek, L.: Robotic bricklaying system. Research reports. Kielce University of Technology, Kielce, Poland (2021). (in Polish)
12. Wos, P.: Robotic bricklaying system. User manual. Kielce University of Technology, Kielce, Poland (2021). (in Polish)
13. Dindorf, R., Takosoglu, J., Wos, P., Chlopek, L.: Utility model Wp.30256. Tracked transporter. Kielce University of Technology, Kielce, Poland (2022)
14. Dindorf, R., Takosoglu, J., Wos, P., Chlopek, L.: Utility model Wp.30764. Tracked transporter housing. Kielce University of Technology, Kielce, Poland (2022)
15. Wos, P., Dindorf, R., Takosoglu, J.: Bricklaying robot lifting and leveling system. Commun. Sci. Lett. Univ. Žilina **23**(4), B257–B264 (2021)

Preliminary Tests for the Use of Hydrostatic Transmission with Oscillating Energy Flow

Kacper Dąbek, Piotr Osiński, and Krzysztof Kędzia(✉)

Wroclaw University of Science and Technology, Wroclaw, Poland
kacper.dabek@pwr.edu.pl

Abstract. The paper presents a simulation model of a hydrostatic transmission consisting of a piston pump and a receiver in the form of a hydraulic motor with an energy compensation system. The current state of research in this area and the patent reservations made by the authors are presented. Possible applications of the solution will be discussed: recovering energy from vehicle suspension and producing electricity in wave power plants, taking into account ecological aspects.

Keywords: Hydrostatic transmission · piston pump · oscillation

1 Introduction

1.1 Ecology in Industry

With the development of civilization, the level of Earth's exploitation is increasing. The economic crisis results in rising commodity prices. The vision of depleting natural resources is becoming an increasingly frequent topic of debate. Continuously introduced new EU climate restrictions serve as a reason to limit current resource utilization methods. Companies are facing pressure related to high energy and resource consumption, climate change, and competition for limited resources. It is necessary to develop new products in accordance with the principles of sustainable development. Legal regulations encourage a closed-loop economy, extending the lifespan of products, and reducing generated waste. The machinery and equipment industry plays a significant role in engineering as even small design changes can have a significant impact on the environment and affect other manufacturing companies [1]. There is an increasing share of so-called green energy in global production. The implementation of new solutions for generating electricity from renewable sources such as water, biomass, or solar power contributes to lower emissions of pollutants, carbon dioxide, and also avoids toxic waste. The use of multi-source systems instead of single-source systems and their appropriate control have a significant impact on energy and ecological efficiency [2, 3]. It is important that implemented power generators exhibit the highest possible energy efficiency. This article will bring the possibilities of using hydrostatic systems to obtain lost energy in nature and industry.

J. Stryczek and U. Warzyńska (Eds.): NSHP 2023, LNME, pp. 184–195, 2024.
https://doi.org/10.1007/978-3-031-43002-2_17

1.2 Hydrostatic Transmission with Oscillating Energy Flow

Many factors contribute to increased pressure on the development of systems that enable energy recovery from existing technological processes or harness energy from renewable sources. One such system is the utilization of hydraulic systems with oscillating energy flow. In such systems, reciprocating pumps are commonly used as hydraulic energy generators. These pumps are employed in various configurations and setups. An example of the application of a reciprocating pump in the configuration of a wave power plant, PELAMIS, is presented in the illustration below [4] (Fig. 1).

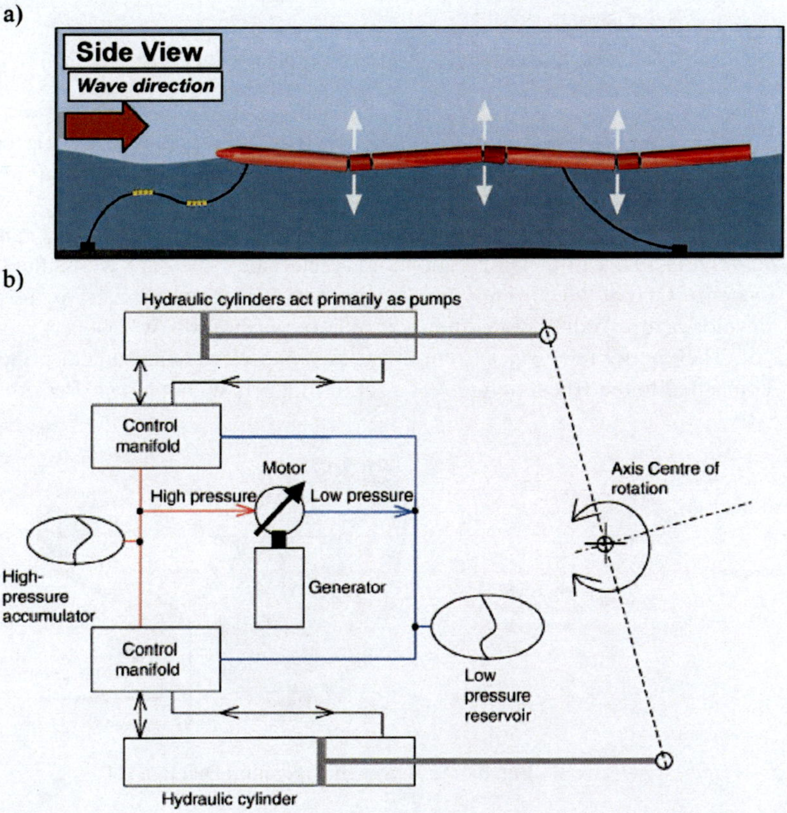

Fig. 1. Diagrams showing the PELAMIS: a) During motion on the water surface, b) a simplified hydraulic scheme [4].

The device operates on the principle of floating members connected by joints, which are in turn attached to the energy-generating system. When the water surface is subjected to waves, the members change their relative angle, thereby driving reciprocating pumps. These pumps, with the assistance of an alternator, convert the oscillatory motion into electrical energy.

Another method of harnessing wave energy is a [5] wave power plant, which is based on a buoy located on the sea surface (Fig. 2).

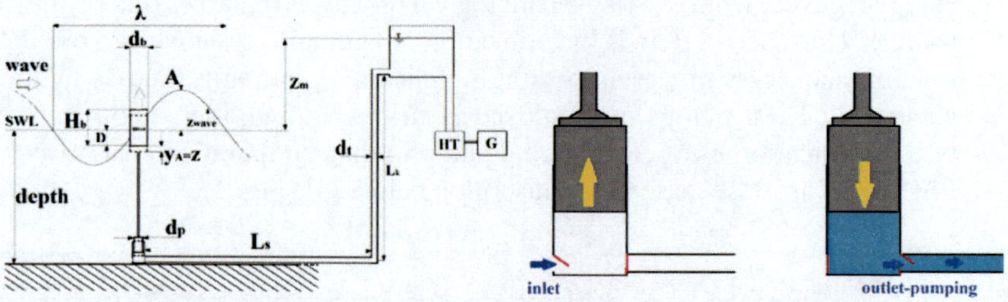

Fig. 2. Diagram of a wave power plant using a piston pump [5].

The wave moves a buoy that is connected to the piston of a reciprocating pump, which in turn pumps seawater into a reservoir. This reservoir serves as a water storage for a pumped-storage hydroelectric power plant.

The application of hydraulic systems with oscillating energy flow can also be used in vehicles. Due to their operational characteristics, they are well-suited for damping systems. One of the currently utilized systems for energy recovery is the eROT system developed by Audi. This system can recover up to 600 W of energy during driving [6]. The energy recovery system operates on an electromechanical principle. An arm connected to the wheel hub moves a set of magnets within a coil (Fig. 3).

Fig. 3. eROT System Operation Diagram [6].

This article will present a hydraulic system similar to the one presented, enabling energy recovery. The following will be presented: mathematical model and computer simulation.

2 Mathematical Model

The analysis of existing solutions for energy recovery does not exhaust all possibilities. As part of our own research work, we have proposed an original solution with oscillating energy flow. The system diagram was revealed in a patent application filed with the Polish Patent Office[7], and the reserved schematic is shown in Fig. 4.

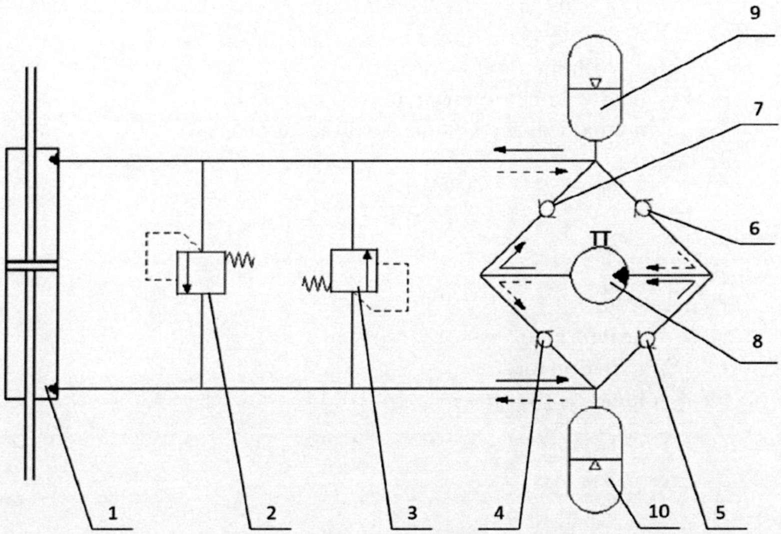

Fig. 4. Hydraulic system scheme: 1 - hydraulic cylinder, 2,3 - pressure relief valve, 4–7 - check valve, 8 - hydraulic motor, 9, 10 – hydraulic accumulator.

The proposed system consists of a double-acting actuator, four check valves, a hydraulic motor, two relief valves, and two compensating devices in the form of hydraulic accumulators. The check valve arrangement forms a hydraulic rectifier, which is capable of converting alternating flow of the working fluid into a flow with a constant direction. In a subsequent step, a numerical model was developed using the MATLAB Simulink software. During the model construction, simplifying assumptions were made: isothermal process, ideal actuator excitation, volumetric leakage occurs between the input and output, mechanical losses are associated with internal friction and infinitely fast pressure propagation in the system. To minimize the possibility of cavitation, the pressure in the entire system has been increased. Below will be presented the equations on the basis of which the simulation model was built.

2.1 Hydraulic Piston

The actuator of the hydraulic system serves as the displacing element, operating in a double-acting manner and alternately supplying both main collectors of the system. Ideal excitation and 100% volumetric efficiency (no internal or external leakage) were assumed. The hydraulic-mechanical efficiency was also assumed to be 100%.

2.2 Hydraulic Motor

The hydraulic motor is responsible for converting the pressure energy of the hydraulic oil into rotary motion. It was assumed that volumetric leakage occurs between the input and output, and mechanical losses are associated with internal friction [8, 9].

Rate flow:

$$q_{Real} = qIdeal + qLeak \tag{1}$$

where:

q_{Real} - real volumetric flow rate,
q_{Ideal} - ideal volumetric flow rate,
q_{Leak} - internal leakage volumetric flow rate.

Torque:

$$\tau = \tau_{Ideal} - \tau_{Friction} \tag{2}$$

where:

τ - net torque,
τ_{Ideal} - ideal torque,
$\tau_{Friction}$ - friction torque.

Ideal volumetric flow rate:

$$q_{Ideal} = D\omega \tag{3}$$

where:

D - the value of the hydraulic motor displacement,
ω - instantaneous angular velocity of the rotary shaft.

Ideal generated torque:

$$\tau_{Ideal} = D\Delta p \tag{4}$$

where:

Δp - instantaneous pressure difference from inlet to outlet.

2.3 Hydraulic Accumulators

The system includes two accumulator devices, one for each collector. The accumulators are designed to smoothen the hydraulic function over time [8, 9].

$$(p_G + p_A)(V_T - V_F)^k = \left(p_{pr} + p_A\right)V_T^k \tag{5}$$

$$p_F = p_G + p_{HS} \tag{6}$$

$$V_C = V_T - V_{dead} \tag{7}$$

$$p_{HS} = \begin{cases} K_S(V_F - V_C) + K_d q_F^+ (V_F - V_C) & \text{if} V_F \geq V_C \\ K_S V_F - K_d q_F^- V_F & \text{if} V_F < 0 \\ 0 & \text{otherewise} \end{cases} \tag{8}$$

$$q_F^+ = q_F \text{if} q_F \geq 0 \tag{9}$$

$$q_F^- = q_F \text{if} q_F < 0 \tag{10}$$

$$q_F = \frac{dV_F}{dt} \tag{11}$$

where:

V_T - Total volume of the accumulator, including the fluid chamber and the gas chamber,

V_F - Volume of fluid in the accumulator,

V_C - Fluid chamber capacity, which is the difference between total accumulator volume and the gas chamber dead volume,

V_{dead} - Gas chamber dead volume, which is a small portion of the gas chamber that remains filled with gas when the fluid chamber is at capacity,

p_F - Fluid pressure (gauge) in the fluid chamber, which is equal to the pressure at the accumulator inlet,

p_{pr} - Pressure (gauge) in the gas chamber when the fluid chamber is empty,

p_A - Atmospheric pressure,

p_G - Gas pressure (gauge) in the gas chamber,

p_{HS} - Hard-stop contact pressure,

K_S - Hard-stop stiffness coefficient,

K_d - Hard-stop damping coefficient,

k - Specific heat ratio (adiabatic index),

q_F - Fluid flow rate into the accumulator, (+) means fluid flows into the accumulator, (−) from accumulator (opposite flow direction).

2.4 Check Valve

The check valve allows the flow of the working fluid in one direction. There is a minimum pressure at which the valve opens. The opening area of the valve changes linearly until reaching the maximum valve area value. The leakage for a closed valve is considered a constant value for the leakage area. The relationships are illustrated in Fig. 5.

Flow area:

$$A(p) = \begin{cases} A_{leak} & \text{for} \, p \leq p_{crack} \\ A_{leak+k\cdot(p-p_{crack})} & \text{for} \, p_{crack} < p < p_{max} \\ A_{max} & \text{for} \, p_{max} \leq p \end{cases} \tag{12}$$

$$k_{cv} = \frac{A_{max} - A_{leak}}{p_{max} - p_{crack}} \tag{13}$$

where:

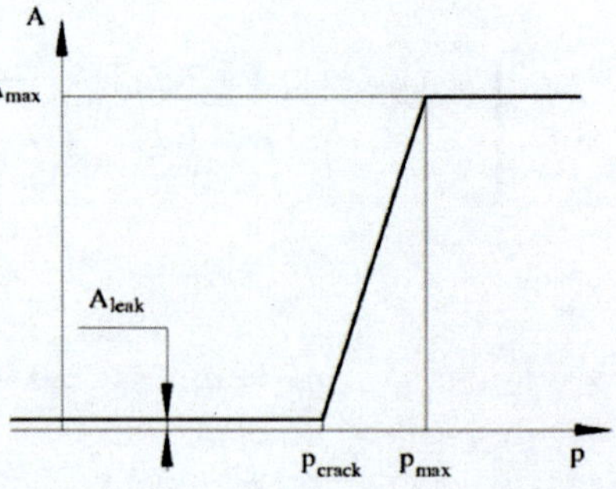

Fig. 5. Dependence of pressure and area [8].

A_{max} – fully open passage area,
A_{leak} – closed valve leakage area,
p_{crack} – valve cracking pressure,
p_{max} - fully open pressure,
k_{cv} – opening coefficient of check valve.

2.5 Pressure Relief Valve

Due to its shape and operational characteristics, the hydraulic system has been protected by two relief valves. Their characteristics are provided below (Fig 6):

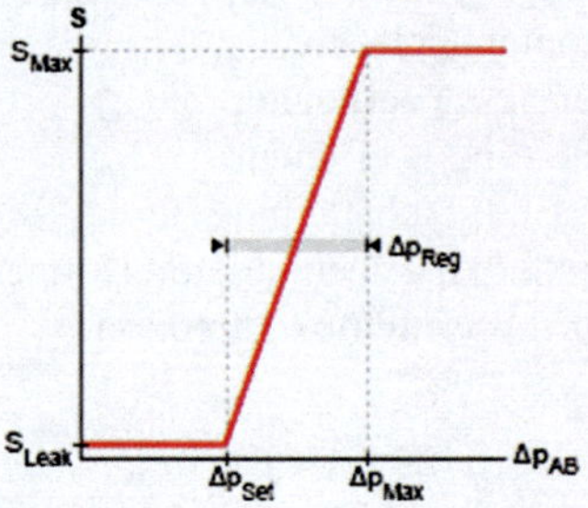

Fig. 6. Dependence of pressure and area in pressure relief valve [8].

$$\Delta p_{max} = \Delta p_{set} + \Delta p_{reg} \quad (14)$$

Δp_{set}– is the value specified in the Valve pressure setting block parameter.
Δp_{reg}– is that specified in the Valve regulation range block parameter.

2.6 Inertia

Inertia has been modeled as a rotating mass attached to the shaft of the hydraulic motor [8]:

$$T = J\frac{d\omega}{dt} \tag{15}$$

where:

T - inertia torque.
J - inertia.
ω - angular velocity.
t - time.

2.7 Summary of Equations

Each of the (1–15) designs describes a particular hydraulic element, together forming a whole. The Simscape program allows to create physical models in a MATLAB Simulink environment.

3 Simulation Results

The simulation was conducted using the MATLAB software with the Simulink extension - Simscape. All simulations involved a single excitation (Fig. 7). It is a sinusoidal excitation that causes ideal actuator motion. During operation, the actuator does not reach extreme positions. The system was initially preloaded with pressure to prevent local pressure drops that could cause cavitation, which is not considered in the simulation system. The simulation focuses on observing the maximum speed in the battery charge function.

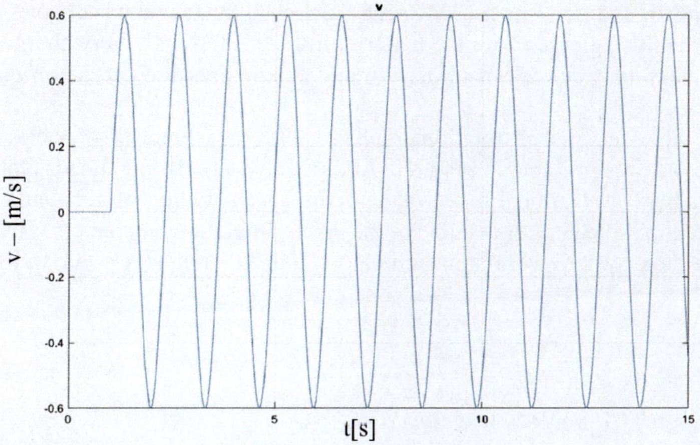

Fig. 7. Excitation signal

The simulation was performed for eleven different pre-fill pressures of the accumulator: 1 MPa, 2.5 MPa, 5 MPa, 7.5 MPa, 10 MPa, 11.25 MPa, 12.5 MPa, 13.75 MPa, 15 MPa, 17.5 MPa, 20 MPa. Figures 8 and 9 present the simulation results for the system without an accumulator and with an accumulator pre-filled at pressure p_{acc}. In each case, the rotational speed of the motor stabilized at a constant level after a certain time.

Simulation Without Accumulator

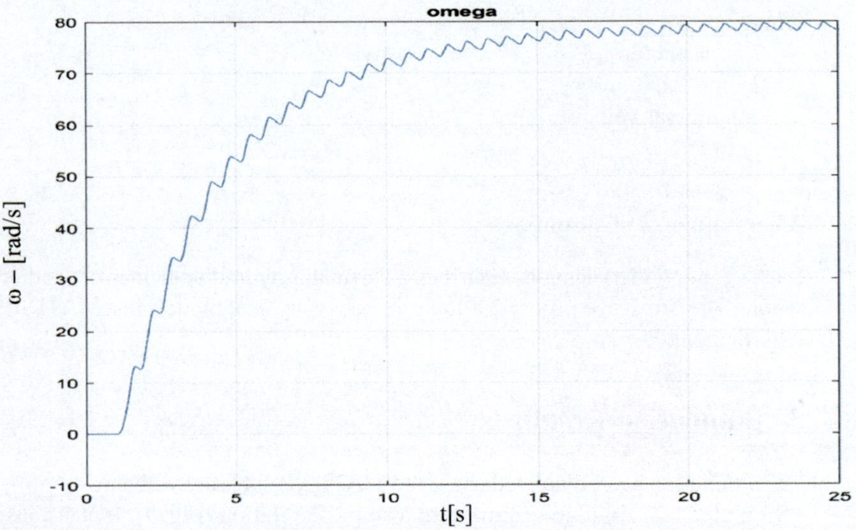

Fig. 8. Simulation without accumulator

Simulation with Accumulator

In the simulation without accumulator chart, a higher speed can be viewed of the hydraulic motor after stabilizing than for simulation with accumulator operated systems. In addition, maximum speed is reached in less time for a case without a battery. By analysing cases with the battery for three different battery charge pressures, a relationship between speed of growth and charge pressure can be observed, namely both increase.

The steady-state rotational speed values as a function of the accumulator pre-fill pressure are presented in Fig. 10. Additionally, a horizontal line is plotted to indicate the steady-state rotational speed value for the system without an accumulator. The maximum rotational speed of the motor for the given simulation parameters is 69.7 rad/s, while for the same system without accumulators, it is 79.0 rad/s (Figs. 11 and 12).

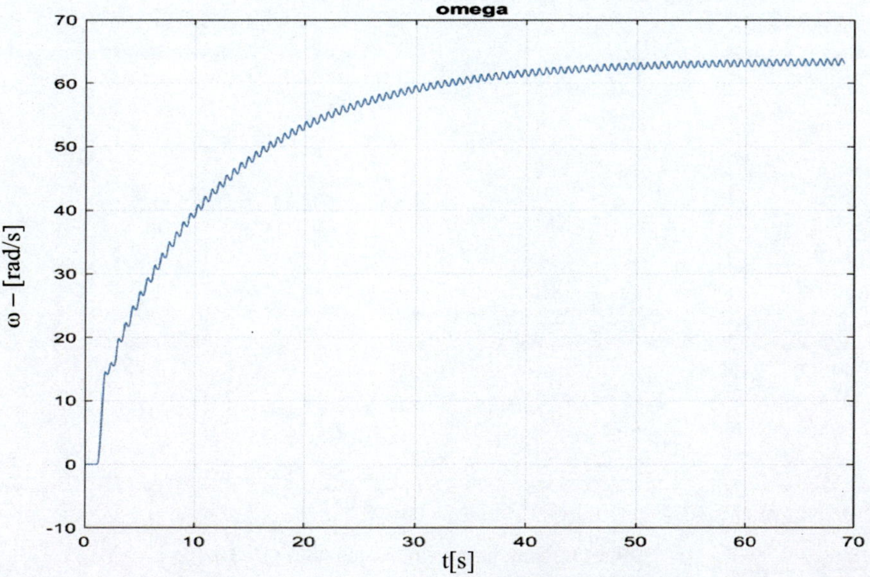

Fig. 9. Simulation with accumulator for $p_{acc} = 10$ MPa

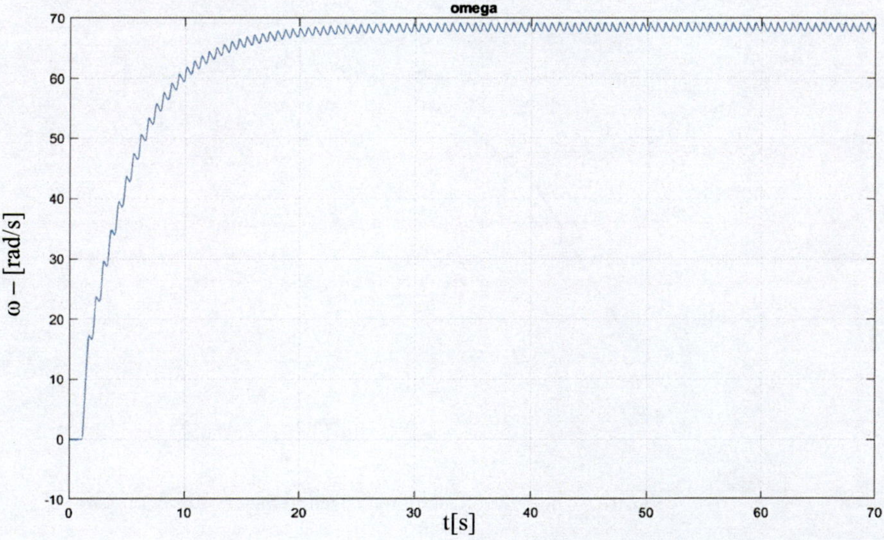

Fig. 10. Simulation with accumulator for $p_{acc} = 15$ MPa

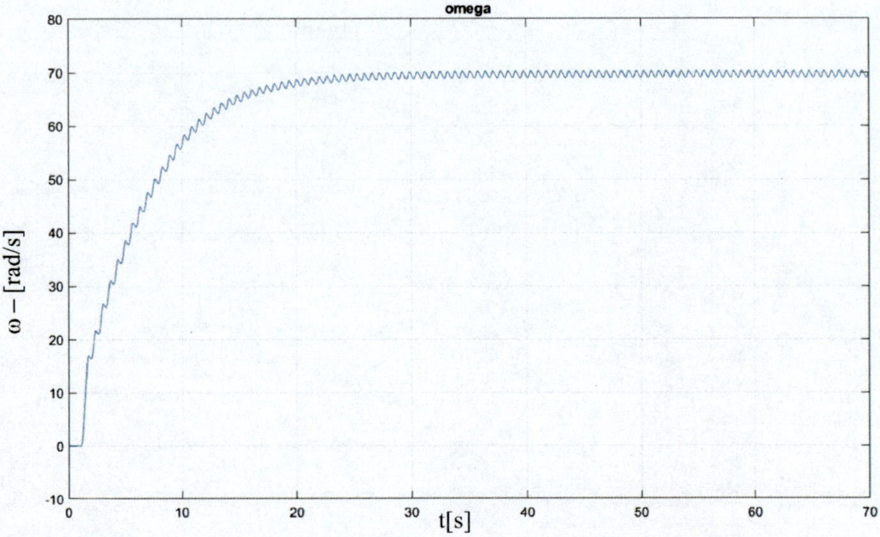

Fig. 11. Simulation with accumulator for p_{acc} = 12,5 MPa

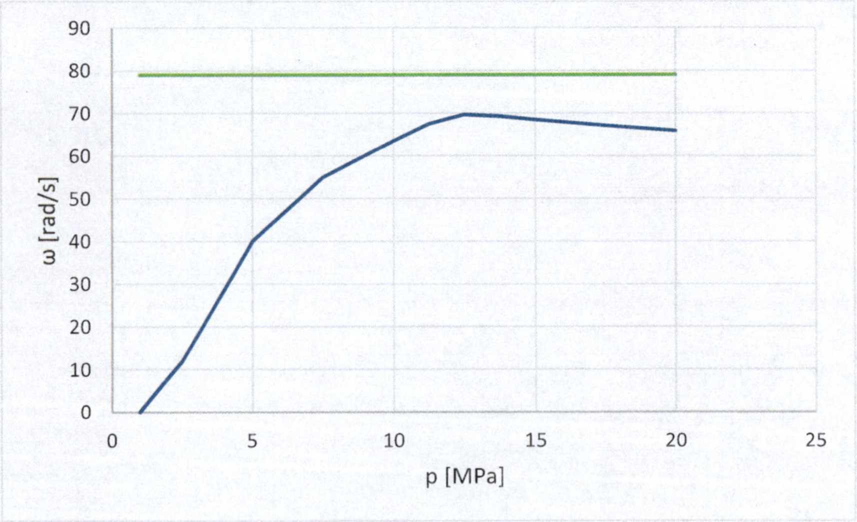

Fig. 12. The ratio of rotational speed to battery filling pressure.

4 Conclusions

In summary, from the presented comparison of rotational speeds, it can be concluded that the use of an accumulator contributes to a reduction in the motor's rotational speed by a minimum of 9.3 rad/s, i.e., at least 11.8% compared to the speed of the motor in the system without an accumulator. The decrease in speed for systems with an accumulator is due to the flow of a portion of the working fluid accumulated in the accumulator back to the pump during the suction cycle. However, this effect is least pronounced for the pre-fill pressure of the accumulator at $p_{acc} = 12.5$ MPa.

The addition of the accumulator prolongs the time required for the rotational speed to stabilize. A long stabilization time indicates a preference for such systems in devices characterized by stable operation (e.g., wave power plants) rather than in rapidly changing processes (e.g., vehicle suspension), where energy efficiency will be lower.

The considerations presented serve as an introduction to further analysis, in which the system efficiency and the impact of the accumulator capacity on pressure pulsations and the stabilization time of the rotational speed will be examined in the next stage.

References

1. Riesener, M., Kuhn, M., Ruschitzka, C., Schuh, G.: Concept for a function-oriented ecology analysis in machinery and plant engineering. Procedia CIRP **116**, 53–58 (2023). https://doi.org/10.1016/j.procir.2023.02.010
2. Kędzia, K.: A method of determining optimal parameters for the secondary energy source of a multisource hydrostatic drive system in machines working in closed spaces. Energies **15**, 5132 (2022). https://doi.org/10.3390/en15145132
3. Kędzia, K.: Wyznacznik Zmienności Cyklu Obciążenia Wieloźródłowego Hydrostatycznego Układu Napędowego; Oficyna Wydawnicza Politechniki Wrocławskiej (2014)
4. Henderson, R.: Design, simulation, and testing of a novel hydraulic power take-off system for the pelamis wave energy converter. Renew. Energy **31**, 271–283 (2006). https://doi.org/10.1016/j.renene.2005.08.021
5. Bonovas, M.I., Anagnostopoulos, I.S.: Modelling of operation and optimum design of a wave power take-off system with energy storage. Renew. Energy **147**, 502–514 (2020). https://doi.org/10.1016/j.renene.2019.08.101
6. Jacota, V., Matache, M.: Calibration of a system capable to recover the dissipated energy into an automobile suspension. U.P.B Sci. Bill. **2018**, 100–101 (2018)
7. Osiński, K., Kędzia, K., Dąbek: Hydraulic Close Transmission with Hydraulic Bridge Rectifier (Przekładnia Hydrostatyczna Zamknięta z Hydraulicznym Mostkiem Prostowniczym), Patent Application: P. 440311 (2022)
8. MatLAB v. R2021b Help on-Line Documentatoion
9. Manring, N.D., Fales, R.C.: Hydraulic Control Systems, University of Missouri–Columbia (2020)

Mobile Machine with Hydrostatic Hybrid Drive Train

Radu-Iulian Rădoi[1], Cătălin Dumitrescu[1(✉)], Bogdan Tudor[1], Ștefan Șefu[1], and Ciprian Culache[2]

[1] National Institute of Research & Development for Optoelectronics/INOE 2000, Subsidiary Hydraulics and Pneumatics Research Institute/IHP, Cuțitul de Argint 14, 040558 Bucharest, Romania
dumitrescu.ihp@fluidas.ro

[2] Hiarom Invest Ltd, Măceșului 8, 077095 Dragomirești-Vale, Romania

Abstract. Reducing the noxious emissions from internal combustion engines is one of the factors that can contribute to amending the quality of the environment in which we live. Even if the efforts and results of the last decades are more visible in the field of road vehicles, the equipment used in the construction field also registers progress in reducing consumption and especially CO_2 emissions. To reduce consumption, equipment manufacturers have applied various hybridization solutions, from the simplest ones that use energy recovery systems to those that use both internal combustion engines and electric motors. A special case is encountered when noxious emissions are completely prohibited, such as work performed in enclosed spaces or which cannot be properly ventilated. In these cases, the only solution is to obtain the necessary energy with an electric motor powered by batteries, in order to drive the transmission and the equipment mounted on the machine. Since hydraulic drive is difficult to replace in machines working in harsh conditions, the hydraulic pumps must be driven by electric motors. The article presents such a machine solution, which uses as the primary source of energy either a regular internal combustion engine or an electric motor powered by batteries.

Keywords: Mobile Machine · Diesel – Electric Hybrid · Hydrostatic Drive

1 Introduction

In the context of the EU directives regarding the reduction of carbon emissions [1], a machine with a diesel-electric hybrid energy source was developed.

A series of authors of some papers studied the trends, the potential, various approaches and challenges related to the hybridization of mobile machinery [2–5].

The developed machine presented in this paper has two independent energy groups, one with an internal combustion engine (ICE) Diesel type and one with an electric motor powered by a battery pack. Depending on the needs, the ICE energy source or the electric motor can be used. The machine is ideal for work in closed halls or tunnels where the electric motor can be used exclusively, and for work outside the ICE can be used as a backup until reaching the garage to recharge the battery.

J. Stryczek and U. Warzyńska (Eds.): NSHP 2023, LNME, pp. 196–203, 2024.
https://doi.org/10.1007/978-3-031-43002-2_18

The transmission of the machine is hydrostatic, the switching of the diesel or electric drive is achieved through hydraulic directional valves. The travel speed is controlled by varying the flow rate of the pump. Hydraulic motors with radial pistons for propulsion are more reliable when used in heavy conditions than electric motors. This type of engine ensures high torque at low revolutions. For propulsion with hydraulic motors, the closed circuit system is used. Closed loop circuit has a number of advantages such as: higher working pressure; it uses less working fluid; the fluid drained from the pump housing is replaced via the auxiliary charge pump, thus ensuring pump cooling and great flexibility in designing system controls [6].

A double gear pump, coupled in tandem with the main pump in a closed circuit, ensures the supply of the power steering and the hydraulic circuits for operating work equipment such as: snow plow, rotary brush, crane arm, etc. The work equipment can be mounted on the machine by means of a fixing plate made according to the EN 15432–1: 2011 standard.

The hydraulic power steering acts on all 4 wheels, ensuring very good and fast maneuverability in tight spaces.

Braking of the car is also done hydrostatically by reducing the flow rate of the pump, the machine being also equipped with a normally closed emergency/parking brake, also operated hydraulically (Table 1).

Table 1. Technical characteristics.

Parameter	Value
Mass	6000 kg
Power of the electric motor	30 kW
Diesel engine power	50 kW
Fuel tank capacity	60 l
Battery pack capacity	30 kWh
Maximum speed – working regime	7 km/h
Maximum speed – travel regime	15 km/h
Maximum traction force	30 kN

2 Configuration of the Mobile Machine

The vehicle (Fig. 1a) is compact and made of a solid steel structure welded from rectangular profiles between 8 and 15 mm thick. A modern cabin is installed in the front of the platform, which combines good visibility with comfort and is equipped with a central dashboard, panel with indicators for various information and control elements. The management of the operation of the machine and the equipment is carried out with a programmable logic controller equipped with a human-machine interface (HMI) installed on the console in the cabin. The energy groups with Diesel engine type Perkins

904J-E28t, N = 50 kW and the one with electric motor with N = 30 kW [7–9] are installed in the rear part of the platform. The engines are coupled with Poclain pumps type PM5052S4Q120840R2516A00VSPA, equipped for closed circuit operation. The hydraulic fluid tank and the blocks with hydraulic control equipment are also installed on the platform. Under the platform there are the two axles with wheels driven by hydraulic motors with radial pistons type MGE05–2-A04–101-1W20-EJ00 and the power steering mechanisms. The machine has the possibility to turn tight by steering all 4 wheels to go sideways or "crab walk" by commanding some directional valves from the hydraulic circuit of the power steering system.

The machine is equipped with a fixing plate, standardized according to EN 15432–1: 2011, for mounting various work equipment: snow plow, crane, etc. (Fig. 1b).

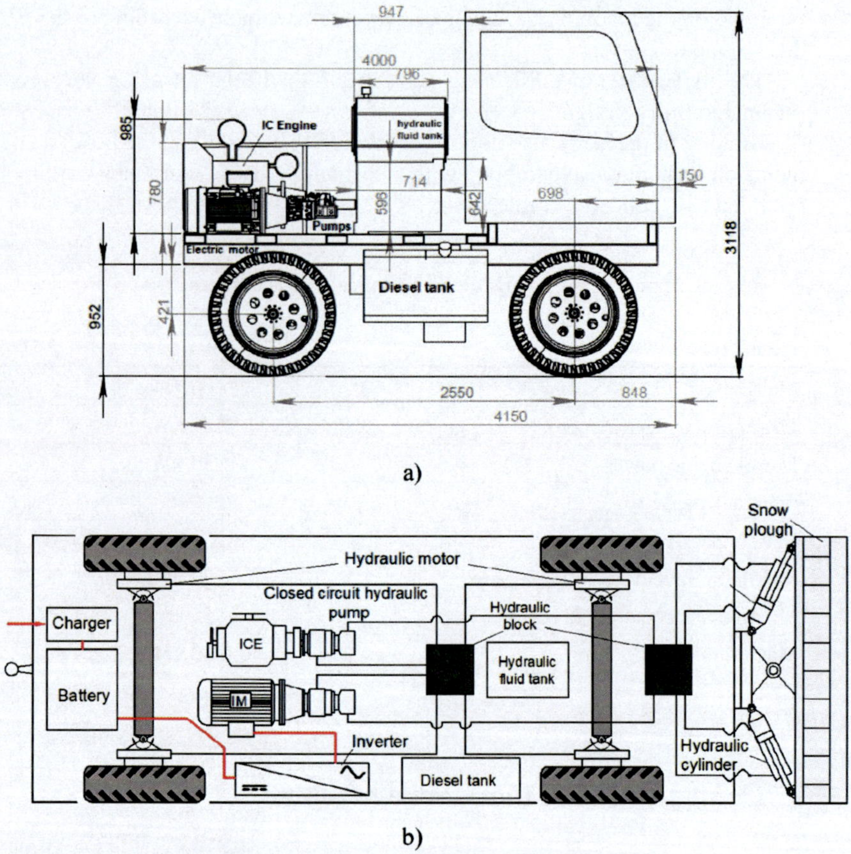

Fig. 1. The sketch with the dimensions (a) and equipment configuration (b) of the mobile machine for utility works.

The configuration of the hydraulic system is based on the working characteristics of the machine. Thus, the working speed is set at 7.5 km/h (2.1 m/s), and the travel speed v_{trav} without performing work will be double – 15 km/h (4.2 m/s).

The wheels of the machine have a diameter of D = 1000 mm, therefore the rotating speed n of one wheel, during the operation of the work equipment, will be:

$$\mathrm{n} = 60\frac{v_{trav}}{\pi D} = 60\frac{2.1}{3.14 \cdot 1} = 39.8 \approx 40\mathrm{rev/min} \tag{1}$$

For fast travel without doing work, the rotating speed will be double: n = 80 rev/min.

These travel speeds will be achieved when using the Diesel engine; by analogy with other machines, a thermal engine with a power of N = 50 kW is chosen.

In the closed hydraulic circuit of these machines, pumps with axial pistons are used, which can achieve pressures of up to 400 bar; therefore the flow rate Q (m^3/s) of such a pump will be:

$$Q = \frac{N \cdot \eta_t}{\Delta P} \tag{2}$$

where power N (*W*) and ΔP (*Pa*).

Given that η_t (the total efficiency of the pump) is 0.9…0.92, and for ΔP the return pressure is neglected, it follows:

$$Q = \frac{50 \cdot 10^3 \cdot 0.92}{400 \cdot 10^5} = 1.15 \cdot 10^{-3}\frac{m^3}{s} = 691/\mathrm{min} \tag{3}$$

At an engine speed of 1500 rev/min, to achieve this flow rate, the pump will have to have a displacement Vg of:

$$V_g = \frac{Q \cdot 60}{n \cdot \eta_v} \tag{4}$$

where Q is flow rate in m^3/s, n is the drive speed (1500 rev/min), and η_v is the volumetric efficiency of the pump, which for this type of pump can have a value of 0.95.

It follows that:

$$V_g = 1.15 \cdot 10^{-3} \cdot \frac{60}{1500 \cdot 0.95} = 4.842 \cdot 10^{-5}\mathrm{m}^3/\mathrm{rev} = 48.4\mathrm{cm}^3/\mathrm{rev} \tag{5}$$

A well-established manufacturer for closed circuit hydrostatic transmissions is Poclain, which produces the PM5052S4Q12 pump that can work at 400 *bar*; the displacement of this pump is 52 cm^3/rev.

At a speed of 3000 rev/min, taking into account the volumetric efficiency above, the pump will give a flow rate of $Q_{max.p} = 2.47 \cdot 10^{-3}$ m^3/s = 148.2 l/min.

Since all 4 wheels driven by a hydraulic motor are used for movement, the flow rate that will reach each will be:

$$Q_{max.m} = Q_{max.p}/4 = 2.47 \cdot 10^{-3}/4 = 6.175 \cdot 10^{-4}\mathrm{m}^3/\mathrm{s} \tag{6}$$

Considering the travel speed, in order to obtain the required speed the hydraulic motor will have a displacement, in cm³/rev:

$$V_{gm} = \frac{Q_{max.m} \cdot 60}{\mathrm{n}} = \frac{6.175 \cdot 10^{-4} \cdot 60}{80} = 4.63 \cdot 10^{-4}\mathrm{m}^3/rev = 463\mathrm{cm}^3/rev \quad (7)$$

Also choosing a Poclain product, we use the largest MGE05 hydraulic motor with 2 geometric volumes (374/749 cm³/rev), which allows the 2 different speeds to be achieved.

Figure 2 shows the structure of the hydraulic installation of the machine.

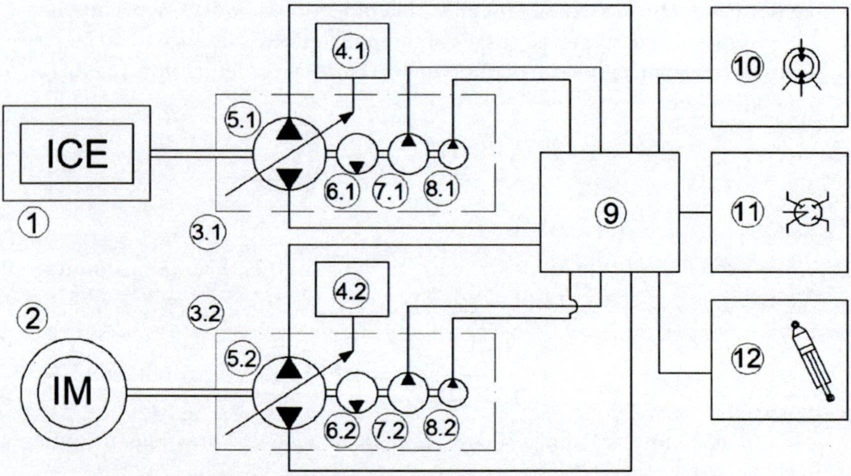

Fig. 2. Block diagram of the hydraulic installation of the machine.

The structure of the hydraulic installation in Fig. 2 is: 1 – internal combustion engine; 2 – induction motor; 3.1, 3.2 – pumping groups; 4.1, 4.2 – pump flow regulation systems; 5.1, 5.2 – closed circuit pumps; 6.1, 6.2 – auxiliary pumps; 7.1, 7.2 – power steering pumps; 8.1, 8.2 – pumps for operating the working mechanisms; 9 - block for directing the hydraulic fluid; 10 – circuit of hydraulic motors with radial pistons for marching; 11 – circuit for power steering mechanisms; 12 – circuit for working mechanisms.

Figure 3 shows a snow plow that can be mounted on the front of the carrier vehicle; the assembly plate has a standardized construction, compliant with the EN 15432 standard.

The machine is made with rigid axles (Fig. 4a) articulated in the middle of the chassis frame, and for this reason it is intended to run only on concrete platforms or paved roads. Any bumps on the roadway are having a mutual influence of the two wheels (Fig. 4b) of a rigid axles, causing the roll movement of front and rear sides of the vehicle.

In Figs. 5a and b, aspects from the testing of the machine running train with the ICE and electric motor energy groups can be noticed.

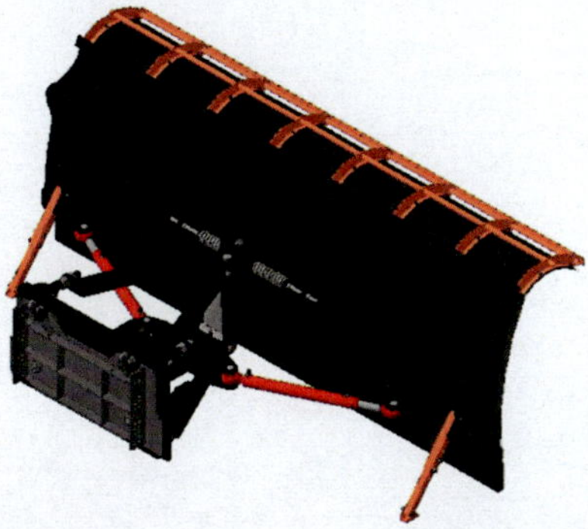

Fig. 3. Snow plow with assembly plate made according to the EN 15432 standard.

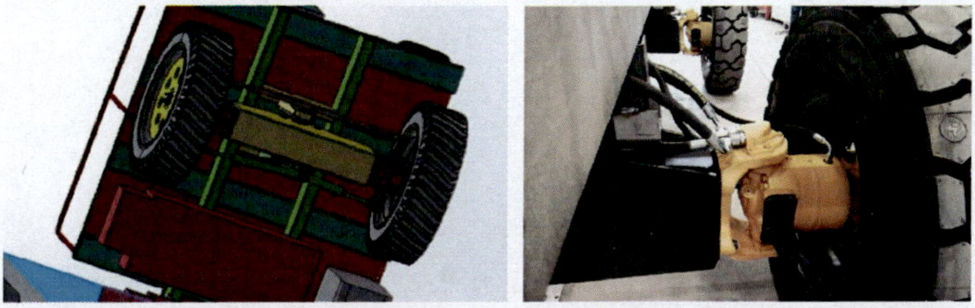

a) b)

Fig. 4. Axle with steering actuators and wheel installed to the hydraulic motor.

a) b)

Fig. 5. Aspects during the tests with the running train of the machine.

3 Electric Drive

In the electric mode, the pumping group is driven by a 30-kW electric motor [7]. The electric motor is powered by means of an inverter that allows: 4-quadrant, synchronous or asynchronous AC motor control, with speed, torque and DC voltage control modes. It

also has an internal common mode and differential mode DC EMC filter which reduces high frequency electromagnetic interference and eliminates DC bus oscillations.

The electrical energy storage system is made with a 30-kWh battery at a voltage of 350 V. The battery is equipped with a battery management system (BMS), which ensures the protection of the cells and provides information on the states of the battery such as: SoC – state of charge; SoH - state of health; I_{max} – maximum current; V_{nom} – nominal voltage. A dedicated on-board charger is used to charge the battery pack.

A future improvement of the machine can be achieved by including systems such as energy recovery and adaptive energy management system [10–13].

The management of the machine's operation is carried out with a programmable logic controller (PLC). The software implemented in the PLC allows the monitoring and control of the functional subsystems of the machine. A PLC control loop activates the cooling of the hydraulic fluid when a temperature threshold is reached, and in case of exceeding preset values for parameters such as pressure, temperature, battery status and others, a series of visual and audible alerts are activated. The driver can interact with the machine through the HMI console and through the dashboard on which information can be viewed regarding the driving speed, the amount of fuel in the tank, the temperature of the engine coolant, the SoC of the battery pack for electric operation, etc. From the control console, the driver can switch between the ICE engine and the electric motor, can control the hydraulically actuated work equipment or turn on the headlights, air conditioning or other equipment on the machine.

4 Summary

The machine is equipped with hydrostatic transmission, and traction is done on all four wheels of the machine, using hydraulic motors with radial pistons embedded in the wheels. This type of motors ensures a high torque at low speeds, necessary for the types of work performed. Moreover, the hydraulic motors ensure safety in operation and high reliability.

The mobile machine offers great mobility; the transmission concept valid on this machine can be very easily adapted to another type of machine because there are no mechanical connections. In addition, the concept can be easily scaled to be used on larger machines.

Being equipped with a hybrid energy source, the machine ensures flexibility in use, being able to tackle works even in closed unventilated spaces [14].

Acknowledgement. This paper has been financed under a project funded by the Ministry of Research, Innovation and Digitalization through Programme 1- Development of the national research & development system, Sub-programme 1.2 - Institutional performance - Projects financing the RDI excellence, Financial Agreement no. 18PFE/30.12.2021.

References

1. Edamura, M., Ishida, S., Imura, S., Izumi, S.: Adoption of electrification and hybrid drive for more energy-efficient construction machinery. Hitachi Rev. **62**(2), 118–122 (2013)

2. Lajunen, A., Suomela, J., Pippuri, J., Tammi, K., Lehmuspelto, T., Sainio, P.: Electric and hybrid electric non-road mobile machinery – present situation and future trends. World Electr. Vehic. J. **8**(1), 172–183 (2016)
3. Ratzinger, J., Buchberger, S., Eichlseder, H.: Electrified powertrains for wheel-driven non-road mobile machinery. Autom. Engine Technol. **6**, 1–13 (2021)
4. Filla, R.: Hybrid power systems for construction machinery: aspects of system design and operability of wheel loaders. In: Proceedings of the ASME 2009 International Mechanical Engineering Congress and Exposition, IMECE, vol. 13, pp. 611–620 (2009)
5. Kumar, L., Jain, S.: Electric propulsion system for electric vehicular technology: a review. Renew. Sustain. Energy Rev. **29**(C), 924–940 (2014)
6. Frank, B.: Using optimal control in concept evaluation and system optimization of diesel-electric hybrid construction machines. In: Proceedings of the 2016 International Conference on Electrical Systems for Aircraft, Railway, Ship Propulsion and Road Vehicles & International Transportation Electrification Conference, ESARS-ITEC, pp. 1–6. Institute of Electrical and Electronics Engineers, Piscataway (2016)
7. Bian, Y., Jiang, J., Shen, R., Jing, Z., Li, A., Wu, H.: Recent patents on hybrid construction machinery. Recent Pat. Mech. Eng. **6**(2), 97–106 (2013)
8. Grammatico, S., Balluchi, A., Cosoli, E.: A series-parallel hybrid electrical powertrain for industrial vehicles. In: 2010 IEEE Vehicle Power and Propulsion Conference, pp. 1–6. Institute of Electrical and Electronics Engineers, Piscataway (2010)
9. Niu, G., Shang, F., Krishnamurthy, M., Garcia, J.M.: Design and analysis of an electric hydraulic hybrid powertrain in electric vehicles. IEEE Trans. Transp. Electrif. **3**(1), 48–57 (2017)
10. Wang, H.: The design of the hybrid energy storage system in hybrid construction machinery. Adv. Mater. Res. **875–877**, 1934–1938 (2014)
11. Li, S., Gu, C., Zhao, P., Cheng, S.: Adaptive energy management for hybrid power system considering fuel economy and battery longevity. Energy Convers. Manage. **235**, 114004 (2021)
12. Woody, M., Arbabzadeh, M., Lewis, G.M., Keoleian, G.A., Stefanopoulou, A.: Strategies to limit degradation and maximize li-ion battery service lifetime – critical review and guidance for stakeholders. J. Energy Storage **28**, 101231 (2020)
13. He, X., Liu, H., He, S., Hu, B., Xiao, G.: Research on the energy efficiency of energy regeneration systems for a battery-powered hydrostatic vehicle. Energy **178**, 400–418 (2019)
14. Hiarom Ltd. product range. https://www.hiarom-railway.com/product-range/. Accessed 30 Mar 2023

Comparative Tests of the Impact of Modifications to the Hydrostatic Drive System of Blasting Utility Vehicle WS-153/173

Tomasz Siwulski[1] (✉) and Mateusz Wolter[1,2]

[1] Wroclaw University of Science and Technology, 27 Wybrzeże Wyspiańskiego st., 50-370 Wrocław, Poland
tomasz.siwulski@pwr.edu.pl

[2] KGHM Zanam S.A., 7 Kopalniana st., 59-101 Polkowice, Poland

Abstract. The article presents the results of a comparative analysis of pressure characteristics in the hydrostatic drive systems of blasting utility vehicles WS-153/173 currently operated and developed for the purpose of loading explosive charges in underground ore mines. The two compared solutions are both designed on the basis of variable displacement pumps with LS (Load Sensing) control. In the first solution, such pumps power all of the hydraulic operating elements, and in the second solution some of the auxiliary elements are powered independently by an added fixed displacement pump, thus reducing both the number of hydraulic energy receivers and the maximum flow rate on the line powered by LS pumps. The article presents and discusses the different time-pressure curves as recorded in typical operating procedures during tests performed on both of the analyzed systems. It also investigates the impact of the introduced modifications on the operation of the analyzed hydrostatic systems. The results of these research and development works seem an interesting material, in which actual devices are used as a direct example of a method for improving the parameters of hydraulic systems with LS pumps by modifying the system architecture.

Keywords: hydrostatic drive · vehicles · hydraulic systems architecture

1 Introduction

Hydrostatic power systems used in mobile machines can be structurally classified into two categories. The first category comprises systems with one central pump, typically a variable displacement pump controlled by the LS system [1–3]. The second category comprises systems in which the individual operating elements or groups of such elements are powered by independent pump systems [4]. This category includes electro-hydraulic actuators, which are a special and increasingly popular solution, extensively described in the literature [5–8]. In addition, recent development works focus on increasing the efficiency of such systems [9–12], in order to reduce the energy consumption of hydrostatic systems installed in battery electric utility vehicles. Leading manufacturers of working machines, including KGHM ZANAM S.A., which specializes in the

J. Stryczek and U. Warzyńska (Eds.): NSHP 2023, LNME, pp. 204–213, 2024.
https://doi.org/10.1007/978-3-031-43002-2_19

production of mining utility vehicles used mostly in underground ore mines, have been making attempts to employ new systems with improvements aimed inter alia at higher efficiencies. The resulting new solutions clearly show a tendency for replacing systems having a central pump with multi-pump systems. Modifications into the architecture of the hydrostatic system cause a change in its dynamic parameters, and the scope of such changes can be defined by comparing the test results. This paper presents the results of a comparative analysis of several time-related parameters of a steering system in the WS-153/173 blasting utility vehicle, experimentally identified for two vehicles having different architectures of the hydrostatic system. The modifications introduced into the systems were limited to the pump and the valve block systems. The results presented in this paper demonstrate that dynamic phenomena have a significant influence on the operating character of the systems and thus the research here described seems an important contribution to the understanding and describing of such dynamic phenomena in hydrostatic systems of working machines.

2 Background

The research and development works aimed at modifying the architecture of the hydrostatic system in the WS-153/173 blasting utility vehicle were initiated in response to reports from the users who observed that the working movement of the steering system in such machines did not reach full velocities. This phenomenon was accompanied by a significant number of defects in the feed pumps of the operated machines. Preliminary tests indicated momentary and significant pressure surges, which were associated with the operating characteristics of the valve system provided with Load Sensing control. The preliminary analysis of the problem resulted in a decision to introduce modifications into the hydrostatic system of the WS-153/173 blasting utility vehicle, i.e. into the architecture of the power-supply system and of the valve control system. The purpose of the modifications was to increase the speed of the working movements in the steering system by employing the entire available pump displacement and to implement such a structure of the system that would allow an increased efficiency. The research and development works resulted in a modified architecture of the hydrostatic system. It was then analyzed by measuring and comparing the time durations of the turning movements in the original steering system and in the modified system. The results of this analysis allowed the identification and description of the influence of adverse dynamic phenomena occurring in the system of the original structure.

3 Problem Analysis

Blasting utility vehicles WS-153/173 are used in underground ore mines (Fig. 1). They are provided with an articulated steering system, in which the actuator is one double-acting cylinder. The steering system is part of the central hydrostatic system of the machine, which also comprises the braking and the auxiliary systems powered independently by separate pumps.

The cylinder of the steering system is powered by a variable displacement pump (A, Fig. 2) and in the original system architecture (system 1, Fig. 2) it powered both the

Fig. 1. Blasting utility vehicles WS-153/173 manufactured by KGHM ZANAM S.A.

steering system (D, Fig. 2) and the motors of the radiator fans (C, Fig. 2). In a scenario when auxiliary electrical power is lost, it could additionally draw power from the internal combustion engine in order to supply a hydraulic motor driving the reel of the emulsion-pumping hose. The system was lubricated with a VG-68 hydraulic oil. The flow from the pump was split with the use of a priority valve which ensured priority flow to the steering system and passed the remaining flow to the radiator fan motors (C, Fig. 2). The system was designed to power the steering cylinder with a flow rate of 130 l/min at the rpm of the diesel engine ranging from 800 to 2400. As a result, the theoretical full extend and retract times of the steering cylinder were 4.4 and 3.3 s, respectively, and the theoretical durations of the full movement of the steering system are shorter, as kinematic constraints are introduced to the vehicle design and the cylinder does not reach its extreme positions.

However, significant differences were reported between the theoretical and the actual extend and retract times of the steering cylinder. Therefore, a thesis was formulated that the direct connection of the LS pump with the priority valve block generates dynamic disturbances in the operation of the valve system and may in effect limit the flow rate on the intake of the steering system. The above thesis and the current directions in the development of such systems motivated a modification of the system architecture (system 2, Fig. 2). The power supply line to the hydraulic motors of the radiator fans was separated and the fans were powered by an additional fixed displacement pump (G, Fig. 2). In addition, the priority block and the valve block controlling the fan motors were removed from the control valve system (B, Fig. 2), as they were no longer required. In effect, the number of elements receiving the flow of the fluid was reduced and the power supply system was significantly simplified (Fig. 2, Table 1).

The modifications to the hydraulic steering system of the machine were advantageously limited in such a manner that they did not require the introduction of new system components. As a result, the modifications could be implemented in the currently used vehicles during standard maintenance downtime.

The actual dynamic parameters of the two steering systems were identified by measuring changes in the pressure values on the high-pressure line of the feed-pump during

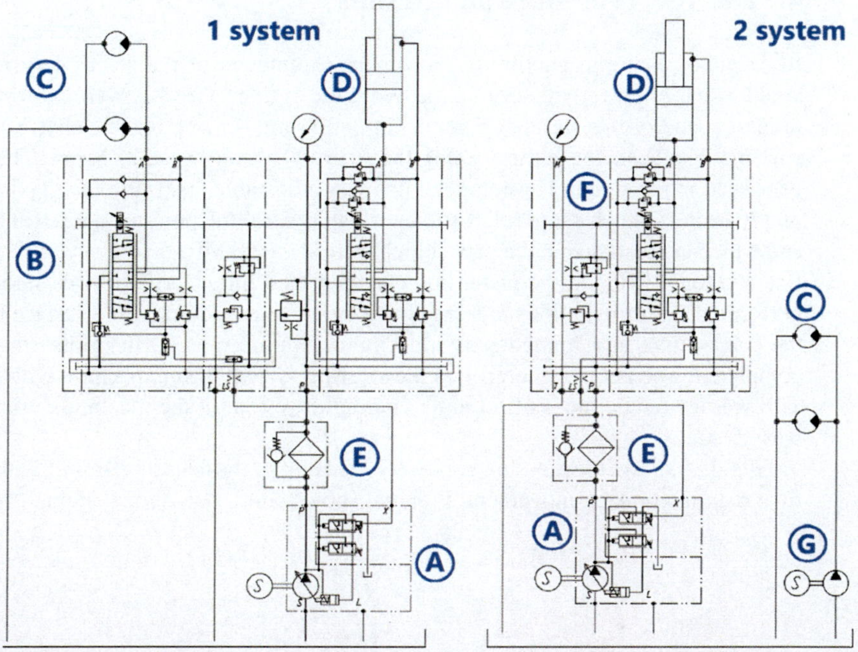

Fig. 2. Schematic diagram of the original (system 1) and the modified (system 2) steering-cylinder power-supply system with indicated locations of the pressure sensor: A – variable displacement LS pump, B – valve system with the priority block, C – radiator fan motors, D – steering cylinder, E – filter, F – valve system without the priority block, G – radiator fan motor feed pump.

Table 1. Basic elements and parameters of the hydrostatic steering system

Element description	Basic parameters
Diameter of steering cylinder piston	$D_p = 140$ mm
Diameter of steering cylinder piston rod	$D_r = 70$ mm
Steering cylinder stroke	$L_s = 592$ mm
Displacement of LS pump	90 cm^3/rev
Maximum pressure of LS pump	35 MPa
Priority valve type (only system 1)	Open/Closed Center PVSPM
Displacement of fixed displacement pump (only system 2)	53 cm^3/rev

the operation of the system. The pressure was measured with the HMG 4000 HYDAC system and an analog pressure transducer with a $0 \div 60$ MPa measurement range. The pressure values were recorded at 1 ms intervals. The results and their analysis are provided further in this paper.

4 The Test Procedure and Results

In order to reach the maximum movement parameters of the steering system and to avoid a lack of power situation in the hydraulic system, the tests were performed at the highest speed of the diesel engine, i.e. at 2400 rpm. As a result, the shaft speed of the variable displacement pump feeding the steering system was 2282 rpm. The flow rate generated in the steering system feed line was adjustable and set to the maximum value of 130 l/min. The pressure relief valves which control the pressure surges in the system and which are installed in the valve block were set to 20 MPa on the piston chamber feed line and to 24 MPa on the piston rod chamber feed line. During the tests, the operator performed full turns of the machine positioned on a non-deformable surface by starting the consecutive turn immediately after the maximum movement in one direction was completed. The operator controlled the steering system of the machine with a joystick. The average temperature of oil in the hydraulic system of the machine during the tests was 25 °C.

Figure 3 shows the pressure curve recorded during the tests of system 1 with indicated time regions for each movement. Figure 4 shows similar data for system 2.

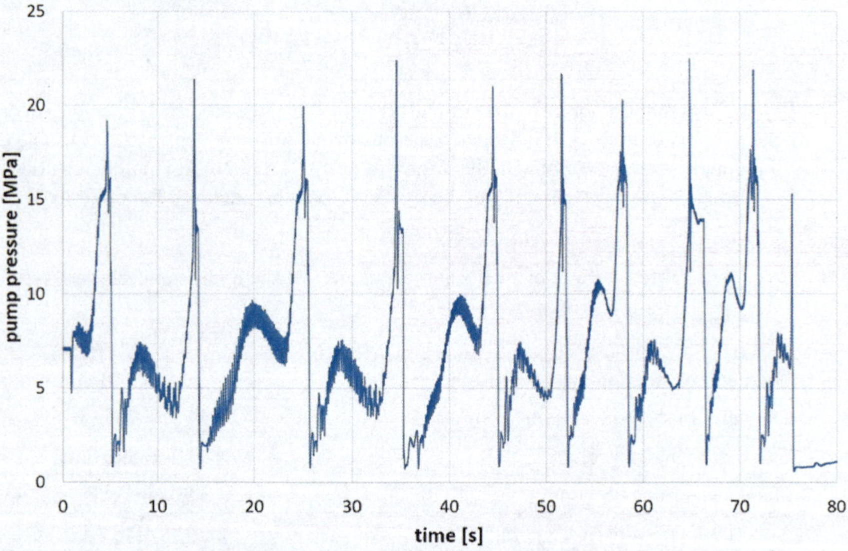

Fig. 3. Time vs pressure on the steering system feed line during the movement of system 1.

The obtained results were expected to serve as a source of information on important operating parameters used in a comparative analysis performed with the use of statistical methods.

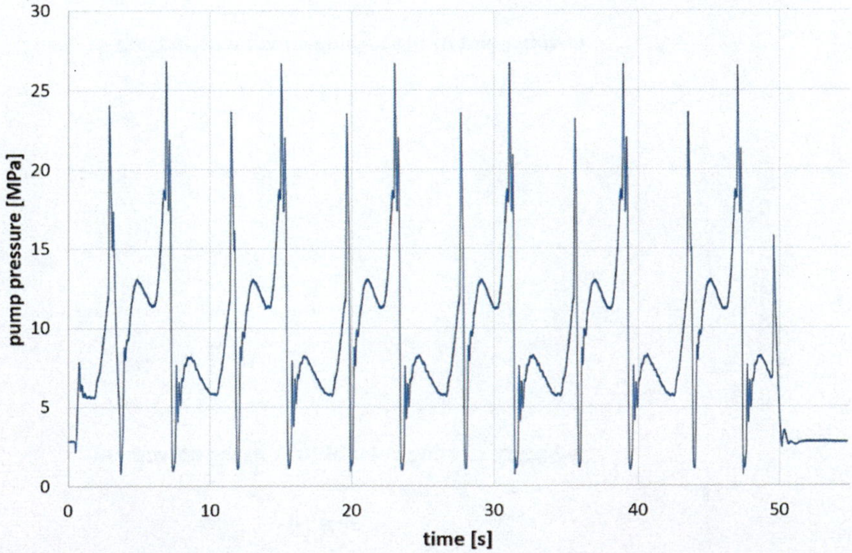

Fig. 4. Time vs pressure on the steering system feed line during the movement of system 2.

5 Discussion

The curves of the pressures on the steering system feed line of the WS-153/173 blasting utility vehicle can be interpreted directly. Their comparative analysis was performed on the basis of the turning movement time of the system. For this purpose, the starting and ending time points of the turning movement were defined from the pressure curves. The movement was assumed to start at the moment when the minimum value of pressure was reached in the system, as analyzed independently for each movement. In turn, the movement was assumed to end at the moment when the maximum pressure value was reached, indicating that the maximum cylinder extension was reached, also as analyzed independently for each movement (Fig. 5).

The results from the obtained measurement data allowed the identification of the time durations between the start and the end of the turning movements. The time durations were subsequently analyzed statistically, separately for the turning movement right and for the turning movement left. The results are shown in Tables 2 and 3. They are also represented graphically in Fig. 6.

The tests of the systems confirm that the mutual interactions between the LS pump and the directly connected priority valve system (system 1) significantly affect the operation of the system and limit the flow rate on the steering system feed line of the machine. The above comparison of the pressure curves recorded during the operation of the steering system demonstrates that the variation amplitudes are greater in system 1 than in system 2. This result is a clear indication that system 1 has a tendency for adverse dynamic phenomena in the valves, resulting in an incomplete and most likely variable opening of the valve. This phenomenon in turn translates directly into limited flow rate

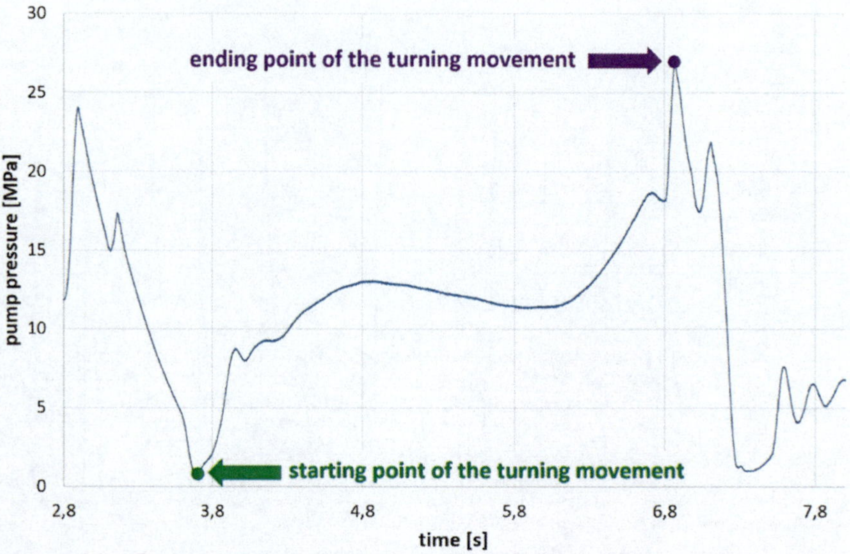

Fig. 5. Representative pressure curve during the movement of the steering system with indicated starting and ending points of the movement.

Table 2. Time durations of individual left-turns together with the Student's t-distribution for the 95% confidence interval, in seconds.

Cycle Number	Time Durations	Average Value	Confidence Interval	Upper Confidence Limit	Lower Confidence Limit
1 system					
1	8,56	7,53	1,86	8,46	6,60
2	8,83				
3	6,49				
4	6,26				
2 system					
1	3,17	3,10	0,09	3,19	3,01
2	3,04				
3	3,13				
4	3,05				

on the steering system feed line of the machine. The above results support the thesis formulated at the earlier theoretical stage of this research.

Table 3. Time durations of individual right-turns together with the Student's t-distribution, in seconds

Cycle Number	Time Durations	Average Value	Confidence Interval	Upper Confidence Limit	Lower Confidence Limit
1 system					
1	10,64	7,54	3,80	9,44	5,64
2	9,05				
3	5,65				
4	4,83				
2 system					
1	4,10	4,11	0,05	4,16	4,06
2	4,09				
3	4,08				
4	4,16				

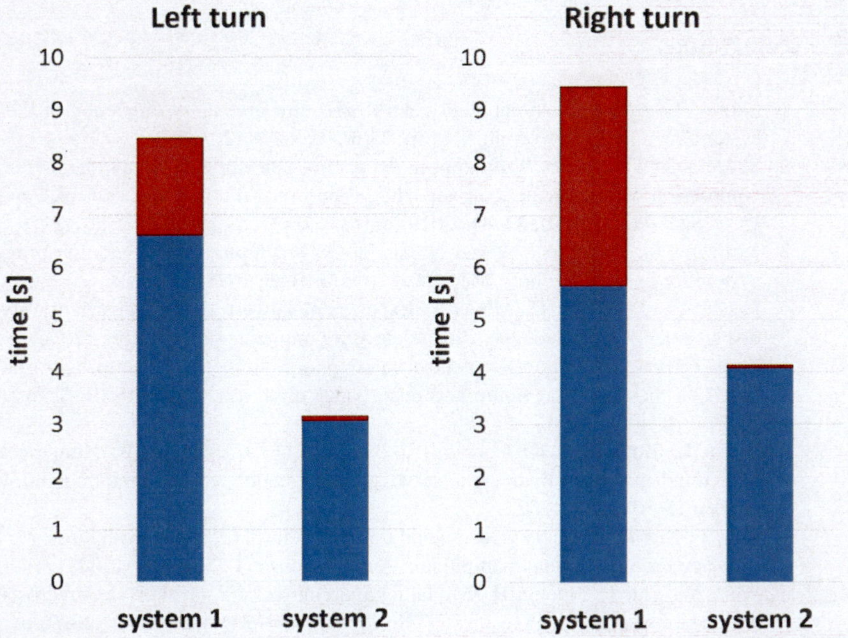

Fig. 6. Comparison of the durations of the turning movements with the Student's t-distribution 95% confidence intervals indicated in red.

6 Conclusions

The modification introduced in accordance with the state-of-art principles into the architecture of the hydrostatic system of the WS-153/173 blasting utility vehicle was advantageous and positively influenced the operating parameters of its steering system. The modification, whose scope was defined on the basis of energy-related assumptions, eliminated the adverse phenomenon which consisted in the partial opening of the valve in the system of the LS pump connected with the priority valve block. Having a negative influence on the operating parameters of the steering system, the phenomenon seems to have a dynamic character, and its explanation requires further research into the mutual interactions of valves and other serially connected elements of the hydrostatic system. The results of the research and development works presented in this paper seem an important contribution to the understanding of these phenomena which are a current and significant problem in the operation of machines with hydrostatic power systems. Moreover, the presented method for limiting the above adverse phenomenon by modifying the architecture of the system may be viewed in itself as an interesting research direction with a significant implementation potential. Importantly, the described method for modifying the system architecture not only limits the adverse dynamic phenomena but also is representative of the development directions in modern high-efficiency hydrostatic systems.

References

1. Skorek, G.: Influence the applied control structure on energy efficiency of the hydrostatic system. J. KONES Powertrain Transp. **25**(3), 411–418 (2018)
2. Skorek, G.: The accuracy of defining the energy efficiency of drive systems exemplified by comparison with hydrostatic drives with proportional motor speed control. Sci. J. Maritime Univ. Szczecin **58**(130), 32–41 (2019)
3. Kittisares S., Hirota, Y., Nabae, H., Endo, G., Suzumori, K.: Alternating pressure control system for hydraulic robots. Mechatron. (Oxford) **85**, 102822 (2022)
4. Fassbender, D., Brach, C., Minav, T.: Experimental investigations of partially valve-, partially displacement-controlled electrified telehandler implements. Actuators **12**(2), 50 (2023)
5. Zhou, F., et al.: High-precision control solution for asymmetrical electro-hydrostatic actuators based on the three-port pump and disturbance observers. IEEE/ASME Trans. Mechatron. **28**(1), 396–406 (2023)
6. Dinca, L., Bogateanu, R., Corcau, J.-I., Dumitrache, A., Suatean, B.: Numerical simulation for redundant electro-hydrostatic servo-actuators under certain special conditions. Energies (Basel) **15**(16), 5906 (2022)
7. Zhu, T., Xie, H., Yang, H.: Design and tracking control of an electro-hydrostatic actuator for a disc cutter replacement manipulator. Autom. Constr. **142**, 104480 (2022)
8. Wang, Y., Guo, S., Dong, H.: Modeling and control of a novel electro-hydrostatic actuator with adaptive pump displacement. Chin. J. Aeronaut. **33**(1), 365–371 (2020)
9. Guo, T., Han, X., Minav, T., Fu, Y.: A preliminary design method of high-power electro-hydrostatic actuators considering design robustness. Actuators **11**(11), 30 (2022)
10. Chao, Q., Zhang, J., Xu, B., Shang, Y., Jiao, Z., Li, Z.: Load-sensing pump design to reduce heat generation of electro-hydrostatic actuator systems. Energies (Basel) **11**(9), 2266 (2018)

11. Leonida, C.: Advancing art of autonomous drilling. Eng. Min. J. (1926) **222**(5), 44–47 (2021)
12. Ge, L., Quan, L., Li, Y., Zhang, X., Yang, J.: A novel hydraulic excavator boom driving system with high efficiency and potential energy regeneration capability. Energy Convers. Manage. **166**, 308–317 (2018)

New Materials and Special Solutions in Fluid Power Technology

Specificity of Designing Hydraulic Cylinders Made of Plastics

Piotr Stryczek(✉)

Mechanical Engineering Faculty, Wroclaw University of Science and Technology, Ignacego Łukasiewicza Street 5, 50-371 Wroclaw, Poland
piotr.stryczek@pwr.edu.pl

Abstract. The principles of designing hydraulic cylinders are generally known, but metals are assumed to be typical design materials for cylinders, which results in the use of standard design solutions and typical manufacturing technologies. However, the use of special design materials, e.g. plastics or composites, may require a departure from traditional, generic methods, or at least an adaptation of them. The article presents the specificity of designing cylinders made of plastics, which have been determined as a result of research presented in the author's earlier publications. The specificity of the designing is presented in an orderly manner, by following the stages of the author's algorithm for designing cylinders made of plastics. The specific challenges connected with this process are presented, which are different for the designing of plastic cylinders and metal cylinders. Some suggested solutions and examples of their practical application are also indicated. In the summary, possibilities offered by the use of plastics in hydraulic cylinders are considered.

Keywords: hydraulic cylinders · plastic materials · design methods · manufacturing technology

1 Introduction

One of the development trends in fluid power is the use of plastics [1, 2]. Plastics are more environmentally friendly than metals because both their production and recycling require less energy and pollute the environment to a less. The use of plastics is beneficial from the design point of view as it is possible to simplify the design solution of fluid power components, reduce friction, energy consumption, as well as the weight of the components. The application of plastics makes it possible to simplify manufacturing methods and reduce manufacturing costs, e.g. by means of the injection molding. The above arguments justify the opinion that fluid power from plastics has significant potential in the scientific and technical area. This is evident in the relevant literature on the subject.

In works [3–6] it was shown that it is possible to build both gear and gerotor pumps made of plastics in the capacity range $q = 1 \div 15\ cm^3$/rev and pressure up to $p = 2$ MPa.

J. Stryczek and U. Warzyńska (Eds.): NSHP 2023, LNME, pp. 217–227, 2024.
https://doi.org/10.1007/978-3-031-43002-2_20

In work [7] it was proved that it is also possible to make simple hydraulic valves (on/off valve and pressure relief valve) of plastics, which can operate at flow rates Q = 15 dm^3/min and pressure p = 3 MPa.

The most intensive works are carried out in the field of plastic hydraulic cylinders [8]. Two directions in their development can be observed. The first is the development of composite cylinders [9, 10], in which thin metal tubes are reinforced with a resin-bonded carbon fiber braid. The second direction is the development of plastic cylinders, in which plastics are used to make cylinder tubes, as well as caps and pistons [11, 12]. So far, piston rods have been made of steel, but, as shown in [13], they can be covered with plastics, e.g. in order to provide protection against aggressive external environment. The cylinders presented in [11, 12] had a piston diameter of D = 30, 40, 50 mm and operated at pressures of p = 6.3 MPa.

Design, manufacturing and experimental research of fluid power cylinders made of plastics has its own specificity and differs from what is used in the case of metal cylinders. Therefore, the aim of this work was to formulate and analyze that design specificity and to present original solutions in the entire process of design, manufacturing and experimental research.

2 Algorithm of Designing Hydraulic Cylinders Made of Plastics

The algorithm for designing cylinders made of plastics, developed on the basis of [14–16], is shown in Fig. 1.

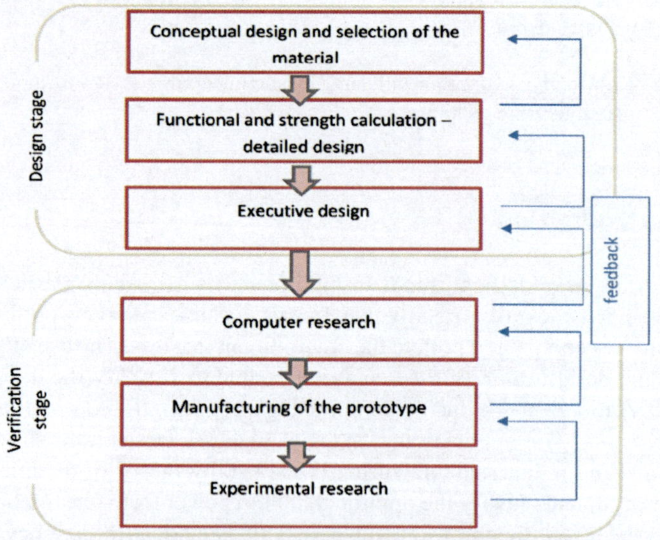

Fig. 1. Design algorithm for plastic hydraulic cylinders made of plastics

The algorithm is divided into two stages: design stage and verification stage. At the design stage, the following are distinguished: conceptual design and selection of the material, functional and strength calculations, and executive design. At the verification stage: computer research using Finite Elements Method, prototype manufacturing and experimental research. At each stage of the process, there are specific aspects and challenges related to the use of plastics, which are discussed below.

3 Design Stage

3.1 Conceptual Design and Material Selection

Typical design materials for the cylinders are metals, which results in the use of standard design solutions. Compared to metals, plastics are characterized by much lower strength, shrinkage during plastic processing, elongation under the influence of temperature during operation, water absorption and creep. The features exert a significant impact at the stage of developing the initial, conceptual solution. Design assumptions should be made considering the impossibility of obtaining such high working parameters as in the case of metal cylinders. Also, some types of connections between components, e.g. direct thread joints between the tube and caps or welded joints, are practically impossible to use. It is also suggested to allow the use of metal for single, most heavily loaded elements of cylinders, such as piston rods. However, in order to talk about cylinders made of plastic, they should remain the dominant material, and allowing the use of metal is only to avoid excessive narrowing of the operating parameters of such cylinders. The fact of using plastics as a design material does not only generate limitations. There are new opportunities related to the possibility of easy processing of plastics, thanks to which it is possible to obtain virtually any shape of plastic parts, e.g. by injection molding. As a result, by designing an appropriate shape, it is possible to develop a solution that will have a much lower weight and will retain sufficient strength. Examples of innovative design solutions developed for the use of plastics for cylinder parts are shown in Fig. 2. Figure 2a shows a tube of a special shape, with ribs increasing stiffness, in which holes for tie rods are made. Figure 2b shows a piston rod in which a plastic tube, ensuring good co-operation with the seal, is filled with a composite inside the hole, what improves stiffness and strength. Figure 2c shows a piston with a built-in guide and sealing ring. Figure 2d and e show slimmed bottom and gland caps with reduced weight resulting from the use of less material. In addition, the gland cap provides for the possibility of using built-in sealing and guiding rings, similarly to the piston in Fig. 2c. Bottom and gland caps modelled on the presented example solutions could have a strength similar to traditional caps, which is to result from the appropriate shaping and use of reinforcements, e.g. ribs, brackets, or locks in which cylinder tubes can be inserted.

Of course, the presented solutions are not the only possible. Some other can be considered, and for that purpose, the methods of generating solutions are very helpful, e.g. morphological tables, solution tree, etc.

The key challenge in developing of a design solution of the structure is the choice of plastic. The material should be selected taking into account numerous factors, such as manufacturing technology, working parameters, operation environment, and work regime. Among the guidelines to pay attention to are:

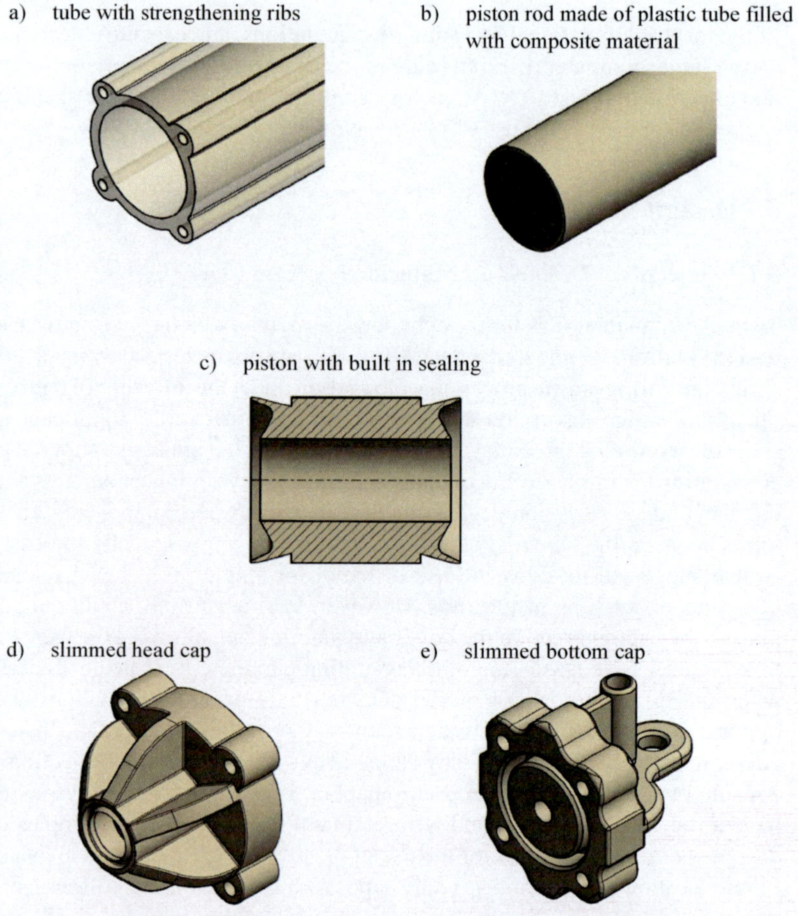

Fig. 2. Conceptual solutions of cylinder parts made of plastics

- mechanical strength (yield point Re, Young's modulus E) and possibly high stability of dimensions which is independent of temperature change (small shrinkage S, small thermal elongation W, low water absorption A)
- possibility to process material by machining and injection molding
- high availability on the market in the form of semi-finished products at the lowest possible price
- low friction coefficient η at the collaboration with seals

After selecting the concept of the solution and the plastic material for the design material, it is necessary to carry out functional and strength calculations.

3.2 Functional and Strength Calculations

Functional calculations are similar to the calculations carried out for metal cylinders. Therefore, it is necessary to determine:

- working pressure p,
- internal diameter of the cylinder tube D,
- piston rod diameter d,
- pushing force F_1 and pulling force F_2,
- velocity of extension u_1 and retraction u_2,
- stroke S and cylinder length,
- diameter of the connection holes (ports)

The main dimensions of the cylinder are largely related to the functional parameters. The scope of the necessary strength calculations may vary depending on the design solution. In general, the calculations boil down to checking the connections between the parts, e.g. connections between cylinder tube and caps, connections between piston and piston rod, as well as checking the cylinder due to the main loads occurring in the structure, e.g. buckling or extension of the tube under the influence of working medium pressure.

In mechanical engineering, it is common practice to use constant strength parameters for a specific material. Plastics as design materials are characterized by the variability of strength and functional parameters during the operation of machines and devices that are made of them. This variability in mechanical engineering is usually not taken into account. The strength of the plastic used is assumed based on the manufacturer's documentation. Changes in the parameters of the material occurring during operation are compensated by appropriately high values of safety factors. Allowable stresses value may be than established using Formula (1), in which σ_{allow}-allowable stress, R_e-yield stress given by the producer of the plastic, x-safety factor.

$$\sigma_{allow} = \frac{R_e}{x} \tag{1}$$

In the case of plastic cylinders, it is necessary to take into account, for example, the purpose of the cylinder's use. If the cylinder is being developed for research purposes, it is desirable to expose the phenomena's related to the use of plastics. The use of lower values of safety factors equal to $x = 1 \div 2$ helps to achieve this goal. In the case of cylinders that are to work at high temperatures, in an intensive work regime or when high reliability is expected, values of the safety factor equal to $x = 3 \div 4$ should be assumed.

3.3 Executive Design

At the stage of developing the executive design, in which a detailed solution is created, special attention should be paid to consider the aspect of manufacturing technology of cylinder parts but also other recommendations should be taken into account:

- manufacturing efforts should be brought to minimum,

- while planning to use traditional machining method it's important to select appropriate semi-finished products, thanks to which it is possible to reduce the number of operations, e.g. use plastic pipes for the cylinder tubes,
- while planning to use the injection molding method detailed shapes of the parts must be adjust to this method, so among others, it's necessary to determine the correct wall thicknesses, choose the location and dimensions, as well as the shape of the reinforcements (ribs),
- if parts do not cooperates as moving pair, it's possible to use composite materials which may allow for gain higher strength or mass reduction,
- special attention should be paid to the surfaces of parts creating moving pairs,
- appropriate tolerances, precision and roughness should be selected and the possibility of obtaining them should be considered

The material of parts creating moving pairs should be characterized by a uniform structure. This requirement is valid regardless of the choice of manufacturing method. Any inclusions, material discontinuities or composite additives in the area of the inner surface of the tube, during operation will have a very negative effect on the service life of the seals. A similar principle should be followed while considering other elements forming moving pairs, consisting of two plastic elements, e.g. a piston if it is to have a built-in guide ring and cooperate with the tube. Than the appropriate tolerances should be selected and the possibility of obtaining them should be considered, as it is assumed, that additional manufacturing efforts should be brought to minimum.

Besides assuming appropriate parameters of the structure, like precision or roughness, a possibility of verifying them should also be considered. For very complex shapes, which are not common for traditional metal cylinders but possible for plastic ones, it may require the use of special measuring machines.

4 Verification Stage

Since the use of plastic causes numerous new challenges in the design process, some of which have been indicated before, the verification of plastic cylinders must be extended in relation to metal cylinders. It includes the following stages: computer research using Finite Elements Method (FEM), manufacturing of the prototype and experimental research. The individual stages are presented below.

4.1 Computer Research

When conducting computer research using FEM analysis, special attention should be paid to the model of restraints and loads of the cylinder. The models should reflect the actual conditions in which the cylinder will work. At the same time, extreme and unfavourable cases for the operation of the cylinder ought to be considered. Such computer research is important source of information about behaviour of plastic cylinders which are new technical objects.

Exemplary results of FEM analysis of a plastic cylinder are shown in Fig. 3.

Presented example was used to obtain information about maximum stresses in the elements and comparing them to allowable stresses, as well as for estimation of direction

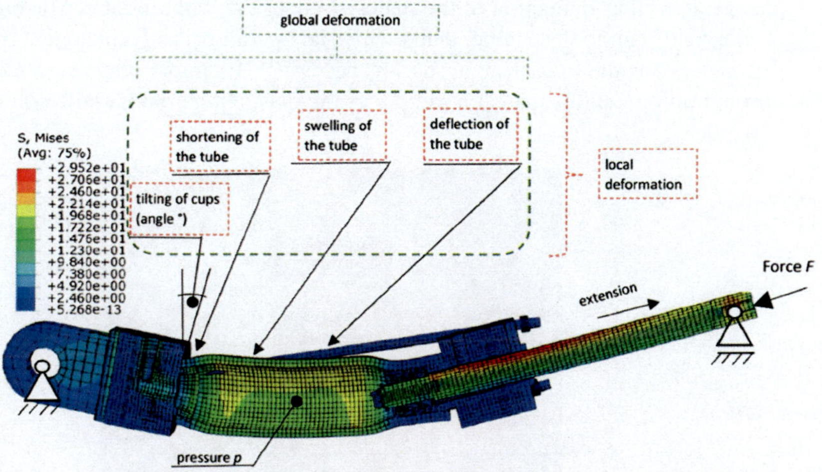

Fig. 3. Results of computer research of plastic cylinder model [11]

and character of deformations. The cylinder was fixed in the mounting eye and at the end of the extended piston rod. The loads were pressure of working fluid and force F acting on the piston rod in axial direction. As a result, model of deformation of plastic cylinder was established, which emerged its characteristic aspects. Local deformations refers mainly to the cylinder's tube and its swelling, what does not appears in metal cylinders. Global deformations are sum of the deformations in all elements, and shows that plastic cylinder may deflect. Even on such basic level of analysis valuable observation may be done, e.g. using appropriate piston sealing is needed to ensure internal tightness of the cylinder. Another example may be recommendation of considering gland cap fixing, e.g. to some supporting structure like frame of a larger machine, which would decrease the risk of buckling.

4.2 Manufacturing of the Prototype

In the process of manufacturing the prototype, special attention should be paid to ensure that the adopted technology of manufacturing elements and subassemblies of the cylinder ensures the accuracy of execution necessary from the design point of view. One of the key aspects is the finish of the inner surface of the cylinder tube. The roundness of the hole, the straightness of the surface along the tube, the surface texture, and especially the smoothness of the surface as well as the orientation of irregularities are of great importance in terms of the sealing durability. Selection of execution methods for this element should be carried out in such a way that the above-mentioned parameters comply with the requirements of the seal manufacturer. Eventual machining traces should be directed along the cylinder axis if possible. This requires the use of special tools. If, for some reason, it is impossible or unjustified, it is suggested at least to perform precise operations with standard tools, e.g. very fine turning, with low feed and small chip thickness. The same applies to other cylinder components made of plastics. Specific aspect of machining of the plastics is change of their dimensions, e.g. as a result of temperature change as well as deflection of the material under the tool pressure. Machining parameters are difficult to determine, that's why trial operations and adjustment of machining parameters for the chosen material are necessary. Example of precise measurements during trail operations made on sample of the plastic material is shown on Fig. 4.

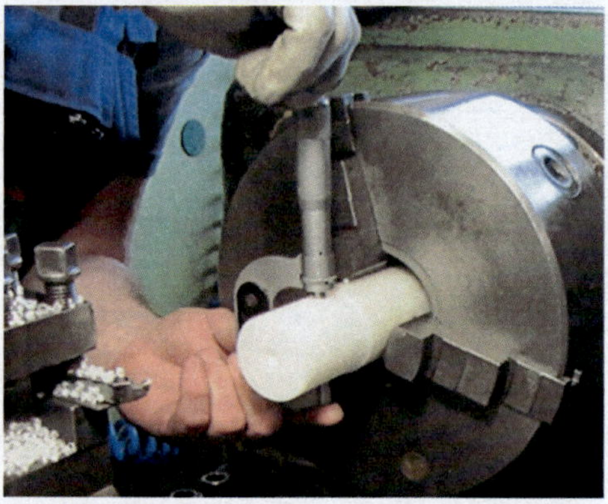

Fig. 4. Measurements after trail operation of high-precision turning of the sample made of the material intended for the plastic cylinder

When technology of injection molding of plastics is used, the precision of the surface of the cylinder elements should be ensured by an appropriate design solution and the quality of the injection molds. Plastic material is pressed into molds in liquid form and then solidify. It is necessary than, to prepare very precise molds which may be very expensive, because the process of designing and producing molds is very complex and time consuming. Such solution is justified if costs of its implementation are spread on large number of produced plastic parts. However, injection molding allows to gain large design advantages, because relatively any shape can be obtained, which makes it specific aspect of plastic cylinders.

The manufacturing and obtaining of a real object is verification step in itself, which allows to make an assessment in relation to design process correctness, correctness of standardized parts selection, manufacturing technology and quality of produced parts as well as chosen design solution.

4.3 Experimental Research

In the case of plastic cylinders, standard and extended experimental research of the cylinder prototype are carried out. Standard examination is carried out according to ISO 101:2001 and partly according to the national standard PN-72/M73202. These stage of examination provide information about correctness of cylinder operation and achieved working parameters e.g. lack of leakages, force at the piston rod, pressure of the working medium, hydraulic-mechanical efficiency. Extended examination concerns the deformation of the structure. On their basis, it is determined whether the cylinder deforms as expected and within the limits acceptable for design reasons. The information obtained as a result of the examination allows for the correction and improvement of the design solution. Recommendations can also be made for the mounting of the cylinders in the machine structure. An example of the results of such examination, in which deformations of the cylinder tube surface were measured at four points located along the axis, is shown in Fig. 5a.

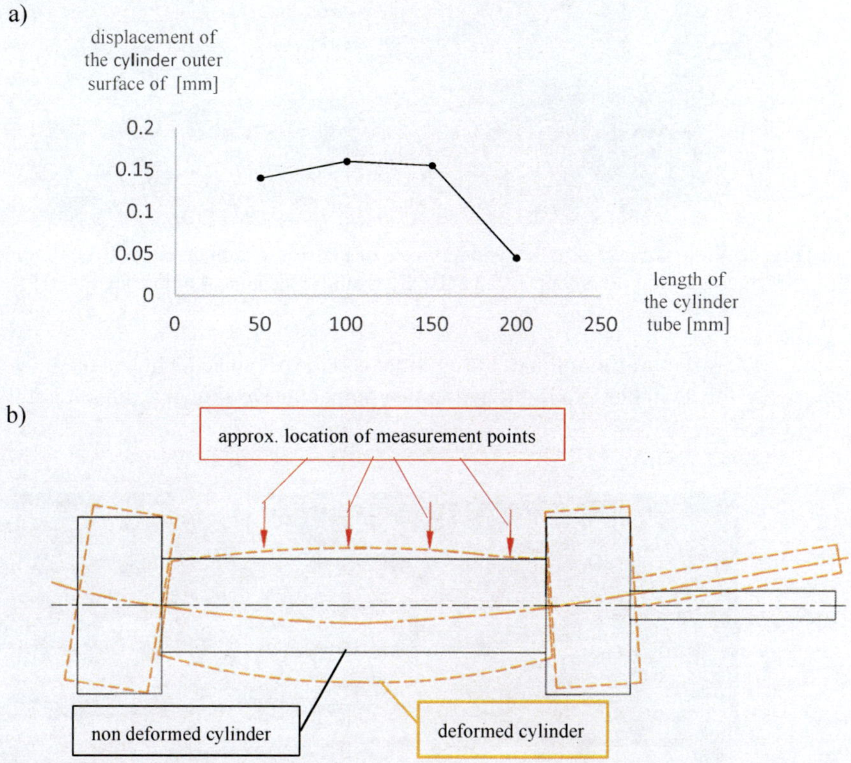

Fig. 5. Hydraulic cylinder deformation test results: a) deformation diagram; b) schematic deformation profile

Figure 5b, on the other hand, supplements the information on measurement results in order to make it more clear. Therefore, schematic image of the deformed cylinder is superimposed on the schematic image of non-deformed cylinder, and red arrows mark the measurements points corresponding to the points indicated on the chart on Fig. 5a.

Standard and exceeded examination results, same as results of fem analysis, provide information about new type of technical object, specificity of its operation and behavior of the structure of the cylinder. Experimental tests performed so far seems to emerge such specific aspect of plastic cylinders behavior, which is correct operation and high working parameters despite deformation of their structures.

5 Summary

The presented design algorithm and the specific principles of designing plastic cylinders contained therein, were used to design and manufacture the cylinder shown in Fig. 6.

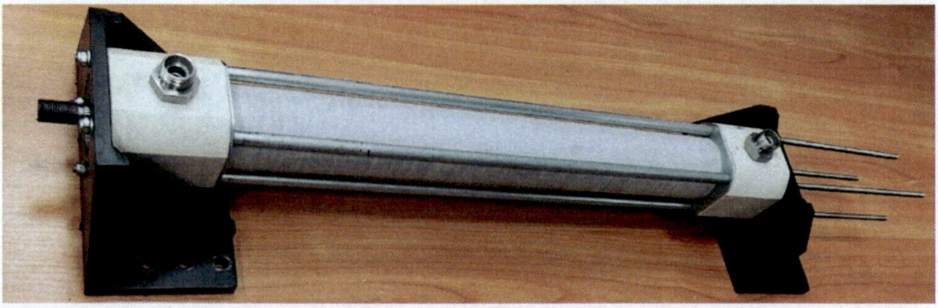

Fig. 6. View of the hydraulic cylinder made of plastics, piston diameter D = 40 mm, stroke s = 400 mm, working pressure p = 6.3 MPa, hydraulic- mechanical efficiency $\eta \approx 0.95$

The structural solution of the cylinder and the obtained technical parameters allowed to use this cylinder in a light, hydraulic compact drive shown as full assembly on Fig. 7 [12].

Fig. 7. Hydraulic compact drive with plastic components

The manufacturing this cylinder and its ability for operating properly, was a confirmation of the correctness and usefulness of the algorithm and the specific design principles contained in it. Against this background, the benefits and new possibilities resulting from the use of plastics are visible. They are related to the fact that:

- plastics are pro-ecological as less energy is used for their production, processing and recycling than in the case of metals,
- the use of plastics makes it possible to simplify the design solution of cylinders, reduce the number of parts, reduce friction between moving elements and reduce the weight of cylinders,

- technology of injection molding of plastics enables simplification and reduction of cylinder manufacturing costs,
- new areas of application for plastic cylinders are emerging, such as water hydraulics, mechatronics, automation and robotics, or medical and chemical appliances.

References

1. Stryczek, J., Banaś, M., Krawczyk, J., Marciniak, L., Stryczek, P.: The fluid power elements and systems made of plastics. Procedia Eng. **176**, 600–609 (2017)
2. Stryczek, J.: Plastics-a new trend in design hydraulic elements and systems. In: Stryczek, J., Warzyńska, U. (eds.) NSHP 2020. LNME, pp. 427–439. Springer, Cham (2021). https://doi.org/10.1007/978-3-030-59509-8_40
3. Rodionov, L., et al.: Challenges in design process of gear micropump from plastics. Arch. Civ. Mech. Eng. **21**, 1 (2021). art. 34
4. Gamez-Montero, P.J., et al.: Magnet-sleeve-sealed mini trochoidal-gear pomp prototype with polymer compozite gear. Energies **10**(10), 1458 (2017)
5. Krawczyk, J., Stryczek, J.: Design and experimental research of plastic gerotor pump. In: Proceedings of Global Fluid Power Society Ph.D. Symposium, Samara (2018)
6. Stryczek, J., Stryczek, P.: Synthetic approach to the design, manufacturing and examination of gerotor and orbital hydraulic machines. Energies **14**(3), 1–31 (2021). art. 624
7. Banaś, M., Antoniak, P., Marciniak, L., Stryczek, J.: Visualisation of flow phenomena in hydraulic throttle valves of plastic. In: Proceedings of the 14th International Conference on Vibration Engineering and Technology of Machinery (VETOMAC XIV), Lisbon (2018)
8. Lubecki, M., Stosiak, M., Skačkauskas, P., Karpenko, M., Deptuła, A., Urbanowicz, K.: Development of composite hydraulic actuators: a review. Actuators **11**(12), 1–15 (2022). art. 365
9. Solazzi, L.: Design and experimental tests on hydraulic actuator made of composite material. Compos. Struct. **232**, 111544 (2020)
10. Lubecki, M., Stosiak, M., Gazińska, M.: Numerical and experimental analysis of the base of a composite hydraulic cylinder made of PET. In: Stryczek, J., Warzyńska, U. (eds.) NSHP 2020. LNME, pp. 396–405. Springer, Cham (2021). https://doi.org/10.1007/978-3-030-59509-8_36
11. Stryczek, P.: Challenges in the designing process of hydraulic cylinders made of plastics. In: Procedings of Global Fluid Power Society Ph.D. Symposium (GFPS), Naples, Italy (2022)
12. Schulze, T., Weber, J. , Banaś, M., Stryczek, P.: Plastic components for electro-hydrostatic drives. In:13th International Fluid Power Conference (13. IFK), Aachen, Germany (2022)
13. Skowrońska, J., Zaczyński, J., Kosucki, A., Stawiński, Ł: Modern materials and surface modification methods used in the manufacture of hydraulic actuators. In: Stryczek, J., Warzyńska, U. (eds.) NSHP 2020. LNME, pp. 427–439. Springer, Cham (2021). https://doi.org/10.1007/978-3-030-59509-8_39
14. Pahl, G., Beitz, W.: Engineering Design. Springer, Heidelberg (1999)
15. Stryczek, S.: Hydrostatic drive. Wydawnictwo Naukowo-Techniczne, Warszawa (1984)
16. Marutov, W.A., Pawłowskij, S.A.: Hydraulic cylinders design and calculations. Wydawnictwo Naukowo-Techniczne, Warszawa (1968)

Computational and Experimental Study of Plastic Valves with 3D Printed Poppets

Urszula Warzyńska(✉), Michał Banaś, and Jarosław Stryczek

Wrocław University of Science and Technology,
ul. Łukasiewicza 5, 50-370 Wrocław, Poland
urszula.warzynska@pwr.edu.pl

Abstract. This research aims to investigate the flow characteristics of plastic valves with 3D-printed poppets using *Computational Fluid Dynamics* (CFD) simulations and experimental validation. Plastic valves have been gaining popularity in various industries due to their lightweight, corrosion resistance, and low cost. However, the use of plastic valves in fluid flow applications is still limited due to concerns about their durability and performance. The poppet, being the main component of a valve, plays a crucial role in determining the valve's performance. In this study, poppets have been 3D printed using *polylactic acid* (PLA) material. Different poppets shapes were tested using CFD simulations and experimental tests to obtain pressure-flow characteristics. The numerical approach to obtain the valves flow characteristics includes the cavitation model to ensure better agreement with the experimental data. The results of this research provide insights into the suitability of plastic valves for various applications and the importance of experimental validation in the design process.

Keywords: Plastic valves · CFD modeling · Valve flow characteristics · 3D printing

1 Introduction

In recent decades, plastics have made the jump from secondary materials used in minor roles to first-choice materials in many technical fields. On the global market, after the COVID-19 pandemic, production increased to over 390 million tons in 2021. The share of Europe in global production, however, fell from 19% in 2017 to 15% in 2021. Many fields of technology, such as IT, electronics, and aviation, would not have a chance to develop without the participation of plastics, although the greatest demand remains in packaging technology (approx. 44%) [1].

Hydraulic drives use plastics still to a limited extent. Advantages of hydraulic energy transfer, e.g. the possibility of obtaining high power with small-size and mass elements is also a challenge when introducing plastics and replacing metal materials. The presence of plastics is most easily seen in composite hydraulic

J. Stryczek and U. Warzyńska (Eds.): NSHP 2023, LNME, pp. 228–237, 2024.
https://doi.org/10.1007/978-3-031-43002-2_21

cylinders, which have found industrial implementation [2] and are still goals of improvement [3]. In other cases, they are rather single, very specialized proposals [4]. Polymers are also suitable for prototypes and small series of components, especially taking into account additive methods [5].

Unlike steel, plastic elements can take almost any shape. This affects the flow conditions in the gaps between the surfaces and gives the possibility of obtaining different flow characteristics in hydraulic components, e.g. valves. The shape of the gap, depending on the function performed by the valve, will be of key importance. The most common method to analyze the flow phenomena is CFD, however, the numerical model needs experimental verification of the assumed simplifications

In the field of hydraulic drives, several papers were presented so far including CFD modelling and research on cavitation phenomena. The research on the flow analysis of an adjustable check valve was analyzed in [6] and in hydraulic cylinders in [7]. The modelling approach on a gerotor pump working in cavitation conditions may be seen in [8], and cavitation modelling in throttle channels were undertaken in several papers [9]. CFD method was also successfully used in leakage flow modelling in hydraulic gear pumps [10].

2 Valve with Printed Poppets

For the tests, a throttle valve made from plastic was selected. Figure 1a shows a simplified cross-sectional view with some important dimensions of the input and output ports. The valve body (1) is made of *poly(methyl methacrylate)* (PMMA) which is transparent and allows observation of the phenomena inside the valve. An adjusting screw (2) moves axially poppet (3) relative to the seat (4) to form a throttle gap with a variable section. Fluid enters the valve through the channel (IN), then flows through the throttle gap into the body chamber and out of the valve through the channel (OUT). Turning the screw (2), we achieve a certain gap with an opening distance of x.

We selected for tests three poppets shown in Fig. 1a–c. The first cone (Fig. 1b) has a typical shape used in hydraulic systems with a vertex angle of 60° [11]. The next solution (Fig. 1c) is based on the poppet (b), however, the side surface of the cone has been machined and the angle changed. For another closing element (Fig. 1d), the angle of 30° has been selected and it was also machined. All poppets have been manufactured using *fused deposition modeling* (FDM) and the surfaces had observable irregularities. Elements (c) and (d) have been machined to increase the smoothness of the surfaces. The process changed the nominal dimensions of the cones and some main dimensions have been presented in Fig. 1.

3 Numerical Simulations

The numerical simulations were performed using CFD by the use of finite volume method in Ansys Fluent® 2021 R2 environment. The geometric models (Fig. 2a)

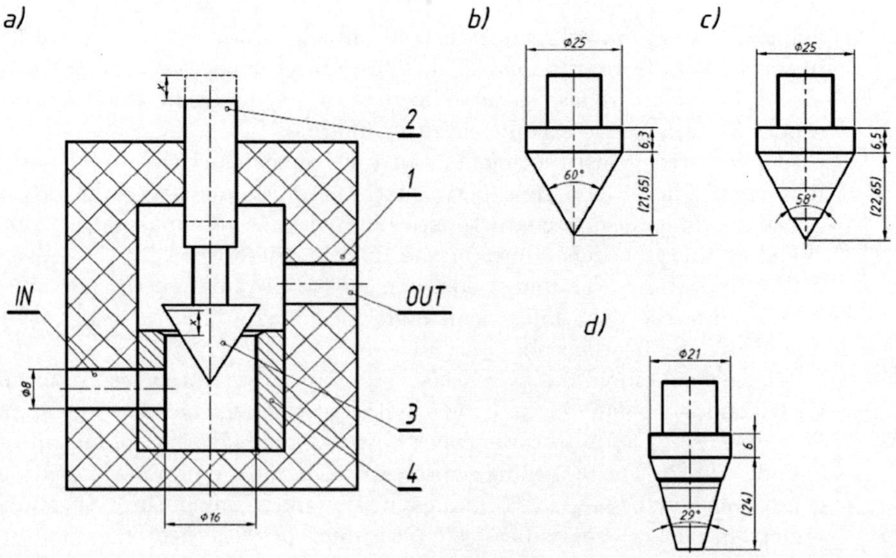

Fig. 1. Plastic valve and tested poppets: a) simplified valve, 1 – valve body, 2 – adjusting screw, 3 – poppet, 4 – seat, b) cone with 60° (*cone60*), c) modified cone with 60° (*cone60mod*), d) cone with 30° (*cone30*).

for the simulations were prepared based on the three CAD models of the conical poppets (Fig. 1). Each simulation model was prepared by the extraction of volume from the inside of the geometrical models, separately for each valve lift (Fig. 2b). The discrete models were prepared by meshing the geometries with tetrahedral elements with a maximum element size of 0.5 mm. The mesh was considerably refined in the flow area between the valve seat and the poppet. Each discrete model was composed of about 2.8 mln elements. The detailed view of the grid is shown in Fig. 2c.

The numerical model included a multiphase mixture model with two Eulerian Phases (hydraulic oil and air) with phase interaction and cavitation mechanism of mass transfer. The cavitation modelling was based on Schnerr-Sauer equations, assuming the vaporization pressure of liquid equal to 20 kPa. The k-omega SST (shear stress transport) model was used for the description of a turbulent flow. The material properties of hydraulic oil HLP68 and air mixture used for the calculations are listed in Table 1.

The boundary conditions included fluid flow velocity at the inlet ($v > 0$ m/s) to the fluid domain and a free flow of fluid at the outlet with pressure ($p = 0$ MPa). The inlet velocity varied accordingly to flow rates Q assumed for simulations in each computational case (see Table 3). All the simulation cases were performed in a steady state, using a double-precision solver, pseudo transient algorithm, and coupled pressure-velocity coupling scheme.

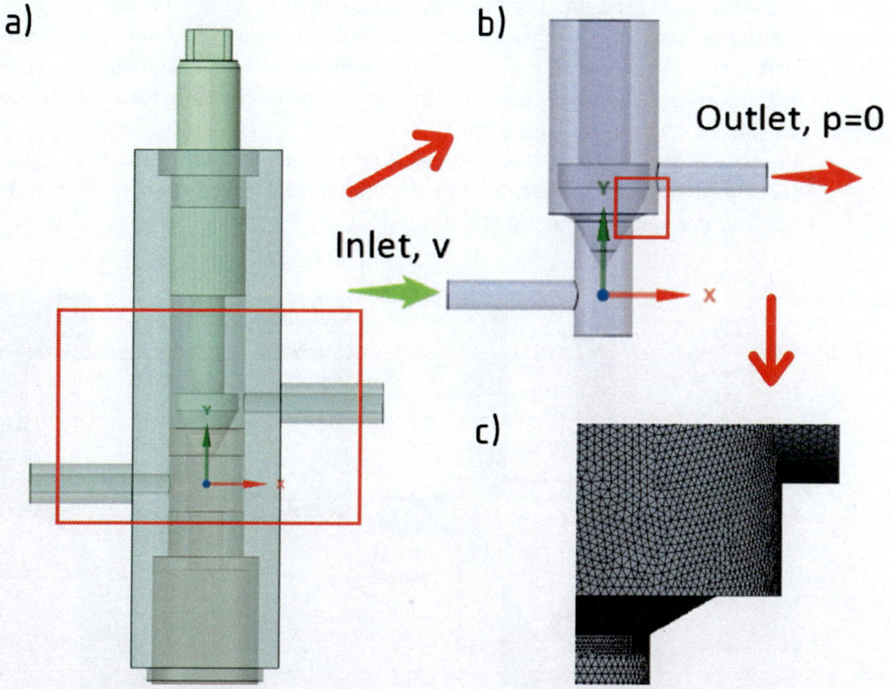

Fig. 2. Numerical model of the valve used in simulations: a) geometrical (CAD) model, b) fluid domain and boundary conditions, c) detailed view of a discrete model.

Table 1. Physical properties of the liquid medium at temperature $\theta = 25$ °C [12].

Phase	Dynamic viscosity (Pa · s)	Density (kg/m^3)
Oil (HLP-68)	0.095	780.0
Air	$1.789 \cdot 10^{-5}$	1.225

The simulation cases included three different poppet shapes (Fig. 1), and different valve lift values x chosen separately for each geometry (Table 3). The pressure drop versus flow characteristics of the valves were obtained by implementing three flow rates Q as an input parameter at each case of a valve lift x and obtaining the pressure difference between the inlet and outlet as a simulation result $\Delta p_{sim} = p_{in} - p_{out}$. The relative difference between simulation (Δp_{sim}) and experimental (Δp_{exp}) pressure drop mean values are calculated as δ:

$$\delta = \frac{|\Delta p_{exp} - \Delta p_{sim}|}{\Delta p_{sim}} \cdot 100\%$$

In Fig. 3 and Fig. 4, the sample results of flow and pressure results obtained by CFD simulations are presented. Based on the analysis of velocity and pressure distribution, it is possible to verify the flow stream distribution in a seat-poppet path, as well as analyse the possible areas of cavitation formation. The visual analysis of the simulation results is an easy way to introduce some geometry changes to obtain more favourable flow conditions. The analysis of cavitation effects and bubble formation will not be further presented in the article, as it will be a concern of a separate study.

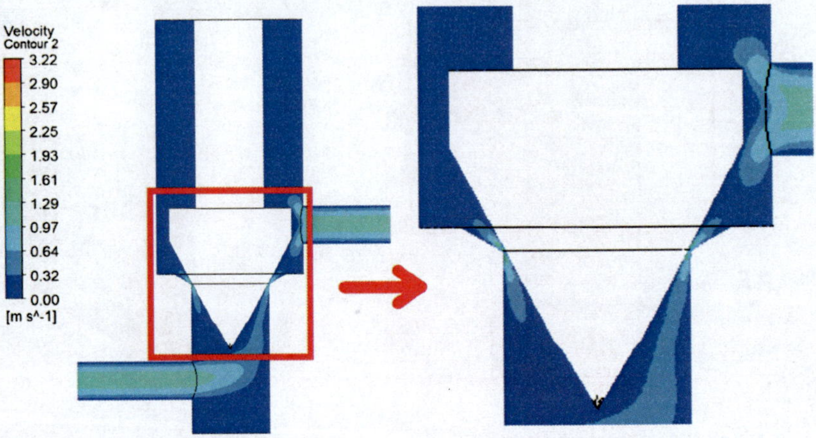

Fig. 3. Flow velocity contour plot – sample results on the exemplary of *cone60mod* poppet with valve opening $x = 0.5$ mm and flow rate $Q = 1.89$ lpm.

4 Experimental Verification

Figure 5 shows the test stand. The pump (2), driven by the electric motor, supplied the tested valve (1) from the tank (3). The valves (4) and (5) helped control the input pressure p_{in} to the plastic valve. Pressure transducers (6) and (7) measured input p_{in} and output p_{out} pressure and flow rate Q was measured with the flow transducer (8). The operating fluid temperature θ was measured by a temperature transducer (9) and maintained in the range of (25 – 35) °C. Parameters of accessible transducers are shown in Table 2.

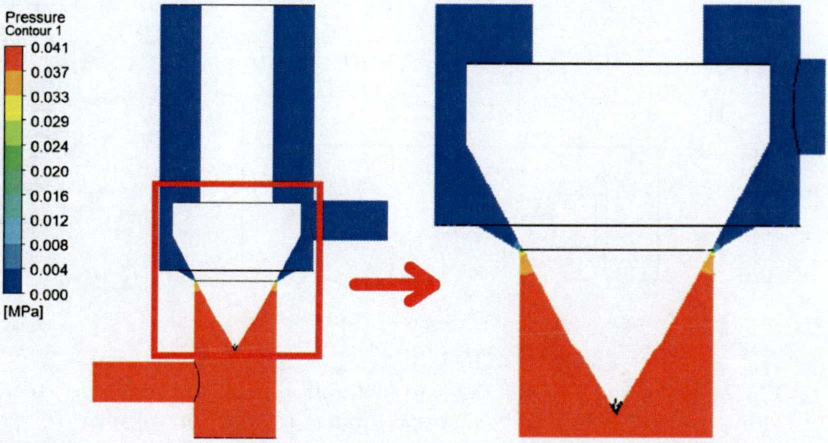

Fig. 4. Pressure contour plot – sample results on the exemplary of *cone60mod* poppet with valve opening $x = 0.5$ mm and flow rate $Q = 1.89$ lpm.

Table 2. Parameters of transducers.

Quantity	Symbol	Accuracy	Full scale
Pressure	p_{in}	±0.05 MPa	10 MPa
Pressure	p_{out}	±0.02 MPa	2.5 MPa
Flow rate	Q	2%	16 lpm
Opening	x	±0.05 mm	–

The research was carried out using the full factorial plan (Table 3). For the opening of the valve x, the flow rate Q was set according to the plan. Three repetitions were made at each measuring point and the mean value was calculated. Due to the low pressure difference Δp or high flow rate Q, not all the assumed measurement points were tested.

Table 3. Measuring points according to the full factorial plan (experimental tests and CFD).

Parameter	Symbol	Measuring points
Flow rate (lpm)	Q	0.70, 1.0, 1.89, 3.54, 5.0, 10
Opening (mm)	x	0.25, 0.50, 0.75, 1.0, 3.0, 12

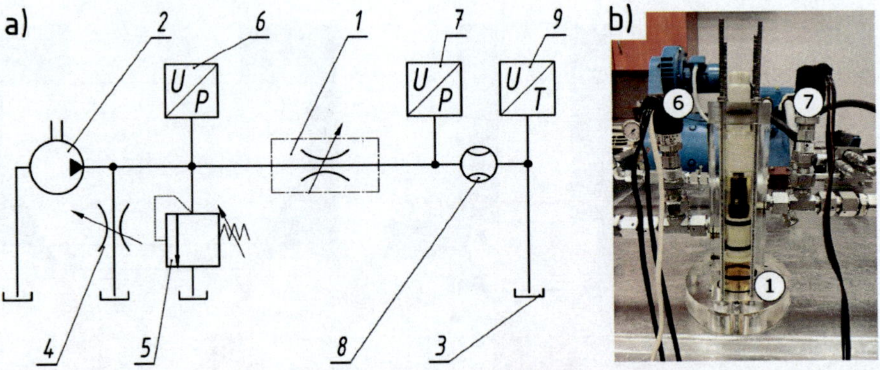

Fig. 5. Test stand with the valve: a) hydraulic circuit; b) the valve on the test stand; 1 – tested valve, 2 – hydraulic pump, 3 – tank, 4 – flow control valve, 5 – pressure relief valve, 6, 7 – pressure transducers, 8 – flow transducer, 9 – temperature transducer.

The pressure-flow characteristics of the throttle valve with analysed poppets *cone60* (Fig. 6), *cone60mod* (Fig. 7), *cone30* (Fig. 8) for various opening values x are depicted.

A good agreement between simulation and experimental curves is visible for higher valve opening values, $x > 0.5$ mm in each valve design. The relative difference is $\delta = (2.5 - 11.9)\%$ for the flow rate $Q > 3$ lpm and $\delta = (11 - 48)\%$ for the flow rate $Q < 3$ lpm. For smaller valve opening values ($x < 0.5$ mm), the difference is significantly higher, $\delta = (36 - 57)\%$.

For *cone60* (Fig. 6), we observe the similarity up to $Q = 10$lpm with small pressure difference $\Delta p < 0.1$ MPa. The difference between $x = 3$ mm and $x = 12$ mm is neglected due to the big gap.

For *cone60mod* (Fig. 7), CFD results are similar to the experimental ones for $x = 0.5$ mm, however, for $x = 0.25$ mm the difference is significant.

For *cone30* (Fig. 8), the difference is observable only for $x = 0.5$ mm. With opening values $x = 0.75$ mm and $x = 1.0$ mm, both methods give comparable results.

The dissimilarity in both methods may result from the geometrical differences between the physical and numerical model, i.e. printing accuracy, surface roughness, inaccuracy of the valve opening setting, etc. An important role plays the accuracy of the measurement instruments, which particularly impacts the observed difference. Another reason for this may be flow phenomena in such small flow surfaces, i.e. additional vortexes, flow turbulences, change in fluid properties not included in simulation settings, etc.

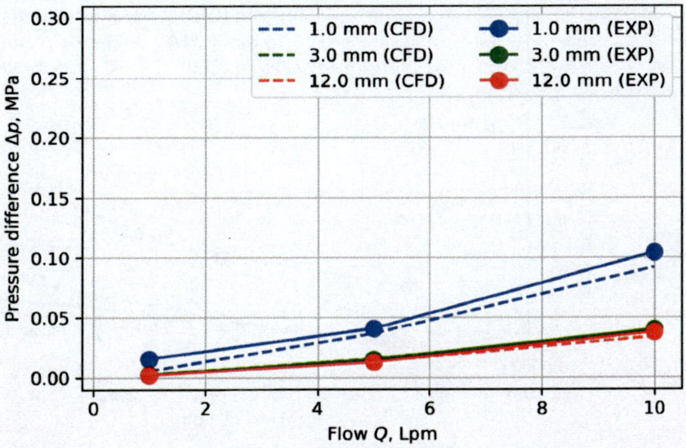

Fig. 6. Pressure-flow characteristics of valve with *cone60* poppet for selected values of valve opening x ($x =$ var); CFD – numerical results, EXP – experimental tests [11].

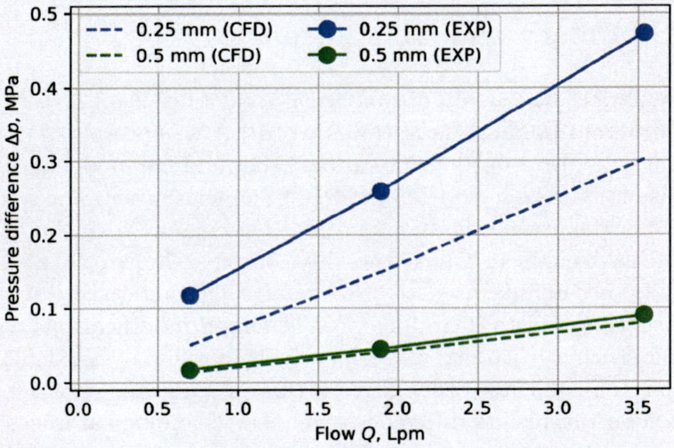

Fig. 7. Pressure-flow characteristics of valve with *cone60mod* poppet for selected values of valve opening x ($x =$ var); CFD – numerical results, EXP – experimental tests.

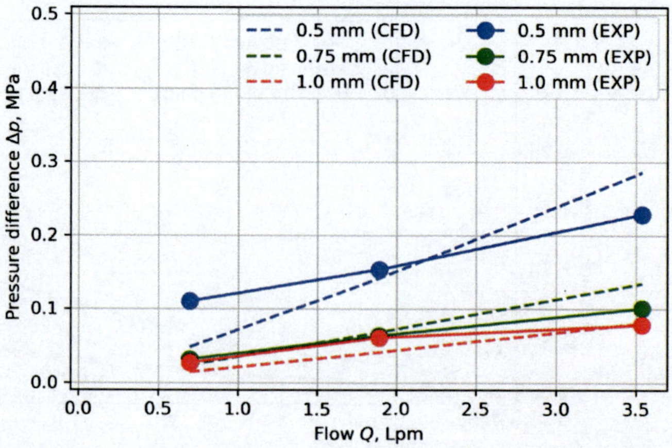

Fig. 8. Pressure-flow characteristics of valve with *cone30* poppet for selected values of valve opening x ($x =\ var$); CFD – numerical results, EXP – experimental tests.

5 Summary and Conclusions

In the article, the study of numerical modeling of oil flow in a throttling valve with different poppets designs made of PLA was presented as well as the experimental validation. In the simulations, the mixture model of hydraulic oil and air was used, and the cavitation model of Schnerr-Sauer. The numerical model presented in the article showed very good agreement with the experimental results for the test cases in which the valve lift was higher than 0.5 mm. The model may be successfully used to predict flow characteristics of valves with similar construction and to introduce poppet shape modifications without the need of manufacturing a variety of poppets with similar dimensions. For the valve lift smaller than 0.5 mm other factors than simulation parameters have the main impact on the results differences, i.e. the technological issues of valve manufacturing or the small gap flow phenomena may influence the flow characteristics. Further research in this area will address those issues as well as further considerations referring to cavitation phenomena.

References

1. PlasticsEurope, *Plastics - the Facts 2022*, Plastics Europe AISBL (2022). https://www.plasticseurope.org/
2. Parker, Lightraulics®Composite Hydraulic Cylinders For working pressures up to 700 bar. Catalogue HY07-1410/UK (2007)
3. Leśniewski, T., Stosiak, M., Lubecki, M., Krawczyk, J.: Wear resistance of selected anti-wear coatings used in multi-material composite hydraulic cylinders. Aviation **26**(3), 153–159 (2022). https://doi.org/10.3846/aviation.2022.17728

4. Stryczek, J., Banaś, M., Krawczyk, J., Marciniak, L., Stryczek, P.: The fluid power elements and systems made of plastics. Procedia Eng. **176**(204), 600–609 (2017). https://doi.org/10.1016/j.proeng.2017.02.303
5. Gibson, I., Rosen, D., Stucker, B.: Additive Manufacturing Technologies. Springer, New York (2015). https://doi.org/10.1007/978-1-4939-2113-3
6. Filo, G., Lisowski, E., Rajda, J.: Design and flow analysis of an adjustable check valve by means of CFD Method. Energies **14**, 2237 (2021). https://doi.org/10.3390/en14082237
7. Siwulski, T., Warzyńska, U.: Numerical investigation of the influence of the inlet nozzle diameter on the degree of fluid exchange process in a hydraulic cylinder. Eng. Appl. Comput. Fluid Mech. **15**(1), 1243–1258 (2021)
8. Buono, D., di Cola, F.D.S., Senatore, A., Frosina, E., Buccilli, G., Harrison, J.: Modelling approach on a Gerotor pump working in cavitation conditions. Energy Procedia **101**, 701–709 (2016)
9. Casoli, P., Scolari, F., Rundo, M.: Modelling and validation of cavitating orifice flow in hydraulic systems. Sustainability **13**(13), 7239 (2021). https://doi.org/10.3390/su13137239
10. Rituraj, R., Vacca, A.: Investigation of flow through curved constrictions for leakage flow modelling in hydraulic gear pumps. Mech. Syst. Signal Process. 153, 107503 (2021). https://doi.org/10.1016/j.ymssp.2020.107503
11. Banaś M., Antoniak P., Marciniak L., Stryczek J.: Visualization of flow phenomena in hydraulic throttle valves of plastics. In: The 14th International Conference on Vibration Engineering and Technology of Machinery (VETOMAC XIV), pp. 1–6. MATEC Web of Conferences, Lisbon, Portugal (2018). https://doi.org/10.1051/matecconf/201821119001
12. Ansys Inc.: Fluid Materials. In: Ansys Fluent Material Database (Software material), Ansys (2021R2)

EHLA-Coated Carbide-Reinforced Control Plates in Axial Piston Pumps First Results from Real-Life Application

Achill Holzer[1](✉), Stephan Koss[2], Jacqueline Rosefort[1], Johannes Henrich Schleifenbaum[2], and Katharina Schmitz[1]

[1] Institute for Fluid Power Drives and Systems, RWTH Aachen University, Aachen, Germany
achill.holzer@ifas.rwth-aachen.de

[2] Digital Additive Production DAP, RWTH Aachen University, Aachen, Germany

Abstract. Hydraulic pumps are a crucial part of every hydraulic system and have been built with similar tribological systems for decades. In a typical axial piston pump, the high surface pressures are hydrostatically balanced, and the residual contact is taken by a steel – special brass contact.

Due to the increasingly complex requirements, such as higher requested speeds and operating pressures, as well as legal constraints, such as the restriction of lead and the use of biodegradable oils, alternatives have to be found for the often-used lead-containing special brass.

Laser coatings applied by means of Extreme High-Speed Laser Application (EHLA) are a possible alternative. These promise almost complete freedom in terms of material selection and impress with a fast and cost-effective coating process.

In preliminary work, the process of material pre-selection by disc-disc tribometer testing up to the first running pump prototype was shown. The results obtained indicated, that with the choice of much harder contact partners, the run-in process also needs to be reconsidered.

In this study, a completely new, laser coated tribological system is tested in a real application. Special focus is set on the operational behavior and efficiency during the first 1000 h of operation, compared to the standard pump.

Keywords: Hydraulics · Axial Piston Pump · Lead-Free · EHLA · Metal Matrix Composition · MMC · 1.4404 · SiC

1 Introduction

Hydraulic systems can be found in a variety of applications, playing an important role in daily life, not only in earth moving and heavy-duty applications, but also in aircrafts and passenger elevators.

Hydraulics has a reputation for being rather conservative and often poorly efficient. To date, there are exceptions for hydraulic applications that exempt them from the REACH lead-free regulations. It is therefore of highest importance to show that

J. Stryczek and U. Warzyńska (Eds.): NSHP 2023, LNME, pp. 238–248, 2024.
https://doi.org/10.1007/978-3-031-43002-2_22

hydraulics with modern circuit designs, as well as new innovative materials do not have to fulfill these prejudices.

Using additive manufacturing processes such as Laser Powder Bed Fusion (LPBF) is not new to hydraulics [1]. Several project have been performed using PVD and other wear resistant coatings [2–4]. Former studies of these authors showed, that EHLA coated surfaces may withstand the tribological loads typically found in hydraulic applications [5, 6]. Consequently, the first functional tests were carried out in real pumps [7].

1.1 Axial Piston Pumps

Axial piston pumps are widely used in stationary, as well as mobile hydraulic applications. Compared to lower-cost gear pumps, they are known for being able to provide higher pressures at better efficiencies.

The Rotary Assembly of pumps consist of a drive shaft, connected to a rotating cylinder block. The cylinder block connects via the control plate to the suction and high-pressure ports, with the control plate acting as a valve. Typically, nine pistons with slipper pads are cinematically driven and supported by the swashplate. A control unit allows the tilting of the swashplate with the stroke piston to a certain angle, adjusting the volume flow.

The tribological contact control plate – cylinder block has been investigated for some years. Important results regarding temperature distribution and wear have been published by [8–10], showing that temperatures of up to 160 °C may be reached in the run-in process. Temperature- and wear distribution seems to be comparable. Both reaching their respective maxima on the high-pressure side of the control plate. Figure 1 shows the open axial piston pump, with the standard special brass (left) and the EHLA coated (right) control plate.

Fig. 1. Standard special brass (left) and EHLA coated control plate (right)

In this work, the conventional hard – soft pairing, consisting of a lead-containing special brass control plate, is compared to a lead free, laser coated control plate.

1.2 Extreme High-Speed Laser Application (EHLA)

The additive manufacturing process EHLA is characterized by a modified energy input compared to the conventional Laser Material Deposition (LMD). In EHLA the powder focus is positioned above the substrate, whereby the laser power is absorbed by the powder particles on their way to the substrate. This results in preheated powder particles which hit the substrate already molten (s. Fig. 2). The offset of the powder focus typically is in a rage of approx. 0.4 to 2 mm. The transmitted laser power, which is not absorbed by the powder particles, generates a small melt pool on the surface of the substrate. The resulting layers are dense and bonded by melting metallurgy. In addition, the ratio of the heat affective zone to the ratio of the substrate is small [11].

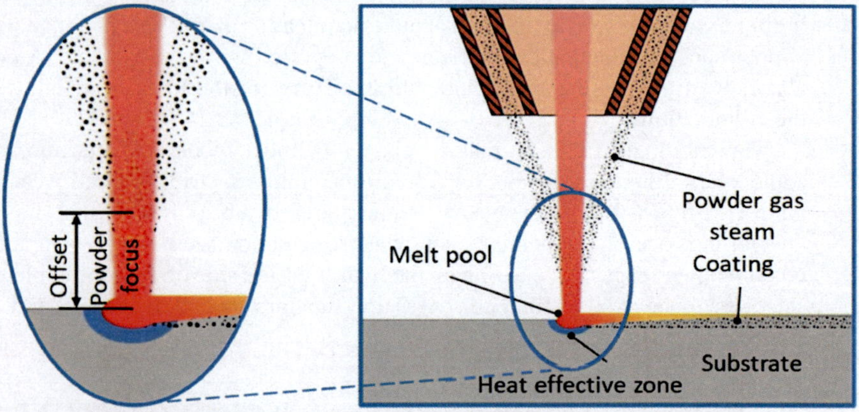

Fig. 2. Schematic of the process Extreme High-Speed Laser Application (EHLA)

Through the energy input into the powder material, heat-sensitive materials as well as materials that are difficult to weld can be coated and unconventional material parings can be produced. The changed energy input to the powder particles allows process speeds of up to 500 m/min and at the same time a powder efficiency of up to 96%, as well as a low surface roughness [1]. These characteristics enable the surface functionalization of hydraulic components by means of adapted coatings.

2 Specimen Preparation

The base body was made of S235JR, a cost-effective structural steel with a yield strength of 235 MPa. It is vital that the steel does not harden too much as a result of thermal processing during the manufacturing process. In a first step, the profile shape was formed by turning on a lathe. Subsequently, the coating was applied. Once the coating process was finished, high- and low-pressure kidneys were added to the base body, by milling.

2.1 Coating Process

The EHLA coating of the control plate for mechanical testing is produced using a Hornet EHLA system (Hornet Laser Cladding BV, Netherlands). The 4-axis handling system has a tiltable turning spindle to process rotationally symmetrical components with dimensions up to Ø 250 × 500 mm. The rotation speed is up to 1300 rpm. The laser beam source is a TruDisc4002 disk laser (Trumpf GmbH + Co. KG., Germany) with a wavelength λ = 1030 nm) and maximum output power of 4 kW.

A schematic figure of the EHLA process with the essential process parameters is shown in Fig. 3. The process parameters that influence the energy input into the powder particles and the substrate material is the particle velocity, which is dependent on the carrier gas flow and the powder mass flow. As the powder mass flow increases, the energy absorption per powder particle decreases. This results in reduced heating of the particles. An increased carrier gas flow leads to an increased particle velocity and reduced interaction time with the laser beam, reducing the heating of individual particles. The energy input into the substrate is mainly influenced by process speed and track displacement, by which the track overlap can be controlled. With increasing process speed, the interaction time between the laser beam and the substrate decreases, reducing energy input into the substrate. This leads to a reduction in melt pool size and the size of the heat-affected zone. Furthermore, the process speed can be used to influence the cooling rate of the layers produced. An increase in the process speed increases the cooling speed [14].

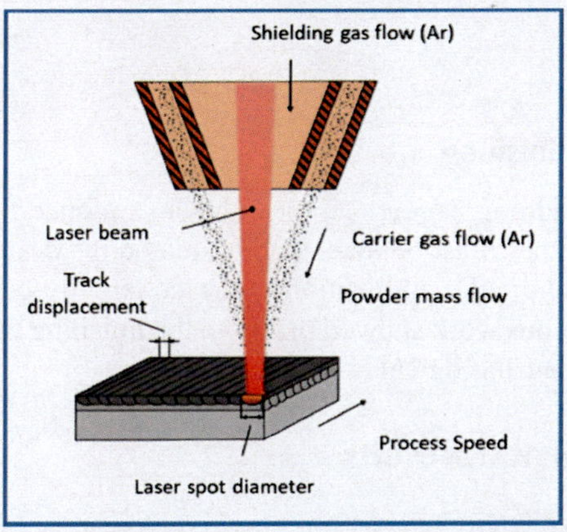

Fig. 3. Schematic of the EHLA including process parameters

For the coating of the control plate, the Metal Matrix Composite (MMC), made of AISI 1.4404 (Oerlikon Diamalloy 1003-1) and SiC (GTV 70.85.8), has been used. The thermal reaction of SiC and 1.4404 requires a high cooling rate in order to keep the reaction as small as possible. Therefore, the process speed of 200 m/min is chosen. An overview of the parameters used to produce the coating is given in Table 1. According to the parameters, a coating with a thickness of approx. 300 μm and a mixing ratio of 75 vol.% 1.4404 and 25 vol.% SiC has been generated (s. Fig. 4). For this purpose, a modified HighNo4.0 nozzle (HD Sonderoptiken) with a larger stand-off has been used for the material feed. The coating is almost defect-free bonded. The coating itself presents some pores and some cracks.

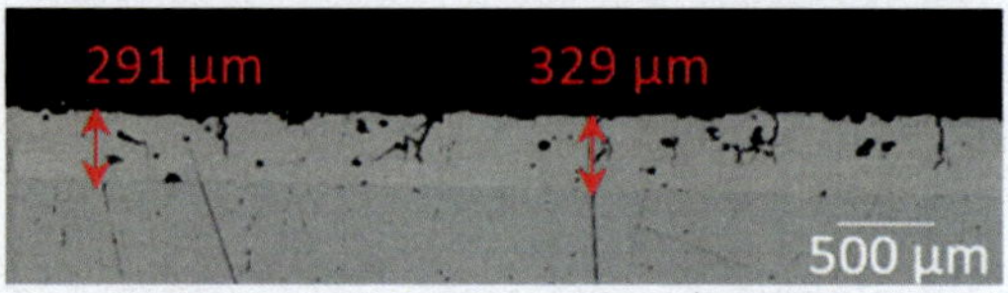

Fig. 4. Cross section of the coating MMC 316L+SiC after deposition

Table 1. EHLA Coating Parameters

Parameters	Values
Laser Power	3400 W
Process Speed	200 m/min
Track Displacement	0.12 mm/rev
Powder Mass Flow	approx. 22.8 g/min
Carrier gas flow	11 l/min
Shielding gas flow	11 l/min
Stand-off Nozzle	15 mm

2.2 Surface Finishing

EHLA, as a welding process, is not able to produce a tribological smooth surface. Therefore, surfaces must be machined, to match the desired geometric dimensions and tolerances. In hydraulic applications, surface lapping or grinding are commonly used processes. Previous work showed that both the finishing method and the roughness, have an impact on the tribological behavior [5].

3 Experimental Results

In order to gain experience with the new type of tribological system, an off-the-shelf pump with a displacement volume of 16cc was modified. There were three main concerns with regards to the general function that could be shown [7]:

- How will the pump behave during long-term operation?
- Does the piston drum wear more quickly as it is now the softer contact partner?
- If the pump can withstand the operation, what efficiency does it achieve compared to a standard pump?

In the following chapter, an endurance test, as well as efficiency measurement procedures and their results are shown.

3.1 Endurance Test

Axial piston pumps are designed for a long service life in rough environments such as construction equipment. Due to the partial hydrostatic relief of the main tribological contacts and the parings consisting of steel against special brass, service lives between 10.000 and 40.000 h can be reached.

However, it is not possible to stress the pump in a test bench for that timespan. In order to have an initial idea on the wear behavior of the coated control, the pump was tested at a constant speed of 1500 rpm, maintaining 200 bar of pressure for a cycle of 1.000 h, which is equivalent to a sliding distance of more than 16.000 km. Following each 100 h of runtime, the pump was disassembled, and both control plate and piston drum, were gravimetrically and geometrically measured.

Wear of Control Plate and Cylinder

As expected, the EHLA coated control plate shows less wear compared to the cylinder block. However, for both surfaces the wear rate can be considered minimal. In Fig. 5 it **can be noted that within the first hours of operation, the wear rate of the cast iron piston** drum, is lower ($0.5 * 10^{-8}\ mm^3/Nm$), reaching a plateau at $5 * 10^6$ m. The wear rate later increases to $1.0 * 10^{-8}\ mm^3/Nm$. **For the coating, the wear rate is significantly** smaller, and remains below $0.2 * 10^{-8}\ mm^3/Nm$.

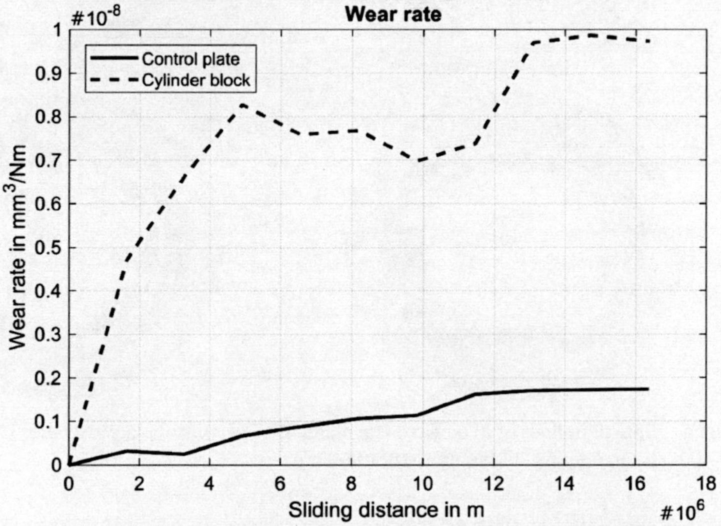

Fig. 5. Wear rate of the coated control plate (solid line) and cylinder block (dotted line)

In the calculation of the normalized wear rate, it must be taken into account that the surface pressure over the control plate was averaged due to the pressure differences.

Surface Analysis

Typical tribological system for the control plate – cylinder contact, consist of a soft contact partner made of special brass and another, much harder, contact partner usually made of nitrated steel or cast iron. For this study, only the control plate was coated, and the cylinder block was kept in the standard, cast iron, version.

In Fig. 6 optical imaging as well as height distribution of the control plate are shown. In the top left, a), the new, grinded control plate is shown. Grinding marks are equally distributed over the whole surface. In the top right, b), the control plate after 1000h of use is shown. In some areas, the original grinding marks can still be seen. Wear tracks can be noticed in the southeastern direction.

The height distribution shows wear in the high-pressure zone located in southeast of the control plate. Subfigure c) shows the control plate after 400h of operation, subfigure d) shows the control plate after 800h. The contact area to the cylinder can be seen clearly. Including the run-in phase, less than 150μm of material was removed from the surface of the coated control plate.

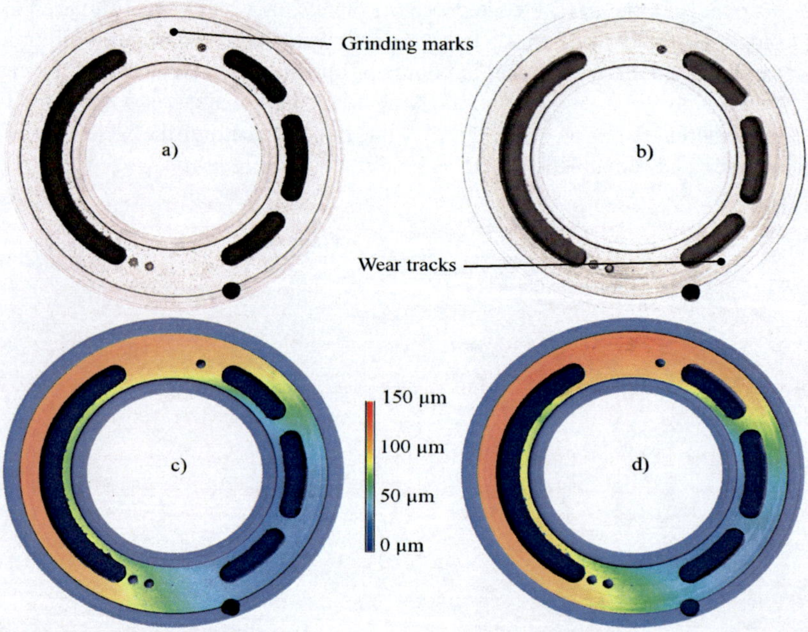

Fig. 6. Optical images of a) new control plate, b) used control plate after 1000 h and height image at 400 h c) and height distribution after 800 h d)

After finishing the 1000 h wear test, the tribological contact was still in good condition and could now be considered ran-in. To get the most relevant results, the efficiency measurement was performed after the endurance test.

3.2 Efficiency Measurement

One of, if not the most, crucial characteristic of a modern axial piston pump is the efficiency. However, this topic will gain importance in the future, as energy consumption and environmental impact play an increasingly important role. For hydraulic pumps and motors, efficiency is measured in accordance with ISO 4409, which specifies the testing environment [12]. In this work, measurements were gathered using one of the institutes standard efficiency test bench. In total, 330 operation points were tested (Table 2). Temperature was kept at 40 ± 1 °C. For the tests, a mineral HVLP oil with a viscosity class of 46 cSt according to DIN 51524-3 was used [13].

Table 2. Efficiency Testing Parameters

Pressure (bar)	Speed (rpm)	Swash Plate Angle (%)	Temperature (°C)
50	500	20	40
75	1000	40	–
100	1500	60	–
125	2000	80	–
150	2500	100	–
175	3000	–	–
200	–	–	–
225	–	–	–
250	–	–	–
275	–	–	–
300	–	–	–

Standard Brass Control Plate

Figure 7 (right) shows the total efficiency of the pump equipped with a standard special brass control plate at a swash plate angle of 100%. A peak efficiency of 90% can be observed at a speed of 1000 rpm and a pressure of 120 bar. The bottom-left part of the diagram (low pressure and speed) shows an efficiency of 88%. The lowest efficiency was 76%, at high speeds with low pressures. There is a drop to 80% efficiency at 300 bar and speeds of around 2000 rpm.

EHLA Coated Control Plate

For the coated control plate, the total efficiency is shown in Fig. 7 (left). The highest efficiency measured is 86%, which can be found at medium speeds and pressures. The lowest efficiency was 74%, at high speeds with low pressures. Compared to the brass control plate, the efficiency is about 2% less. In general, the efficiency is lower at low rotational speeds.

In order to investigate the behavior of both control plates in more detail, the volumetric and hydro mechanical efficiency maps were plotted in Fig. 8.

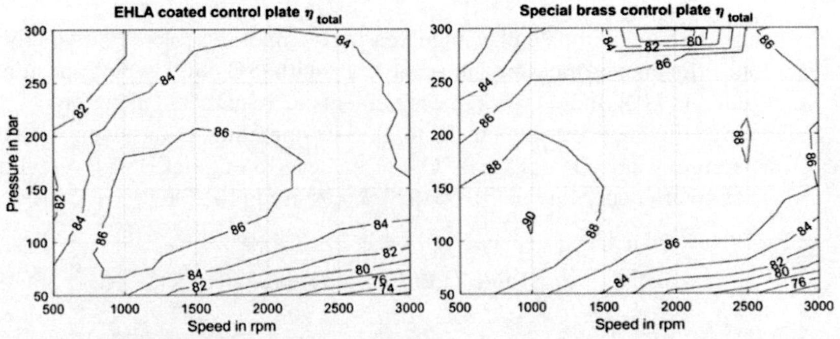

Fig. 7. Total efficiency of EHLA coated (left) and Special brass control plate (right)

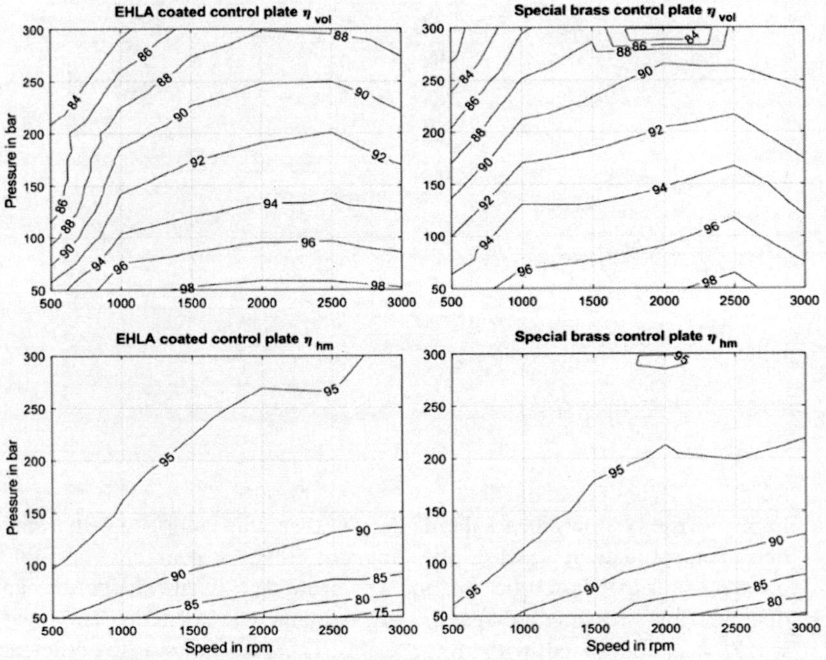

Fig. 8. Volumetric- and hydro mechanical efficiency of EHLA and special brass control plate

4 Summary and Outlook

With these results, both control plates reach industrial-like efficiencies and could be used in standard applications. The coated prototype of the control plate underwent an endurance test with constant parameters, confirming that the tribological pairing EHLA coating – cast iron can be used also over longer time periods. Compared to the standard pump, run-in times were much higher (600 h) than the special brass (1 h) which may be explained with the increased hardness level of the used materials.

To summarize, it was possible to manufacture the prototype of a coated control plate, which can be used in a state-of-the-art hydraulic pump for at least 1000 h at constant operation parameters. Compared to the standard special brass control plate, the run-in took significantly longer. A final efficiency test showed almost the same efficiencies. Overall, there is about 2% less efficiency. This could also be due to the fact that the prototype was manufactured as a single piece. It is certain that in a fine-tuned series production even better efficiencies would be possible.

As a possible next step, more detailed investigations, including the behavior at several operating points, including lower, rotational speeds and tests with special fluids such as biodegradable oils or water-based fluids (HFC) should give important feedback and may open new applications where the anticorrosion features of the coating may be used.

References

1. Matthiesen, G., Merget, D., Pietrzyk, T., Ziegler, S., Schleifenbaum, J.H., Schmitz, K.: Design and experimental investigation of an additive manufactured compact drive. In: International Fluid Power Conference, pp. 137–147 (2020). https://doi.org/10.25368/2020.21
2. Equey, S.: Investigation of the interaction between diamondlike carbon coatings and lubricant additives. Ph. D. thesis, ETH Zuerich (2008). https://doi.org/10.3929/ethz-a-005730458
3. Kalin, M., Majdicˇ, F., Vizˇintin, J., Pezdirnik, J., Velkavrh, I.: Analyses of the long-term performance and tribological behavior of an axial piston pump using diamondlike-carbon-coated piston shoes and biodegradable oil. ASMEJ Tribol. **130**(1), 011013 (2008). https://doi.org/10.1115/1.2805442
4. Schuhler, G., Jourani, A., Bouvier, S., Perrochat, J.-M.: Efficacy of coatings and thermochemical treatments to improve wear resistance of axial piston pumps. Tribol. Int. **126**, 376–385 (2018). https://doi.org/10.1016/j.triboint.2018.05.007
5. Holzer, A., Koß, S., Ziegler, S., Schleifenbaum, J.H., Schmitz, K.: Extreme high-speed laser material deposition (EHLA) as high-potential coating method for tribological contacts in hydraulic applications. In: Reisgen, U., Drummer, D., Marschall, H. (eds.) EMPOrIA 2020. LNME, pp. 153–167. Springer, Cham (2021). https://doi.org/10.1007/978-3-030-70332-5_14
6. Koß, S., Holzer, A., Megahed, S., Ziegler, S., Schleifenbaum, J.H., Schmitz, K.: Investigation of the coating of hydrodynamic plain bearing contact surfaces by means of Extreme High-Speed Laser Material Deposition (EHLA). In: IOP Conference Series: Materials Science and Engineering, vol. 1097, no. 1, p. 12016 (2021). https://doi.org/10.1088/1757-899X/1097/1/012016
7. Holzer, A., Koss, S., Schleifenbaum, J.H., Schmitz, K.: Extreme high-speed laser material application-coated carbide-reinforced control plates and conventional lead-based plates in axial piston machines. Chem. Eng. Technol. **46**(1), 110–117 (2023). https://doi.org/10.1002/ceat.202200402
8. Wegner, S., Löschner, F., Gels, S., Murrenhoff, H.: Validation of the physical effect implementation in a Simulation model for the cylinder block/valve plate contact supported by experimental investigations. In: 10th International Fluid Power Conference, Dresden, 8–10 March 2016, vol. 1, pp. 269–281 (2016)
9. Shorbagy, A., Ivantysyn, R., Weber, J.: Holistic analysis of the tribological interfaces of an axial piston pump – focusing on the pump efficiency. Chem. Eng. Technol. **46**, 5–13 (2023). https://doi.org/10.1002/ceat.202200450
10. Ivantysyn, R., Shorbagy, A., Weber, J.: An approach to visualize lifetime limiting factors in the cylinder block/valve plate gap in axial piston pumps. In: ASME/BATHSymposium on Fluid Power and Motion Control (2017). https://doi.org/10.1115/FPMC2017-4327

11. Schmitz, K., Murrenhoff, H.: Grundlagen der Fluidtechnik, Teil 1: Hydraulik. RWTH Aachen University, Shaker, Aachen (2020)
12. ISO4409:2019-07: Information Technology—Hydraulic fluid power - Positive-displacement pumps, motors and integral transmissions - Methods of testing and presenting basic steady state performance. International Organization for Standardization Geneva, Switzerland
13. DIN 51524-3:2017-06: Pressure fluids - Hydraulic oils - Part 3: HVLP hydraulic oils, Minimum requirements. International Organization for Standardization Geneva, Switzerland
14. Koß, S., et al.: Comparison of the EHLA and LPBF process in context of new alloy design methods for LPBF. In: AMR, vol. 1161, pp. 13–25 (2021). https://doi.org/10.4028/www.scientific.net/AMR.1161.13. 2021

Experimental Study of the Phenomenon of Luminescence in Electrodynamic Processes During Hydrodynamic Cavitation

Ihor Nochnichenko(✉), Vladyslav Kryvosheiev, Oleg Yakhno, and Dmytro Kostiuk

National Technical University of Ukraine «Igor Sikorsky Kyiv Polytechnic Institute», Kyiv, Ukraine
nochnichenko.ihor@lll.kpi.ua

Abstract. The issue of the physics process of hydrodynamic luminescence formation requires careful research. In scientific and technical literature, two terms denoting the glow of a liquid are widely used: sonoluminescence and light emission. The first term is directly related to ultrasound as the cause the liquid to glow; the second - provides a wider range of reasons that cause the glow. One way or another, both processes have two main theories of their occurrence – "thermal" and "electric". The "thermal" theory assumes that when an external influence is applied to a cavitating bubble, high temperatures are formed inside it, which stimulate the emission of radiation by the bubble. The "electrical" theory is based on electrical phenomena inside the bubble itself or interaction with nearby cavitating bubbles.

The practical value of the work results is determined by the search for a reliable theory of the phenomenon of hydrodynamic luminescence. The obtained results contribute to a deeper understanding of the studied phenomena and create a basis for the development of an experimentally based theory of the origin of the phenomena of sonoluminescence and hydroluminescence, which has a positive effect on the introduction of these phenomena into technological processes.

The relevance of the study is also due to the wide use of dielectric pipelines when supplying hydrocarbon and other liquids to various technical devices. In the future, it is possible to use these processes in the chemistry, pharmaceutical industry, and the creation of effective methods of surface treatment with liquids.

Keywords: Cavitation · Sonoluminescence · Hydroluminescence · Triboluminescence · Visualization of liquid flow

1 Introduction

In hydraulic equipment, various types of throttle elements are used as shut-off and control elements, in which the flow of the working fluid has a turbulent character, and in a narrow section, high speed causes cavitation and the associated active release of bubbles of undissolved air and steam. In the zone of increased pressure, the bubbles instantly collapse, which causes erosive destruction of the material of the channel walls

J. Stryczek and U. Warzyńska (Eds.): NSHP 2023, LNME, pp. 249–257, 2024.
https://doi.org/10.1007/978-3-031-43002-2_23

and active acoustic processes. The rapid closing of steam-gas caverns, following the laws of thermodynamics, can cause a local increase in temperature, and under certain conditions, the glow of the liquid. The issue of the physics process of hydrodynamic luminescence formation requires in-depth research [1].

In scientific and technical literature, two terms denoting the glow of a liquid are widely used: sonoluminescence and light emission. The first term is directly related to the ultrasound as the cause the liquid to glow; the second - provides a wider range of reasons that cause the glow. One way or another, both processes have two main theories of their occurrence - "thermal" and "electric" [2].

The thermal theory assumes that when the cavitation bubble is affected, high temperatures are formed inside it, which stimulates the emission of the bubble. The thermal theory of hydrodynamic luminescence was proposed by Koldamasov [3], indicating in his article that the source of luminescence is a plasma clot formed at an average temperature. 10^4 K. Proponents of the thermal theory of hydrodynamic luminescence are Gordeev and Serbinov [4, 5]. In a series of experiments to study the excitation of an explosion in liquid explosives in [5], it was found that the initiation of an explosion by cavitation does not occur during expansion, but during the closure of the cavitation cavity in the explosive liquid. Based on the obtained results, the authors [4, 5] concluded that the cause of light emission during hydrodynamic luminescence is the thermal processes that occur during cavitation.

The "electrical" theory is based on electrical phenomena inside the bubble itself or interaction with nearby cavitating bubbles. Reports on the correspondence of the continuous spectrum observed during sonoluminescence to the emission spectrum of a black body play a certain role as a justification for this theory.

Also [6], light radiation was associated with electric discharges in a liquid. Electrical effects that accompany liquid flow at high pressure were directly observed in [7, 8], in [3] a similar conclusion was made as a result of indirect measurements. In general, the connection between cavitation and electrical discharges can be important for understanding the physical nature of the breakdown of liquid dielectrics.

However, the detailed understanding of the mechanism of hydroluminescence is still unclear. What phases - gas (bubble), liquid (fluid itself), or even solid (channel wall) - are responsible for light radiation, to what extent can this physical mechanism be applied to the phenomenon in general?

2 Method

For experimental research, a model of a hydrodynamic cavitator was made, with a nozzle in the form of a transparent plexiglass cylinder, Fig. 1. The flowing part of the cavitator contains an ejector at the entrance, which allows to introduce air or another phase into the flow of liquid in the appropriate concentration. The main part of the cavitator consists of an adjustable channel in the form of a throttle of variable cross-section (the active passage area is set by turning the screw) [2].

Also, the basic hydraulic scheme of the test bench was developed, Fig. 2 and the experimental setup in Fig. 3.

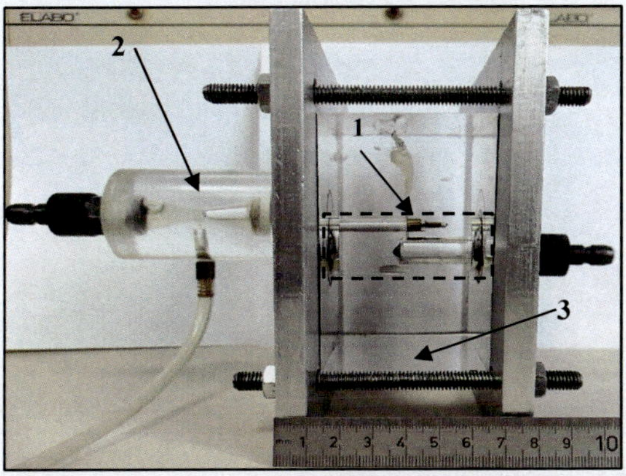

Fig. 1. Hydrodynamic cavitator with a cylindrical nozzle and an ejector at the entrance (1 - reactor, 2 – ejector, 3 – body)

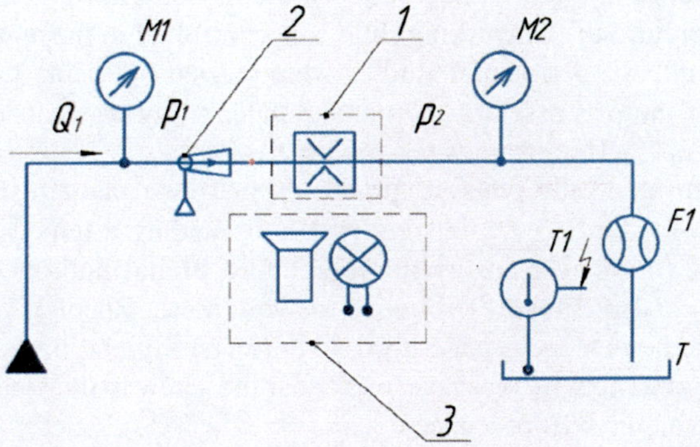

Fig. 2. Basic hydraulic diagram of the test bench (Q1 – flow rate through the cavitator, p1, p2 – pressure at the cavitator inlet and outlet, M1, M2 – pressure gage, T1 – temperature sensor, F1 – flowmeter, T – tank reservoir, 1-cavitator, 2-ejector, 3-high-speed video camera)

The principle of operation of the experimental setup is as follows. The liquid under study passes through a rough oil filter and enters the high-pressure pump. The high-pressure pump is driven by a three-phase asynchronous electric motor through a belt drive. A small-sized vector frequency converter was used to regulate the frequency of rotation of the three-phase asynchronous motor and, as a result, the frequency of rotation of the gear pump, which in turn made it possible to create different pressures of the investigated liquid. At the output of the gear pump, the liquid enters the working area through a high-pressure hose, after the working area, the liquid flows back into the filling tank reservoir. The temperature and pressure sensors are calibrated with a reference thermocouple and pressure gauge, and the digital flow meter is calibrated with a reference volumetric flow meter. The appearance of the experimental stand is shown in Fig. 3.

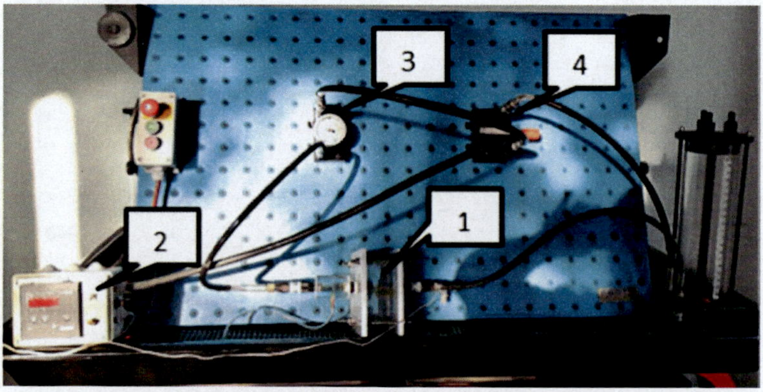

Fig. 3. Experimental setup (1-cavitator, 2-temperature sensors, 3-manometer 4-safety valve)

The main parameters that were monitored: the pressure level at the inlet, the pressure difference at the inlet and outlet of the device, flow and temperature at the outlet. The temperature of the working fluid was controlled in the apparatus using a reference thermocouple. Experimental studies were carried out using the "H-L" working fluid. Filming of the processes of the nozzle under study was carried out by a camera at a speed of 120 … 1000 frames per second [9–11].

To investigate the physical processes that take place inside the cavitator, it was decided to modernize the existing nozzle by adding a tinned copper wire to which it is possible to connect an oscilloscope and a digital multimeter Fig. 4. The existing stand and equipment were grounded to avoid noise. Rheological tests of model liquids were carried out, which made it possible to classify and investigate the impact of this phenomenon. The wire was "cleaned" near the screw in the area where the phenomenon of hydrodynamic luminescence occurs.

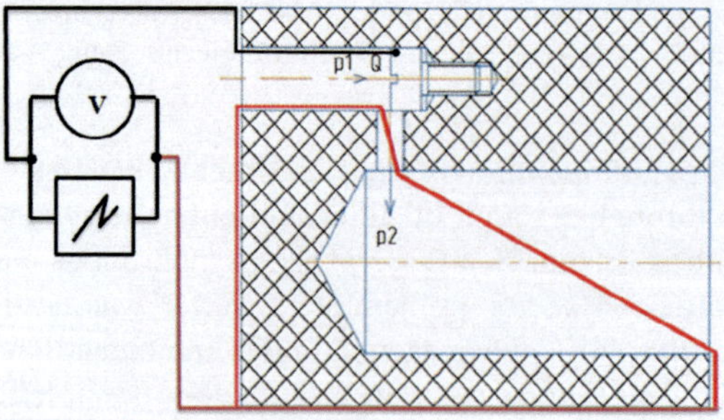

Fig. 4. Hydrodynamic cavitation nozzle under investigation

The experiment was conducted at a pressure drop of up to 5 MPa and a flow rate of up to 10^{-5} m^3/s. Additional input parameters are listed in Table 1.

Table 1. Initial parameters of the experiment

Parameters	Value
Pressure difference	$\Delta p = 5$ MPa
Temperature	$t_o = 33$ °C
Flow rate	$Q = 0.0000004$ m^3/s
The velocity of the liquid	$v = 100$ m/s

3 Results

1. When an oil flow under a pressure of up to 1 MPa is passed through the cavitator, no visible changes occur and the liquid remains transparent.
2. When the pressure in front of the cavitator is increased to 1.5 MPa, a fairly thin layer of microbubbles localized on the inner surface of the passage section is formed directly near the entrance to the through hole (attached cavitation).
3. When the pressure in front of the cavitator is increased to 2 MPa, short-term (lasting 1–2 s) flashes of the blue-violet glow of hydroluminescence are noted directly near the entrance inside the passage hole. At the exit from the opening, a torch of cavitation bubbles develops, gradually filling the entire exit chamber, but this part of the cavitation bubbles never initiated hydroluminescence. The beginning of deviations of the curve was noticed on the oscillogram of Fig. 5. The indicators of the digital multimeter remain at the 0 V mark.

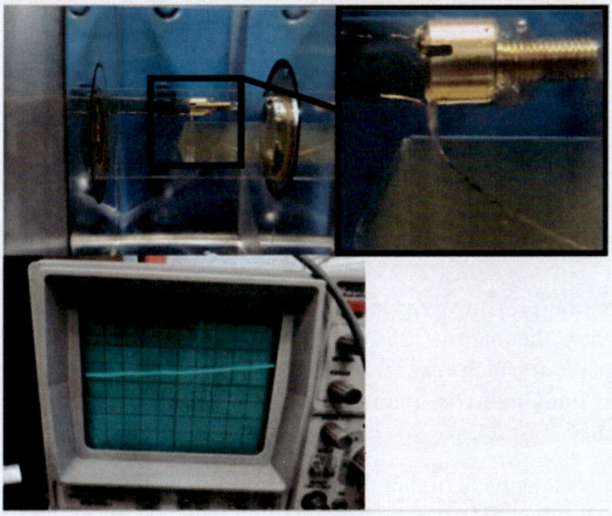

Fig. 5. Oscillogram of the initial cavitation with short-term flashes of hydroluminescence

4. As the pressure increases to 2.5 MPa, the frequency of hydroluminescence flashes increases, and upon reaching a pressure of 3…3.5 MPa, the hydroluminescence glow becomes continuous, stable, and a further increase in pressure only slightly increases its intensity. When observing the process in slow motion (images obtained during high-speed video recording), it can be seen that hydroluminescence is not a continuous light, but occurs with a certain frequency in frequent, unsynchronized, random flashes.
5. After reaching a pressure of 3 MPa, there is a gradual convergence of attached cavitation (in narrowing) with the torch of cavitation bubbles in the outlet part of the cavitator (development of supercavitation), and the attached cavitation and the torch of bubbles are spatially separated. When setting the supercavitation mode, when the torch of cavitation bubbles fills the entire outlet area of the throttling device and closes with the area of attached cavitation, abnormal "noises" were noticed on the oscillogram and the multimeter readings were recorded at 1.904 V (Fig. 6).

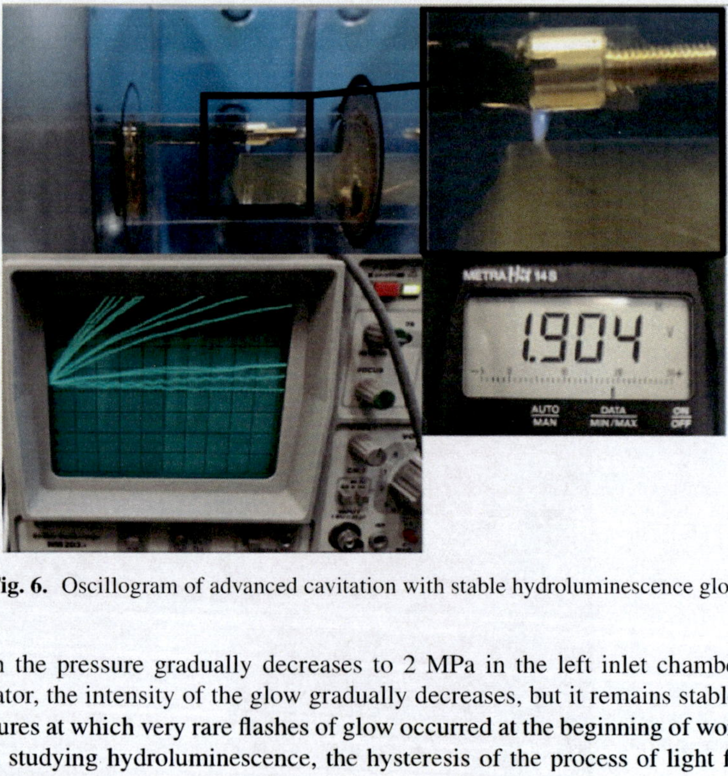

Fig. 6. Oscillogram of advanced cavitation with stable hydroluminescence glow

6. When the pressure gradually decreases to 2 MPa in the left inlet chamber of the cavitator, the intensity of the glow gradually decreases, but it remains stable even at pressures at which very rare flashes of glow occurred at the beginning of work. Thus, when studying hydroluminescence, the hysteresis of the process of light flashes is revealed (Fig. 7).

The glow starts at higher pressure drops in the throttle device and remains stable at lower pressures. This can be explained by the fact that in the process of oil pumping, microbubbles formed as a result of cavitation, which are the seeds of cavitation, do not have time to be completely removed from the system. Oil containing a large number of fairly stable microbubbles is naturally characterized by lower cavitation strength. During

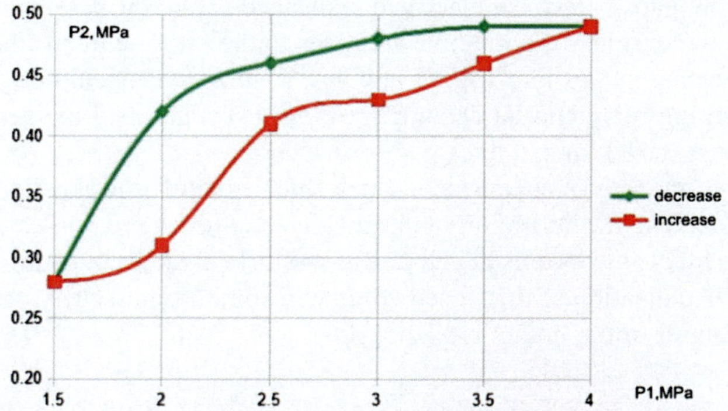

Fig. 7. Hysteresis of the process of light flashes

the circulation of oil, as a result of its impact on the inlet section of the through hole, hydrodynamic cavitation occurs at lower tensile stresses.

Based on the results of the processing and analysis of the conducted experimental studies, a conceptual model of the stages of the occurrence and development of the cavitation process and associated effects was built (Fig. 8).

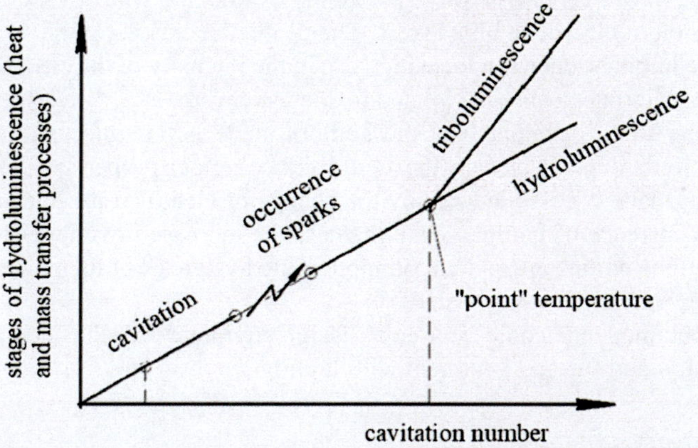

Fig. 8. Occurrence of cavitation and accompanying physical processes in the fluid flow

According to our observations, when the flow speed increases in the narrowing zone, cavitation first occurs, then with a further increase in the flow speed, single sparks begin to appear, and from a certain point - a "shot" and there is a stable glow. When the "spot temperature" is reached and in the presence of materials with cavitation radiation, the intensity of the glow increases due to triboluminescence; in the absence of the corresponding materials, the process proceeds less intensively.

The study of the nature of hydrodynamic glow can be compared to fire: it starts with a weak spark, then a flickering flame, and then a powerful flame endowed with speed and power. It can be assumed that the process of cavitation in the flow of mineral oil in narrow gaps can cause electrical resistance of the oil base and channels, as well as lead to the burning of corrective additives to the oil. The main source of light radiation is the release of gas into vapour bubbles, such as hydrogen. For example, the water in the working body can be decomposed into hydrogen and oxygen under the action of cavitation, which then ignites with a discharge.

The conducted research made it possible to identify the cavitation zones arising in the cavitator. With the help of high-speed video recording, areas of cavitation were detected, and the mechanism of its development was investigated. In addition, the characteristics of the visualization of the closed volume to some extent clarify the existing ideas about the behavior of the liquid in the nozzle.

4 Conclusions

Thus, it was possible to assume that the phenomenon of hydroluminescence has an electrical character under certain conditions. A review of studies has shown that light emission depends on the rheological properties of the liquid. It is more intense where the viscosity is lower. The obtained experimental data were approximated for the possibility of using a mathematical model of a hydrodynamic cavitation generator and allowed to take into account the properties of the working body.

Based on the work performed, we can conclude that the cause of hydrodynamic luminescence is the friction of the liquid against the walls of the channel and the light emission of electric double layers. Therefore, the reason for the occurrence of hydrodynamic luminescence is a local increase in the intensity of the electric field, which occurs when electroneutrality is violated inside the cavern.

The stand developed by the authors made it possible to test the theory of light radiation's dependence on the liquid's rheological properties. The proposed original design made it possible to study the change of electrical parameters in the flow during the occurrence of luminescence, namely the increase in voltage and the appearance of pulsations in the supercavitation mode. The hysteresis of the process of the appearance of light radiation was revealed.

Obtained data could be used to design equipment for the chemistry, pharmaceutical industry, and surface treatment with liquids.

References

1. Kryvoshejev, V.J.: Eksperymentaljne doslidzhennja ghidrodynamichnoji kavitaciji, shho suprovodzhujetjsja javyshhem ljuminescenciji. MS thesis. KPI im. Ighorja Sikorsjkogho (2020)
2. Nochnichenko, I., et al.: Experimental study of cavitation-hydrodynamic luminescence in gas-liquid environment. Bull. Natl. Tech. Univ. "KhPI". Ser.: Hydraul. Mach. Hydraul. Units: zb. nauk. pr. – Kharkiv: NTU "KhPI" **1**, 32–39 (2021)
3. Farhat, M., Chakravarty, A., Field, J.E.: Luminescence from hydrodynamic cavitation. Proc. R. Soc. A **467**, 591–606 (2011). https://doi.org/10.1098/rspa.2010.0134
4. Eberlein, C.: Theory of quantum radiation observed as sonoluminescence. Phys. Rev. A **53**(4), 2772–2787 (1996). arXiv:quant-ph/9506024. Bib-code: 1996PhRvA..53.2772E

5. Milton, K.A.: Dimensional and dynamical aspects of the casimir effect: understanding the reality and significance of vacuum energy. preprint hep-th/0009173 (2000). arXiv:hep-th/0009173
6. Gertsenshteyn, S.Ya., Monakhov, A.A.: Elektrizatsiya i svechenie zhidkosti v koaksialnom kanale s dielektricheskimi stenkami. Izv. RAN. Mekhanika zhidkosti i gaza. **3**, 114–119 (2009)
7. Leighton, T.G., Farhat, M., Field, J.E., Avellan, F.: Cavitation luminescence from flow over a hydrofoil in a cavitation tunnel. J. Fluid Mech. **480**, 43–60 (2003). https://doi.org/10.1017/S0022112003003732
8. Konstantinov, V.A.: DAN SSSR, vol. 56, no. 3, pp. 259–260 (1947)
9. Nochnichenko, I.V., Luhovskyi, O.F., Kostiuk, D.V.: Study of hydrodynamic luminescence in a cavitation liquid medium. Naukovo-tekhnichnyi zhurnal «Problemy tertia ta znoshuvannia» **3**(84), 57–62 (2019). https://doi.org/10.18372/0370-2197.3(84).13853. Index Copernicus - ISSN: 0370-2197
10. Nochnichenko, I., Luhovskyi, O., Kostiuk, D., Jakhno, O.: Research of the influence of hydraulic orifice material on the hydrodynamic cavitation processes accompanied by luminescence. In: Stryczek, J., Warzyńska, U. (eds.) NSHP 2020. LNME, pp. 293–300. Springer, Cham (2020). https://doi.org/10.1007/978-3-030-59509-8_26
11. Nochnichenko, I.V., Luhovskyi, A.F., Jakhno, O.M., Kostiuk, D.V., Komada, P., Kozbakova, A.: Experimental research of hydroluminescence in the cavitating flow of mineral oil. In: Proceedings of the SPIE 11176, Photonics Applications in Astronomy, Communications, Industry, and High-Energy Physics Experiments, 6 November 2019, vol. 1117615 (2019). https://doi.org/10.1117/12.2536946. PDF: 8 pages (Scopus -ISSN: 2577–5421)

Stand for Testing Hydraulic Actuators with Modified Working Surfaces

Justyna Skowrońska(✉), Andrzej Kosucki, and Łukasz Stawiński

Institute of Machine Tools and Production Engineering, Lodz University of Technology, ul. Stefanowskiego 1/15, 90-924 Lodz, Poland
justyna.skowronska@dokt.p.lodz.pl

Abstract. Increasing production costs require the search for new cheaper or more durable solutions for hydraulic components, such as actuators. Direction of the development of actuator constructions is to apply new methods of modifying surfaces of working elements, for example, pistons and piston rods. Operation of actuators under high pressure and variable temperatures, additionally in a hydraulic oil environment, requires special properties of the surfaces, in particular, low friction coefficients, high hardness, and resistance to abrasion, corrosion, or aggressive environments. This paper presents a test stand for actuators with various working surfaces. The test stand can be used for long-term tests (fatigue tests) performed on four actuators simultaneously, as well as for testing selected parameters, such as mechanical-hydraulic efficiency. One of the actuators is a reference actuator made of typical material and standard surface modification, while the others are manufactured using modern surface modification methods, such as nitriding, nitrosulphurization, or application of a DLC carbon coating enriched with tungsten carbide particles. The design of the workstations makes it possible to set different load levels on all cylinders simultaneously, as well as different operating speeds. All important parameters of the working cycles and the number of cycles performed are measured. The article presents examples of measurement cycles.

Keywords: Power Hydraulics · Hydraulic Actuators · Surface Modifications · Test Stand · Actuators Research

1 Introduction

Hydraulic actuators, as executive elements in power hydraulics, are often exposed to damage caused by operation and harmful external factors. Any failures may result in damage of the entire hydraulic system, which in extreme cases may even endanger human health and life. The choice of material is a very important point in the design of hydraulic components such as pistons, cylinders, or piston rods [1].

Actuators' failures are caused by many factors. Too high pressure can strain the material. High axial forces during operation can cause buckling and therefore loss of stability [2]. Actuator components in contact with the external environment are often exposed to fresh water, salt water, and sand. Insufficient smoothness of the surfaces

J. Stryczek and U. Warzyńska (Eds.): NSHP 2023, LNME, pp. 258–268, 2024.
https://doi.org/10.1007/978-3-031-43002-2_24

cooperating with the seals, or any scratches, may reduce the adhesion degree of the seals, which in turn will result in insufficient tightness and leakage of working liquid. Nowadays, more and more emphasis is placed on ecological issues. The aim of the research is to ensure the highest energy efficiency of production and operation. The durability of the actuators is a very important aspect as fewer replacements and repairs are associated with less energy and material consumption [3]. Preventing breakdowns is cheaper and much more environmentally friendly.

Actuators often are covered with various materials or their surfaces are modified by thermochemical treatment. That is related to the desire to reduce the impact of manufacturing processes on the natural environment and reduce production costs while maintaining good mechanical and operational properties. Currently, galvanic chromium coatings are most often used for piston rods, which allow to improve wear resistance and improved tribological and anti-corrosion parameters. Unfortunately, recently a lot has been said about the harmfulness of chromium to employees during the application process and to the environment - during operation or disposal [4]. Therefore, it is especially important to find a replacement for the hexavalent chrome coatings applied on piston rods.

The topic of research on actuators on specially designed stands has been mentioned in many articles in the field of power hydraulics. The work [5] discusses the testing of actuators on the stand in order to determine the dynamic phenomena occurring during operation. In the article [6], a stand was designed for the diagnosis of hydraulic cylinder failures in environments with limited power. In [7], the authors proposed a hydraulic system to test a hydraulic active heave compensation system. In [8], the stand for testing actuators in the context of the simulation of aerodynamic forces is described. However, most of the works concern the research of the drive system or control system, while the stand described in this article also raises the issue of materials from which the actuators are made.

This article describes the design of the stand for testing hydraulic actuators during its operation under the variable load conditions. Thanks to the use of a contradictory system, it is possible to examine four actuators at the same time, ensuring identical operating conditions. Each of the actuators is made in a different material configuration and surface modification of its main elements.

Section 2 presents a description of the tested elements, including the materials used. Section 3 discusses the test stand with a description of the main components. Section 4 describes the feasibility of the research performed and the expected results. Finally, Sect. 5 contains conclusions about the possibilities of the test stand.

2 Materials

The surfaces of the piston rods and pistons should be as smooth as possible in order to ensure appropriate tribological properties. The contact of the components with water-containing fluids determines the corrosion resistance requirement. Moreover, the cooperating surfaces should be characterized by high resistance to fatigue and abrasion. The improvement of the operational parameters is applied both through the use of innovative materials and the modification of the surfaces of the working elements. On the stand presented in this paper, 4 cylinders in various variants of their piston rods and pistons will be tested. One of them will be a typical actuator made of the most common materials, i.e. C45 steel with chrome plating. This will be the reference actuator. The configurations of the proposed materials of the tested actuators are shown in Table 1.

Table 1. Configuration of the tested actuators.

Material/Surface Modification	Actuator A	Actuator B	Actuator C	Actuator D
Piston	C45	38H2MUA/DLC	38H2MUA/nitriding	38H2MUA/nitrosulphurization
Piston rod	C45/chrome-plate	38H2MUA	38H2MUA/nitriding	38H2MUA/nitrosulphurization

Nitriding and nitrosulphurization were selected from the thermo-chemical treatments, as both treatments lead to the creation of durable layers resistant to corrosion and fatigue. In turn, by using DLC coatings on the piston, they wanted to obtain a smooth, durable surface, which is important due to the contact with the seal.

The selected 38H2MUA steel is the typical steel for nitriding. The key to this issue was to select a material whose surface could be subjected to all three modifications, which is important in terms of the repeatability of experimental tests. The selection of surface modification methods was in turn based on the characteristics of the layers that can be obtained with a given method to meet the requirements for pistons and piston rods.

3 Experimental Test Stand

The test stand for actuators consists of a specially designed frame and a hydraulic system. The frame is made of steel square section profiles. Its relatively small dimensions (930 mm × 500 mm × 212 mm) allow the mobility of the workstation and allow it to be extended with additional elements, such as wheels. A set of four hydraulic actuators is attached to the frame with flanges. All four actuators operate at the same time under at least comparable if not identical conditions. This makes it easier to compare all four versions and reduces testing time and costs. Photos of the stand with markings of driving, control, and measuring elements are presented in Fig. 1. Table 2 presents individual elements of the stand with a description of the most important parameters and Fig. 2 shows the hydraulic scheme of the test stand. Table 3 shows the designations of signals sent to the computer.

The test stand is powered by a hydraulic unit consisting of an asynchronous motor (a) and a fixed displacement pump (b) with a pressure relief valve (c). The working liquid is taken from tank (d). It is equipped with a filling cap with a vent with a filter net (e). An oil-air cooler (f) is installed on the fluid reservoir to reduce the temperature and prevents the deterioration of the parameters of the working liquid. To the outlet of the oil-air cooler is mounted a return filter (g). Thanks to the use of a proportional directional valve (h), it is possible to smoothly change the speed and the direction of movement of the piston rods. Next to the pairs of actuators (Fig. 2), there are two throttle-check valves (i) for manual adjustment of the oil flow between the two sides of the system. These valves are used to change the load on hydraulic cylinders. The working fluid used in the described hydraulic system is HL – 46.

The main part of the power supply control system is the Mitsubishi FX5U controller (j). The use of such a solution allows for long-term tests without the connection of computer equipment. The hydraulic system can be controlled using the GT2107-WTSD touchscreen (k). The stand is also equipped with a set of sensors that enable real-time measurement of selected parameters or their recording using a data acquisition card (l) and a PC. Pressure transmitters (m) provide pressure control at key points in the hydraulic system to change the position of the directional valve. The use of additional manometers (n) enables visual control of the pressure level in three lines, independently.

The applied force sensor (o) enables the measurement of the compressive force between alternating pairs of actuators. The force sensor is attached on one side to an intermediate element, which is screwed to two piston rods moving in the same direction. On the other hand, the intermediate element remains free and rests only on the piston rods of the two actuators, which move counter-moving to the former ones. This is possible because the force between the counter-moving piston rods is always positive, compressing the sensor in both directions. Free mounting of the sensor on one side prevents the occurrence of additional side loads and stiffening of the system, which is an unfavorable phenomenon and could interfere with the measurement. The temperature sensor (p) measures the temperature of the oil in the tank. The use of an inductive sensor (r) allows to control the number of cycles of the system operation. An additional element is also a position sensor (a set of a enco-meter with an incremental encoder) (s), whose main task is to control the position of the piston rod.

Due to the presence of both current and voltage outputs, current-to-voltage converters were used to connect the sensors with the measurement card.

By a slight modification of the working table, it is possible to test new types of actuators in any configuration, provided that the diameter of the mounting flange does not exceed 75 mm and the stroke does not exceed 150 mm. The system generates a flow of 1.7 lpm (resulting from the parameters of the pump used). The designed system can perform about 6, 7 cycles per minute, while the maximum pressure is 12 MPa.

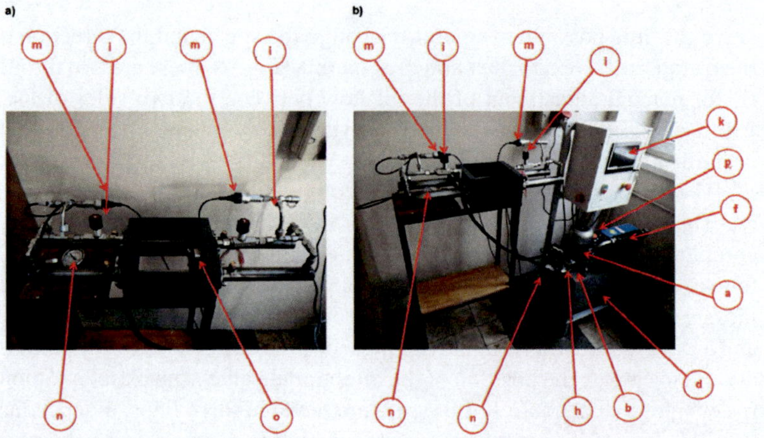

Fig. 1. Photo of the test stand [a) set of actuators, b) test stand with control system].

Table 2. Main elements of the test stand.

Symbol	Element	Type/Model	Main parameters
a	asynchronous motor	T0710402	power: 0,75 kW rotational speed: 1390 rpm (50 Hz)
b	gear pump	MC4	unit capacity: 1,2 cm^3 working pressure: 250 bar
c	pressure relief valve	V1B	maximum pressure: 190 bar
d	oil tank	BEK 20	volume: 26 l
e	filler cap with vent and filter net	SY-06S	diameter of the required filling hole: 32 cm
f	oil-air cooler	HY01001	capacity: 0,4 l flow rate: 10 ÷ 40 lpm
g	return filter	SH 63061	flow rate: 60 lpm
h	proportional directional valve	XD3C03T2G	flow rate: 10 lpm voltage: 24 V maximum current: 0,88 A
i	throttle check valve	DRV-08-01.5/0	maximum flow rate: 50 lpm maximum pressure: 350 bar
j	controller PLC	FX5U	digital inputs: 16 digital outputs: 16 analog inputs: 4
k	touch screen	GT2107-WTSD	screen size: 7 in. screen resolution: 800 × 480
l	data acquisition card	USB-6212	analog inputs: 16 analog outputs: 2 digital outputs: up to 32

(*continued*)

Table 2. (*continued*)

Symbol	Element	Type/Model	Main parameters
m	pressure transmitter	A-10	maximum pressure: 250 bar output signal: 4 ÷ 20 mA accuracy: ± 0,5%
n	manometer	–	maximum pressure: 160 bar
o	force sensor	KMM40E	maximum force: 2 kN output signal: 0,5 ÷ 4,5 V accuracy: ±0,5%
p	temperature sensor	22DT-12P	temperature range: 0 ÷ 160 °C output signal: 0 ÷ 10 V accuracy: ±0.5 °C
r	proximity inductive sensor	LM30-34025PA-L	measurement range: 25 mm output signal: PNP NP 200 mA
s	position sensor	90.1404.FX+58-11112-2000	measurement range: 4 m accuracy: ±0,06 mm/rev number of pulses of encoder: 2000 output signal of encoder: A/A-B/B-0/0 resolution of encoder: 5

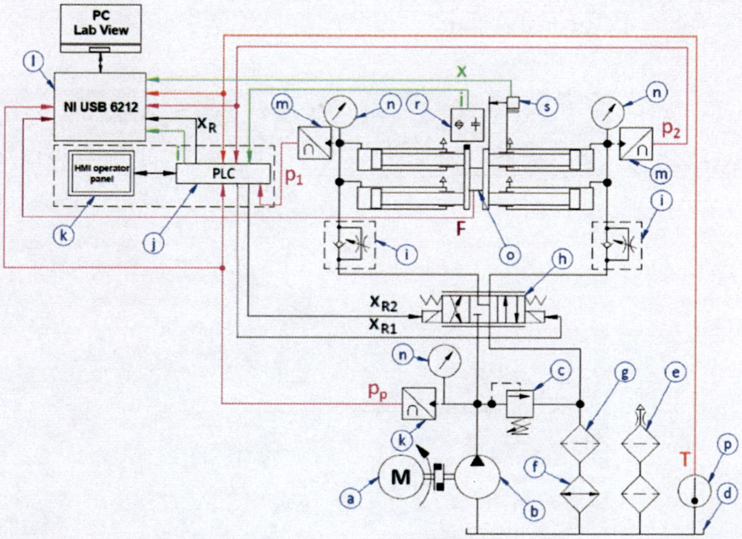

Fig. 2. Hydraulic scheme of the test stand.

4 Possibilities of Research

For research purposes, 5 sets of actuators were prepared. Several cycles must be run for each set. Before the actual start-up of the stand, material tests of the obtained layers must be performed. The layers were examined using a microscope, mechanical parameters

Table 3. Designations of signals sent to the computer on the hydraulic scheme.

Symbol	Designation
T	temperature
p_1, p_2	pressures in lines 1, 2
p_p	pump pressure
x	displacement of the actuators
i	number of cycles
x_{R1}, x_{R2}	repositioning of proportional directional valve
x_R	repositioning of proportional directional valve - transferred to the data acquisition card

(e.g. roughness) were determined and specially prepared test discs were tested on a tribometer, and their wear profile was determined. In addition, the surface profile of all pistons and piston rods used in the tests should be also checked on the profilometer (Fig. 3). The test consisted of moving the measuring tip in contact with the sample and determining a graph corresponding to the surface profile, including parameters such as surface roughness. The path of the tip was 4 mm long and several measurements were taken for each sample. Figure 4 presents an example of the roughness plot for the end face of a DLC-coated piston.

Fig. 3. The method of measurement on the profilometer.

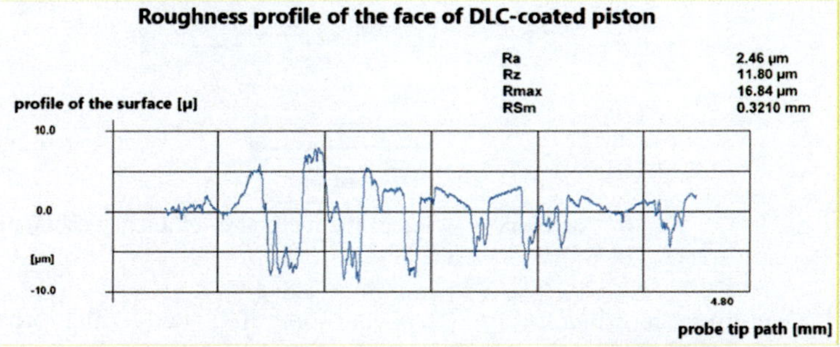

Fig. 4. Exemplary roughness profile of DLC-coated piston.

After a specified number of work cycles, the mechanical and material properties of the actuator components should be tested again. It is necessary to re-examine the modified surfaces under the microscope and on the profilometer. Carrying out material tests will help to investigate how surface properties have changed after exposure to external factors (for example hydraulic oil, high pressures or various temperatures). Therefore, it will be possible to compare it with the condition of the surface before the station was launched - determining whether there is corrosion wear, significant deterioration of tribological parameters, or damage to the modified layers.

For performance measurements, the stand has been slightly modified. The efficiency will be determined for each actuator under both passive and active load using pressure and force sensors. These modifications to the test stand concern the separation of a system containing only two actuators (Fig. 5). After this change, the system works identically to the previous one, only for two actuators. This is possible thanks to the use of a different connector between the actuator piston rod and the force sensor. Additionally, two cut-off valves are installed, which close the liquid supply to the non-working actuators. This doubles the speed of the actuators because the same amount of fluid generated from the pump flows into one actuator instead of two.

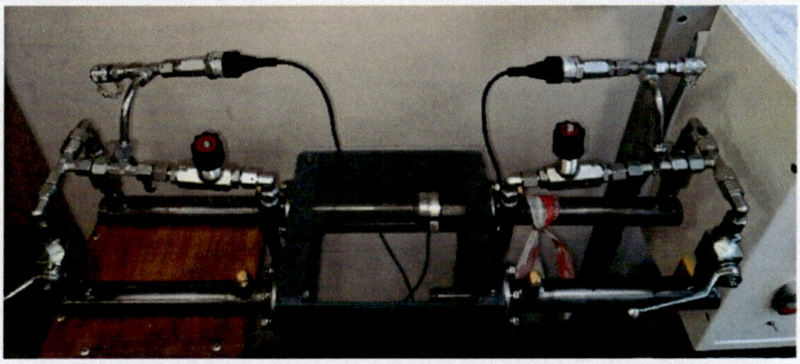

Fig. 5. Modified test stand used for efficiency measurements.

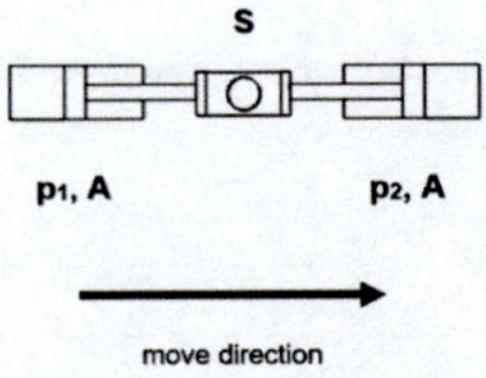

Fig. 6. Auxiliary drawing of the method of efficiency calculation

Denoting, according to Fig. 6, by *S* the force indicated by the force sensor between the piston rods, by p_1, p_2 the pressures in the piston chamber of actuator respectively and by *A* the working area of the actuator, the efficiency under passive load η_p and active load η_a can be determined from the formula (1).

$$\eta_p = \frac{S}{p_1 \cdot A}, \quad \eta_a = \frac{p_2 \cdot A}{S} \tag{1}$$

The exemplary runs of pressures during efficiency tests are presented in Fig. 7 and Fig. 8 (the areas of pressure peaks when reaching the endpoints of the piston rod have been omitted). Figure 9 and Fig. 10 show the efficiency diagrams for the pressures obtained.

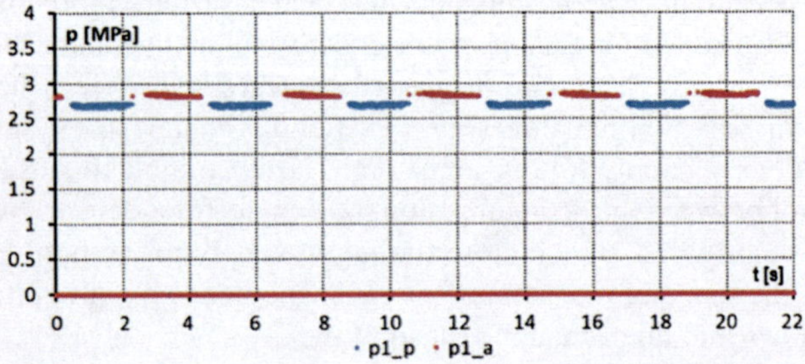

Fig. 7. Exemplary pressures (*p1_p* - pressure during passive operation, *p1_a* - pressure during active operation) courses.

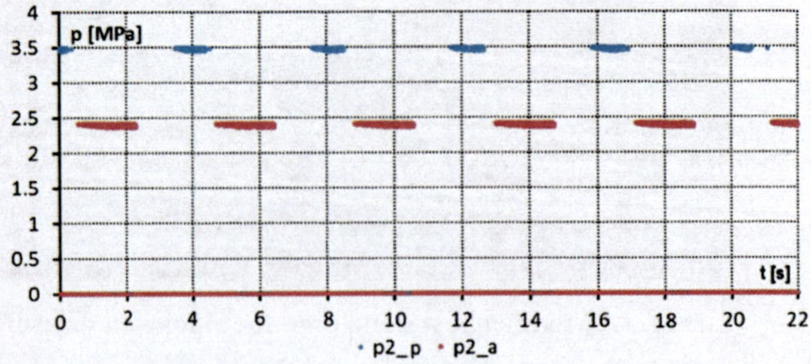

Fig. 8. Exemplary pressures (*p2_p* - pressure during passive operation, *p2_a* - pressure during active operation) courses.

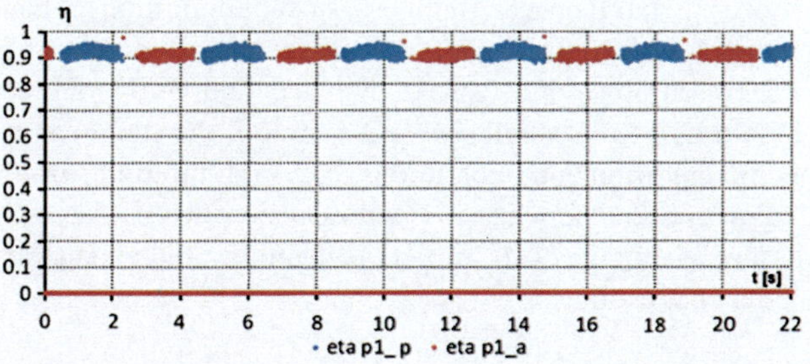

Fig. 9. Exemplary efficience values (*eta p1_p* – efficiency during passive operation, *eta p1_a* - efficiency during active operation) courses.

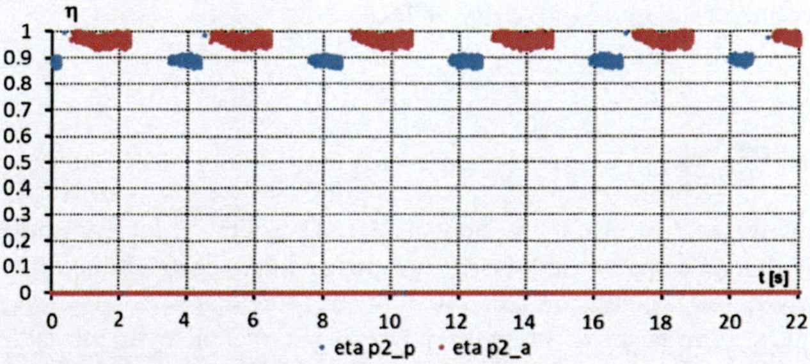

Fig. 10. Exemplary efficience values (*eta p2_p* – efficiency during passive operation, *eta p2_a* - efficiency during active operation) courses.

The above graphs will be the basis for answering the question of which actuator modification will be better - however, this requires long-term trials and the results will be published in future papers.

After a series of work cycles, fatigue wear tests of the actuators can also be performed. The expected results are to obtain layers performed with at least the same good parameters as those used in the reference actuator. It is important to maintain good tightness of the actuators and favorable tribological parameters. Additionally, a low wear ratio is important.

5 Conclusion

The presented test stand was built in order to examine the legitimacy of the use of selected modern modifications of the surface of pistons and piston rods in hydraulic systems. The main goal is to find a manufacturing technology that allows the elimination of coatings that are harmful to the environment and people used on the elements of hydraulic actuators. The stand enables simultaneous testing of four hydraulic cylinders and provides them with identical working conditions. The scope of tests of actuators working at the specially designed stand is very wide. Nowadays it is possible to test different mechanical or operational properties both during the system operation (e.g. pressure, temperature, load conditions, efficiency) and after a specified number of cycles of operation (e.g. fatigue wear, tightness, surface condition). Thanks to the use of devices counting the number of cycles, it is possible to test the actuators following the standards in variable conditions of load, speed, displacement, temperature, etc. [e.g. standard ISO 10100:1990]. Due to its small size and mobility, the stand can be transported to places with specific environmental conditions, e.g. high humidity or low temperature. This is of particular importance when it is necessary to test actuators in special applications. An important advantage of the project is the unlimited expansion possibilities with new components or sensors.

Acknowledgement. The first author would like to acknowledge the financial support from the project co-financed by the European Union under the European Social Fund as a part of the Operational Program: Knowledge, Education, Development, project No. POWR.03.02.00-0-I042/16-00 and from internal grant from the Young Researchers' Fund of the Faculty of Mechanical Engineering of the Lodz University of Technology.

References

1. Skowrońska, J., Kosucki, A., Stawiński, Ł.: Overview of materials used for the basic elements of hydraulic actuators and sealing systems and their surfaces modification methods. Materials **14**, 1422 (2021). https://doi.org/10.3390/ma14061422
2. Uzny, S., Kutrowski, Ł.: Strength analysis of a telescopic hydraulic cylinder elastically mounted on both ends. J. Appl. Math. Comput. Mech. **18**, 89–96 (2019). https://doi.org/10.17512/jamcm.2019.1.08
3. Solazzi, L.: Stress variability in multilayer composite hydraulic cylinder. Compos. Struct. **259**, 113249 (2021). https://doi.org/10.1016/j.compstruct.2020.113249
4. Flitney, B.: Alternatives to chrome for hydraulic actuators. Seal. Technol. **10**, 8–12 (2007). https://doi.org/10.1016/s1350-4789(07)70460-9
5. Siwulski, T.: Comparative studies of the dynamic response of hydraulic cylinders with different hydraulic supply systems design. In: Stryczek, J., Warzyńska, U. (eds.) Advances in Hydraulic and Pneumatic Drives and Control 2020. NSHP 2020. LNME, pp. 301–310. Springer, Cham (2021). https://doi.org/10.1007/978-3-030-59509-8_27;
6. Adams, S., Beling, P., Farinholt, K., Brown, N., Polter, S., Dong, Q.: Condition based monitoring for a hydraulic actuator. In: Annual Conference of the Prognostics and Health Management Society 2016 (2016)
7. Jakubowski, A., Kubacki, A., Rybarczyk, D.: Design of the test stand for hydraulic active heave compensation system. Arch. Mech. Technol. Mater. **37** (2017). https://doi.org/10.1515/amtm-2017-0012;
8. Zhou, Y., Zhou, X.: Modeling and controller design for an experimental test bench for aircraft actuators. Adv. Mech. Eng. **10**(12) (2018). https://doi.org/10.1177/1687814018815362

Energy–Saving in Hydraulic Drives in Experimental Approach and Biomimetric Similarity

Mykola Karpenko[1(✉)], Michał Stosiak[2], Olegas Prentkovskis[1], and Paulius Skačkauskas[1]

[1] Vilnius Gediminas Technical University, Plytinės st. 27, 10105 Vilnius, Lithuania
mykola.karpenko@vilniustech.lt

[2] Wrocław University of Science and Technology, Łukasiewicza st. 7/9, 50-371 Wrocław, Poland

Abstract. The current article presents the research results of the proposed energy-saving way in hydraulic drives based on a biomimetic approach. Experimental measuring for investigation energy consumption by different types of high-pressure hoses and its influence on the inside fluid flow are presented. The methodology presented in this paper for experimental measurement is based on fluid pressure measuring and high-pressure hose vibration measuring analysis. In the research, by experimental measuring, different braided high-pressure hoses, used in modern hydraulic drives, were compared. The research was performed to describe a high-pressure hose behavior dependency on the fluid flow inside and its influence on the energy parameters of the system. The proposed concept for energy-saving in a hydraulic drive is based on biomimetic approaches. In the research it is disclosed that, by using different types of high-pressure hoses, a reduce of power losses and an increasing hydraulic drive efficiency can be achieved, by replacing one type of a high-pressure hose with another type, during machinery hydraulic drive maintenance or on design stage.

Keywords: Energy-saving · pressure pulsation · composite · high-pressure hose

1 Introduction

Based on studies in the industry, fluid power drives play a crucial role in the acquisition of raw materials for the global economy. This is due to the fact that a majority of mobile machinery used in sectors such as mining, transportation, agriculture, and aviation employ these drives. [1]. Machines' operational equipment is powered by hydraulic drives through power transmission, which can use up between 30% to 90% of the primary combustion engine or electric motor's energy during the working cycle, as noted by [1]. However, as highlighted by [2], the requirements for hydraulic drives are continually growing, with productivity, environmental friendliness, reliability, and low energy intensity being the main ones. The hydraulic drives' efficiency value is low at around 60%, according to [3], and the European Commission strategy emphasizes the use of

J. Stryczek and U. Warzyńska (Eds.): NSHP 2023, LNME, pp. 269–279, 2024.
https://doi.org/10.1007/978-3-031-43002-2_25

energy-saving technology in all types of transport and machines (urban and industrial) applications [4]. Consequently, investigating ways to enhance the efficiency of hydraulic drives using energy-saving technology is both pertinent and essential. As pointed out in [5], energy-saving technologies not only increase efficiency and reduce energy consumption but also increase reliability, simplify the design, and significantly increase the service life of components, working fluids, and seals, which is beneficial for hydraulic drives.

Reviewing modern energy-saving systems and technologies in hydraulic drive, [6] indicates that reducing dynamic loads (pressure pulsations) in a system is one of the most feasible ways to reduce power consumption and increase hydraulic drive efficiency. According to [7], solutions for reducing dynamic loads typically involve installing active damping elements into the hydraulic drive. [8] notes that hydraulic accumulators and different fluid flow dampers are commonly used to reduce dynamic loads in hydraulic drives. However, in some cases, integrating damping elements into the hydraulic drive may not be feasible or rational, as indicated by [9]. For example, hydraulic accumulators and other pulsation dampers may be too large to integrate into mobile machine hydraulic systems, may not work over the required frequency range or even provide negative effect, according to [9]. Pumps with integrated pulsation dampers, as suggested by [10], are suitable for use in mobile systems, but their integration may prove difficult due to the dampers requiring a relatively large space. [11] also points out that active vibration compensation systems' disadvantage is that they add energy to the hydraulic system via the actuator, which can worsen the system's stability. Therefore, there is a need to damp pressure pulsation inside the hydraulic drive by finding other passive damping components, as [11] suggests.

According to [12] and [13], by adopting biomimetic approaches, new strategies for dampening pressure pulsations in hydraulic drives can be discovered. The authors suggest that since the delivered flows by pumps found in nature are mostly pulsatile, there is potential in applying the nonlinear viscoelasticity of vertebrates' circulatory system to improve existing technical dampers or find novel biomimetic damping solutions. [12] compared the damping principle of vertebrates' circulatory system with existing technical solutions and suggested that the nonlinear viscoelasticity of the vessel walls of arteries offers a high potential for improving existing technical dampers or finding novel biomimetic damping solutions, according to [14]. According to [15], the arterial walls of the major vessels passively expand and elastically recoil, resulting in a reduction in the magnitude of pressure pulsations in the circulatory system and a smoothing of blood flow throughout the system. The vessel walls are composed of elastin and collagen fibers and smooth muscle cells, representing a composite material, as noted by [16]. According to [15], part of the pulse energy is dissipated by the viscoelastic component of the vessel walls, about 15% to 20% strain energy lost during each cycle. The underlying damping principle is comparable to one of the expansion hoses and inline hydraulic accumulators.

Nonetheless, achieving precise dynamic analysis of hydraulic driving processes can be challenging due to the nonlinear nature caused by the compressibility of oil and the nonlinear characteristics of the pump and hydraulic high-pressure hoses. Hence, it is crucial to conduct experimental investigations to enhance the efficiency of hydraulic drives using various types of high-pressure hoses. In this paper, the possibility of limiting the amplitude of the pressure pulsation and energy loss in flexible hoses depending on the structure of the hose wall is indicated.

2 Concept Idea

Biomimetics is the process of using biological systems as models for human-made technologies. In hydraulic drive science, biomimetics is being applied to develop new designs, materials, and techniques that are more environmentally friendly and sustainable (Fig. 1), [17]. Figure 1 shows the analogy between biological and hydraulic components. In developing their concepts, designers look to the most efficient natural solutions. Just as in a hydraulic system the source of pulsatile flow is the positive displacement pump, in the blood circulation system of many living organisms the source of pulsatile flow is the heart. For typical hydrostatic systems, the basic component of the pulsation spectrum is in the range 150–350 Hz, while for living organisms it depends mainly on the size of the organism, with the general rule being that the smaller the organism, the higher the blood pressure pulsation frequency (e.g. for a human about 1 Hz, for an elephant about 0.4 Hz, for a mouse about 12 Hz). This pulsating flow of the working fluid (hydraulic fluid, blood) and the impedance of the hydraulic or blood system contributes to the pressure pulsation and therefore to the time-varying, harmonic loading of the walls of the hoses or veins. An analogy is also noticeable in the area of the elements transporting the working fluid in the system from the flow source to the consumers. In the hydraulic system, these are the pipes (rigid or flexible), and in the circulatory system, the veins, aorta and arteries. Just as in the circulatory systems of living organism's veins are made up of three layers (outer membrane, endothelium, muscular layer), designers of high-pressure hydraulic hoses are turning to newer and newer materials and conductor designs. The new hydraulic lines are made of composite materials and often have three layers. They transmit high pressures while at the same time ensuring that fluid pressure pulsations are reduced compared to rigid hoses. While in hydraulic systems, fluid is transported to hydraulic receivers (motors, cylinders), in biological systems, the analogy is with muscles.

The concept aims to enhance the efficiency of hydraulic drives through experimental research and a biomimetic approach that involves utilizing various types of high-pressure hoses. The concept also involves employing a biology analogy to dampen pressure pulsations in hydraulic drives.

This concept of utilizing the nonlinear viscoelasticity of the high-pressure hose walls for improving the dampers in hydraulic drives is suggested to have a high potential based on biomimetic arterial and venous damping solutions. The passive expansion and elastic recoil of the high-pressure hose walls should reduce the magnitude of pressure pulsations and smooth fluid flow throughout the hydraulic system. Like vessel walls, high-pressure hose walls are composed of multiple layers, including rubber and fabric or steel braid. The proposed concept suggests that the viscoelastic component of the high-pressure hose walls should dissipate part of the fluid pulse energy. By using flexible instead of rigid hoses in hydraulic systems, the capacitance c_K of the system (hoses) is increased by reducing the value of the equivalent bulk modulus of elasticity B_Z. The relationship

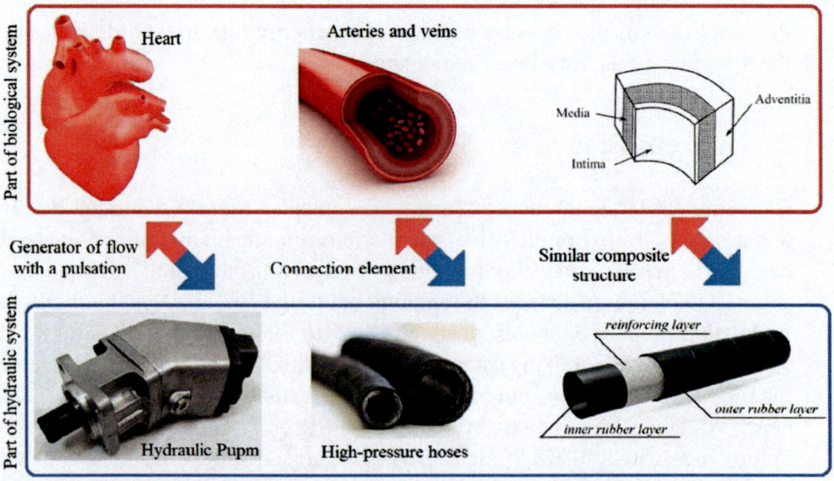

Fig. 1. Hydraulic drive energy-saving concept based on biomimetic approach.

between these parameters is expressed by linearized relationships:

$$B_z = \frac{\Delta p}{\frac{\Delta V}{V_0}}[Pa] \quad (1)$$

$$c_k = \frac{V}{B_z}\left[\frac{m^5}{N}\right] \quad (2)$$

where: Δp - change in pressure, ΔV - change in volume due to pressure increment, V_0 - initial volume of liquid enclosed in the hose.

However, due to the strong nonlinearity of the hydraulic driving process caused by the compressibility of oil and the nonlinear characteristics of the pump and hydraulic pipeline, an accurate dynamic analysis may be challenging. Therefore, experimental investigation is suggested to increase the efficiency of hydraulic drives using different types of high-pressure hoses.

3 Describing of Research Objects Parameters and Test Bench

One option for the high-pressure hoses in this concept is to use a composite material made from rubber (Fig. 2), it's an important structural part of any hydraulic drive for connection all elements in one operation system. In current research, for proving concept idea a comparing analysis between using one braid high-pressure hoses of different types was used: one braid fabric reinforced hydraulic hose (*TE*) [18]; standard one steel braid reinforced hydraulic hose (*SN*) [19]; compact one steel braid reinforced hydraulic hose (*SC*) [20]. *SC* steel wire reinforced hydraulic hose is almost similar to *SN* steel wire reinforced hydraulic hose in structure, but difference is that it has smaller outside diameters, which solves the problem of the compact and narrow installation space of some equipment piping.

The high-pressure hose with a diameter of the conditional pass $1/2''$ was accepted for research. The material of high-pressure hoses layers presented in Fig. 2. The property of each hoses layer can be found in [22]. Main geometrical parameters of used high-pressure hoses in the research is presented in Fig. 3.

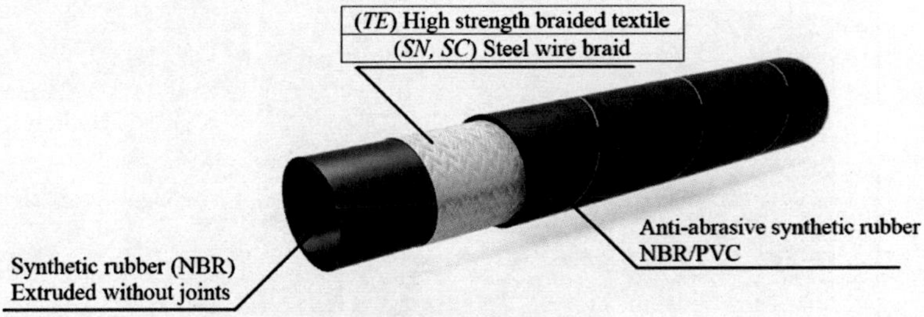

Fig. 2. Construction and layers material of high-pressure hoses

Type	*ID*	*OD*
TE	1/2" or 12.7 mm	17.7 mm
SC	1/2" or 12.7 mm	18.9 mm
SN	1/2" or 12.7 mm	20.6 mm

Fig. 3. Main geometrical parameters of used hoses in research

The first step in establishing the evidence of the proposed concept of energy-saving by using different types of high-pressure hoses was to conduct experimental research. The experimental research involved measuring and analysing the fluid pressure drop inside the high-pressure hoses, as well as conducting vibration analysis of the hoses depending on the fluid flow. The research was conducted using a two-sample measurement design and based on the single-sample statistical method for estimating uncertainty in repeated measurements of data processing, as stated in [21]. Test bench for the experimental research of fluid pulsation inside a high-pressure hose and its deformation shown in Fig. 4.

The high-pressure hoses standards presented are internationally certified and commonly utilized in hydraulic drives for mobile machinery. In the current research, high-pressure hoses with a length of 1 m were utilized, the nominal pressure on the inlet of each high-pressure hose ~3.45 MPa with a fluid flow rate ~40 l/min produced by hydraulic pump. The temperature of the working fluid was stabilised by the cooler.

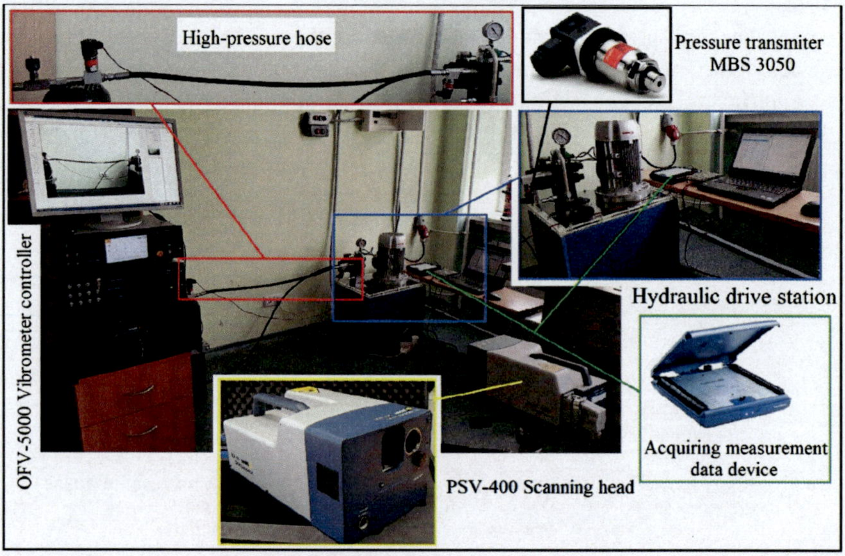

a

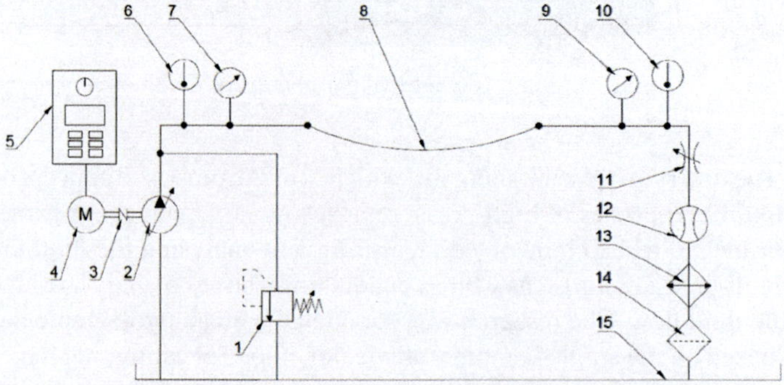

b

Fig. 4. The performing of tests: a – view of test bench for measurements; b – scheme of hydraulic test bench: 1 - safety valve; 2 – pump; 3 - flexible coupling; 4 - electric motor; 5 - electric motor control cabinet; 6 - temperature sensor on inlet; 7 - pressure sensor on inlet; 8 - tested hose; 9 - pressure sensor on outlet; 10 - temperature sensor on outlet; 11 - adjustable throttle valve; 12 - flow meter; 13 – cooler; 14 - return filter; 15 - oil tank.

4 Experimental Tests Result

The experimental measurement included several measuring of fluid pressure at the inlet and outlet of the high-pressure hose, as well as several readings of the displacement and velocity of the high-pressure hose surface and its frequency response. In order to reduce measurement errors, the average results of multiple readings were presented. The velocity was measured using a PSV Sensor Head on the surface of the high-pressure hose, as shown in Fig. 5. The fluid pressure pulsation was measured at the inlet and outlet of the high-pressure hoses, as shown in Fig. 6. The course of the pressure pulsation at the beginning of each test hose was identical and is shown in Fig. 6 - green line.

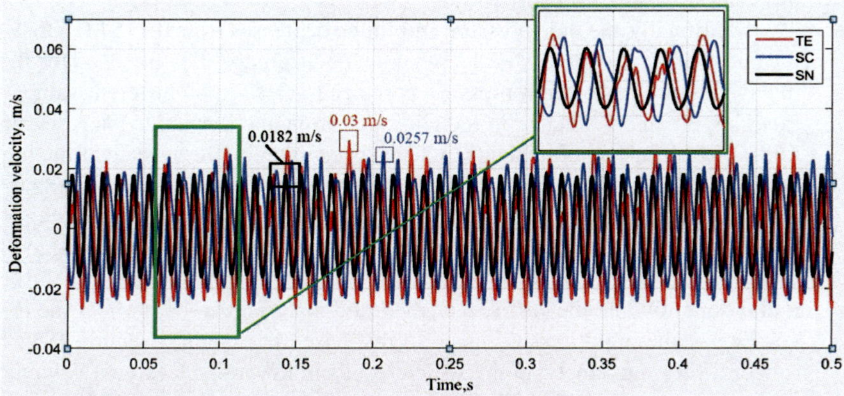

Fig. 5. The measurement of hoses surface deformation velocity (in a middle of hose): TE - one braid fabric reinforced hydraulic hose, SN - standard one steel braid reinforced hydraulic hose, SC - compact one steel braid reinforced hydraulic hose.

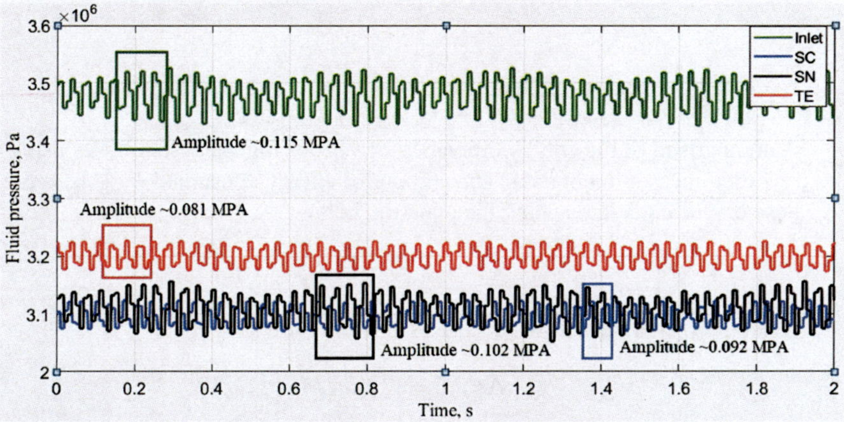

Fig. 6. The measurement of fluid pressure pulsation (in the inlet and outlets of hoses): TE - one braid fabric reinforced hydraulic hose, SN - standard one steel braid reinforced hydraulic hose, SC - compact one steel braid reinforced hydraulic hose.

Measurements indicate that there is a nominal fluid pressure drop of approximately 0.332 MPa for *SN* and 0.381 MPa for *SC* and 0.25 MPA for *TE* between the inlet and outlet of the high-pressure hoses and the calculations indicates laminar flow inside tested hoses, $R_e \approx 256$ and can be calculated using Hagen-Poiseuille formula:

$$\Delta p = \frac{128 \cdot Q \cdot \mu \cdot l}{\pi \cdot d^4} \tag{3}$$

where: Q – flow rate [m^3/s], μ – fluid dynamic viscosity [Ns/m^2], l – hose length [m], d – inner hose diameter [m].

Additionally, the fluid pressure amplitude decreases from 0.115 MPa to 0.102 MPa at the outlet of *SN*, to 0.092 MPA at *SC* outlet and from to 0.081 for *TE*. This demonstrates that different types of high-pressure hoses characterised by different values of the bulk modulus B can be effective in reducing pulsation and dynamics loads within hydraulic lines. Measurements obtained from a laser scanning on the surface of the high-pressure hoses reveals an amplitude of displacement velocity of 0.0182; 0.0257 and 0.0182 (m/s) on *SN*, *SC* and *TE* hoses, respectively. The graph illustrates that the standard version *SN* of the high-pressure hose has lower displacement velocity compared to the compact version *SC* and version with textile cord *TE*. The thickness of the rubber layers in the radial direction and the stiffness of the braid material can both affect the displacement velocity of the high-pressure hose. In this case, the standard version (*SN*) has thicker rubber layers and a metal braid, which results in lower displacement velocity compared to the compact version (*SC*) and the version with a textile cord (*TE*) that have thinner rubber layers and a more stiffening braid material.

It's important to note that the pressure drop and fluid pressure amplitude values may vary depending on the specific hydraulic system and application. These values should be considered as indicative results based on the measurements performed in the presented study.

5 Energy-Saving Research

A comparison of the energy efficiency of various high-pressure hoses using previously recorded measurement data. The amount of energy consumed $E_{con.}$ by a high-pressure hose can be calculated using the equation below:

$$E_{con.} = \Delta P_i(t) \cdot Q_i(t). \tag{4}$$

where, $Q_i(t)$ – a fluid flow, m^3/s; $\Delta P_i(t)$ – pressure losses, Pa:

$$\Delta P_i(t) = P_{inlet}(t) - P_{outlet}(t). \tag{5}$$

where $P_{inlet}(t)$, $P_{outlet}(t)$ – fluid pressure at the inlet and outlet hose, Pa.

Energy chart have been created based on the archived results and are presented in Fig. 7. These charts illustrate the energy transformation visually and quantitatively during the use or replacement of high-pressure hoses in mobile machinery hydraulic drives. It should be noted that these results were obtained for a high-pressure hose length of 1 m. The values of the energy consumed in each hose were determined by multiplying the flow rate and the measured value of the pressure difference at the beginning and end of the hose. Relationships (4) and (5) were used for the detailed descriptions in Fig. 7. The

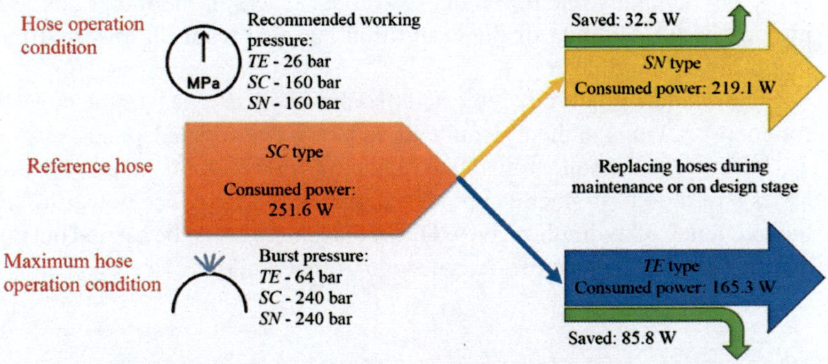

Fig. 7. The energy flow chart from comparing analysis: SC hose type – reference, SN and TE hose type – comparison.

pressure loss values for each hose can be read off the pressure time courses shown in Fig. 6.

Based on the experimental results, it was found that the most efficient option for energy saving in the presented hydraulic drive is using TE hose type. This type of high-pressure hose can save around 53.8 W and 85.8 W compared to SN and SC types, respectively (by 1-m hose length). This translates to savings of 24.5% and 34.3%, respectively. If the pressure in the hydraulic drive is over 26 bar, using SN type can save around 32.5 W or 12.91% on a length of one meter, compared to using SC hose type.

6 Conclusions

This study involved conducting experimental measurements to compare the performance of a textile braid high-pressure hose with one or two metal braid high-pressure hoses. The methodology used in this paper relied on fluid pressure pulsation measurements, vibration measurements of the high-pressure hose. The proposed concept for energy-saving in a hydraulic drive is based on biomimetic approaches. It was shown that by using flexible hoses with different wall structures, the amplitude of pressure pulsations in the hose can be influenced. Changing the wall structure of a hose leads to a change in the value of the bulk modulus of elasticity, which results in a change in the value of capacitance. It should be noted that a decrease in the value of the bulk modulus of elasticity leads to an increase in capacitance and, at the same time, to a decrease in the amplitude of pressure pulsation in the duct. However, an increase in capacitance makes the hydraulic system less rigid, 'spongy', which leads to a reduction in the natural frequency of the hydraulic receiver. This must be taken into account when analysing the dynamic states of a hydraulic system with a loaded hydraulic receiver.

Based on the measurements, it seems that the standard version (*SN*) of the high-pressure hose has better performance in reducing fluid pressure pulsation and dynamics loads within hydraulic lines, as it shows a smaller fluid pressure drop and lower fluid pressure amplitude at the outlet compared to the compact version (*SC*). At the same time the best energy-saving characteristics can be achieved by using version with textile cord

(*TE*). This is likely due to the thicker rubber layers in the radial direction, which may provide better damping of fluid pulsation but also result in lower stiffness and higher flexibility of braid layer.

The results of the study indicated that the *TE* hose type was the most effective option for energy savings in the hydraulic drive being investigated.

Ultimately, the study found that replacing high-pressure hoses with different types of hoses based on biomimetic approaches could help to reduce power losses and increase the efficiency of hydraulic drives. This replacement could be carried out during hydraulic drive maintenance or at the design stage of machinery.

References

1. Patrosz, P., Bąk, M., Załuski, P., Śliwiński, P., Karpenko, M.: Methodology of experimental research on efficiency of hydro-mechanical automatic gearbox. In: Prentkovskis, O., Yatskiv (Jackiva), I., Skačkauskas, P., Maruschak, P., Karpenko, M. (eds.) TRANSBALTICA XIII: Transportation Science and Technology. TRANSBALTICA 2022. LNITI, pp. 393–403. Springer, Cham (2023). https://doi.org/10.1007/978-3-031-25863-3_37
2. Kanezawa, Y., Daisho, Y., Kawaguchi, T.: Increasing efficiency of construction machine by hybrid system. JSAE Annu. Congress 17–20 (2001)
3. Minav, T., Heikkien, J., Pietola, M.: Direct driven hydraulic drive for new powertrain topologies for non-road mobile machinery. Electr. Power Syst. Res. **152**, 390–400 (2017)
4. European Commission. European Energy and Transport Trends to 2030, Luxemburg (2003)
5. Zhao, K., Liu, Z., Yu, S., Li, X., Huang, H., Li, B.: Analytical energy dissipation in large and medium-sized hydraulic press. J. Clean. Prod. **103**, 908–915 (2015)
6. Karpenko, M., Bogdevičius, M.: Review of energy-saving technologies in modern hydraulic drives. Sci. Future Lith. **9**(5), 553–558 (2017)
7. Stosiak, M., et al.: Research of vibration effects on a hydraulic valve in the pressure pulsation spectrum analysis. J. Mar. Sci. Eng. **11**(2), 1–15 (2023)
8. Amirante, R., Distaso, E., Tamburrano, P.: Experimental and numerical analysis of cavitation in hydraulic proportional directional valves. Energy Convers. Manag. **87**, 208–219 (2014)
9. Kitajima, D., Machimura, H., Munakata, A., Nemoto, M., Yamauchi, H.: Fluid pressure pulsation damper mechanism and high-pressure fuel pump equipped with fluid pressure pulsation damper mechanism. US patent, US8366421B2 (2013)
10. Chai, L., Jiao, Z., Xu, Y., Zheng, H.: A compact design of pulsation attenuator for hydraulic pumps. In: Proceedings of IEEE International Conference on Aircraft Utility Systems (AUS), pp. 1111–1116 (2016)
11. Shen, W., Jiang, J., Su, X., Karimi, H.: Control strategy analysis of the hydraulic hybrid excavator. J. Franklin Inst. **352**(2), 541–561 (2015)
12. Bach, D., Schmich, F., Masselter, T., Speck, T.: A review of selected pumping systems in nature and engineering — potential biomimetic concepts for improving displacement pumps and pulsation damping. Bioinspir. Biomim. **10**(4), 1–28 (2015)
13. Bach, D., Masselter, T., Speck, T.: Damping of pressure pulsations in mobile hydraulic applications by the use of closed cell cellular rubbers integrated into a vane pump. J. Bionic Eng. **14**(4), 791–803 (2017)
14. Camasão, D., Mantovani, D.: The mechanical characterization of blood vessels and their substitutes in the continuous quest for physiological-relevant performances. A critical review. Mater. Today Bio **10**, 1–18 (2021)
15. Shadwick, R.: Mechanical design in arteries. J. Exp. Biol. **202**(23), 3305–3313 (1999)

16. Wolinsky, H., Glagov, S.: Structural basis for the static mechanical properties of the aortic media. Circ. Res. **14**(5), 400–413 (1964)
17. Barış, G., et al.: Investigation on flow and heat transfer of compact brazed plate heat exchanger with lung pattern. Appl. Therm. Eng. **175**, 115309 (2020)
18. European standard. EN 854:2015. Rubber hoses and hose assemblies. Textile reinforced hydraulic type. Specification (2015)
19. European standard. EN 853 1SN:2015. Rubber hoses and hose assemblies. Wire braid reinforced hydraulic standard type. Specification (2015)
20. European standard. EN 857 1SC:2015. Rubber hoses and hose assemblies. Wire braid reinforced hydraulic slimline type. Specification (2015)
21. Karpenko, M., Nugaras, J.: Vibration damping characteristics of the cork-based composite material in line to frequency analysis. J. Theor. Appl. Mech. **60**(4), 593–602 (2022)
22. Karpenko, M.: Investigation of energy efficiency of mobile machinery hydraulic drives. PhD dissertation, Vilniaus Gedimino technikos universitetas, p. 164 (2021)

Check for updates

Research on a Composite Micro-hose Subjected to Harmonic Loading. Experimental Approach

Michał Stosiak[1(✉)], Łukasz Przybylak[1], Adam Deptuła[2], Mykola Karpenko[3], Kamil Urbanowicz[4], and Paulius Skačkauskas[3]

[1] Wrocław University of Science and Technology, Łukasiewicza st. 7/9, 50-371 Wrocław, Poland
michal.stosiak@pwr.edu.pl
[2] Opole University of Technology, Prószkowska 76, 45-758 Opole, Poland
[3] Vilnius Gediminas Technical University, Plytinės st. 27, 10105 Vilnius, Lithuania
[4] West Pomeranian University of Technology in Szczecin, 70-310 Szczecin, Poland

Abstract. The article presents an original test rig for investigating the dynamic and static properties of hydraulic hoses. A composite micro-hydraulic hose with an inner diameter of 2 mm was taken as the test object. To carry out hydrodynamic tests in a composite hydraulic micro-hose, a test rig was built. The test stand consists of a hydraulic system whose purpose is to supply the test object with working fluid, and measurement paths that allow observation of hydrodynamic phenomena occurring in the tested hose. In addition, the test stand was equipped with a working fluid temperature stabilization system. The measured values during the measurement series were: pressure at the beginning and at the end of the tested hose (AC and DC components), temperature of the fluid inside the micro-hose at the beginning and at the end, vibrations in three directions at a selected point of the micro-hose, tension force of the micro-hose (DC and AC component), reaction in the supports (DC and AC component), fluid flow rate in the micro-hose, displacement of the movable support differential with the axial deformation of the micro-hose. In the research was determined the influence of the mean pressure value on the amplitude of micro-hose vibrations. Amplitude-frequency spectra were used to analyses the results obtained.

Keywords: Composite · micro-hose · pulsation · frequency · experimental approach

1 Introduction

Hydraulic systems are commonly found across a wide range of industries. They also play a special role 'in the mechanical engineering industry. Due to their numerous advantages, they are used wherever there is a need to use high power with a compact design [1]. At the same time, new trends in the construction of hydraulic components and systems have been recently noticed. One of the directions of development is the pursuit of miniaturization of hydraulic components. To adapt to the current market demand, hydraulics has formed a new sub-area, which is micro-hydraulics [2] As the name suggests, it is

J. Stryczek and U. Warzyńska (Eds.): NSHP 2023, LNME, pp. 280–290, 2024.
https://doi.org/10.1007/978-3-031-43002-2_26

characterized by the significantly small size of the components involved, compared to their conventional counterparts. Generally accepted that microhydraulics are characterized by small and very small flows, i.e. flows of less than 50 cm^3/s (3 dm^3/min). In the case of valves, the criterion size is the seat dimension, which is for micro-hydraulic components WN < 6. For pumps and displacement motors, the concept of displacement chamber volume per one full revolution of the shaft is used, which for micro-hydraulic displacement units should be less than 1.2 cm^3/rev. Micro-hydraulic systems are slowly replacing pneumatic and electromechanical systems, which can largely be explained by the fact that they have a large number of useful features, where some of the most important are [3]: small system dimensions; high power transmission and operational reliability; smoothness of movement and cleanliness of the propulsion; low noise level.

Another direction of development is the use of new materials that make it possible to reduce the weight of hydraulic components without changing values of operating parameters (operating pressures). New materials based on plastics and composites are used for the bodies of pumps, valves [4], on hydraulic cylinders [5, 6] or hydraulic lines. In addition to durability, the new materials are also required to be resistant to environmental factors. This is particular important for offshore platforms where hydraulic components are subjected to the action of salt sea water or water mist. [7]. Composite materials are also characterized by very high specific strength, in this respect the most basic composites outclass good steels and their alloys, in addition, their low fiber density contributes into low component weight [8, 9]. Other features of composite components that are also used in hydraulics are: high chemical insensitivity; damage to the element in the form of cracks in the matrix; high strength at low weight; the limit between brittle and plastic states is gradual at lower temperatures; often lower production costs, which results into lower component costs; easy manipulation of physical and mechanical properties; high corrosion resistance; excellent fatigue stability; vibration damping by composite fibers.

The high strength of the new materials and their lightness make them increasingly used for hydraulic components [10]. Composite micro hoses have appeared on the market for use as power hoses in lubrication systems, signal hoses in load-sensing systems or in control and measurement systems [11]. Changing the size of a hydraulic component is often not compatible with maintaining its physical-mechanical properties, and the next problems may involve phenomena that can occur at the microscale but are not noticeable at the macroscale. Therefore, to effectively exploit the capabilities of composite micro-hoses, it is necessary to know their properties and behavior during operation under quasi-steady and variable conditions. Typically, a composite hose is constructed of several layers, 3–6 layers of materials, using a composite fabric layer (such as aramid fibers) to transfer the load. The composite fabric layer is the most important structural factor of composite hoses because it is the layer where the hose gains strength while maintaining high flexibility.

The paper presents a unique experimental rig for studying hydrodynamic phenomena in composite hoses. Experimental results of flexural vibrations of a composite microhose for various average pressures inside the pipe are presented. The analysis of the dynamic behaviour of hydraulic composite components can be useful for designers and users of these components.

2 Object Under Study. Test Site

2.1 Description of the Test Object

The test object is a meter long 2020N-012V30 hydraulic microhose from Parker [12], it is a thermoplastic, three-layer composite hose. The hose consists of three layers. The inner layer is made of thermoplastic polyamide, which is a polymer with better physical and physicochemical properties than other materials used for the base layers of hydraulic lines. Polyamide is characterized by high mechanical strength, resistance to a wide temperature range, abrasion and tensile resistance, low coefficient of friction, UV resistance, broad chemical resistance, light weight, lack of electrical conductivity and thermoplastic. Another, middle layer is a high-strength synthetic composite fiber, which acts as a reinforcement for the hose. The layer of composite fiber significantly improves the physical and mechanical properties of the hose, so that the maximum working pressure of the hose is 50 MPa, and the minimum burst pressure oscillates around 200 MPa.

The last layer of the hose is made again of polyamide and is the outer layer of the hose, tasked to protect the hose from mechanical factors such as abrasion. Unlike the base layer of the hose, the polyamide surrounding the outer layer is punctured, resulting in the presence of a significant number of micro-holes. Water vapor, air and other gaseous products can get inside the tube and accumulate in the reinforcement area, creating areas of lower strength resistance. This phenomenon shows itself by means of the appearance of bubbles or blisters in the area. Puncturing the tube layer is very important, because puncturing the tube cover allows trapped gases and vapors to get out.

2.2 Description of the Test Stand

A test stand was configured for carrying out preliminary vibration tests of a composite hydraulic micro-hose (Fig. 1). This test stand consists of a hydraulic system designed to generate fluid flow rate and pressure in the tested object, and measurement tracks that allow observation hydrodynamic phenomena occurring in the tested tube.

Hydraulic system

The task of the hydraulic system, which was made specifically for the purpose of the conducted scientific tests, was to enable fully controlled flow of the working fluid through the test object. The hydraulic station makes it possible to control the value of the average pressure, flow rate and temperature of the working fluid. The hydraulic stand is shown in Fig. 2.

Driven by an electric motor (1), a multi-piston pump (2) feeds the system and the tested hose with mineral oil. Tested hose (7) is placed in the test stand (Fig. 2). The working fluid is supplied to the test hose by the overdriving the directional control valve (5). Adjustable throttling valves (6) and (8) are used to set the flow rate and average pressure in the test line in accordance with the test program. In addition, the system was equipped with a safety valve (3), oil filter (11). An important component of the system is the oil cooler (10), as it allowed to stabilize the temperature of the working fluid during the tests. The Manometer (4), located in the hydraulic power unit, was used to roughly control of the average pressure.

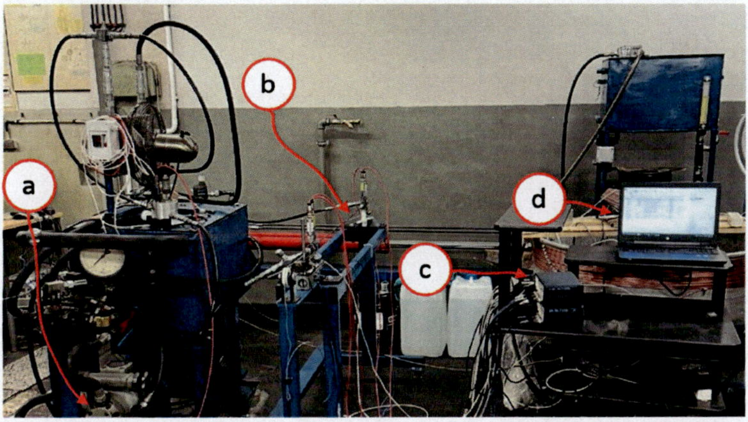

Fig. 1. Test stand: a - piston pump; b - tested hose; c – conditioners; d - measurement devices.

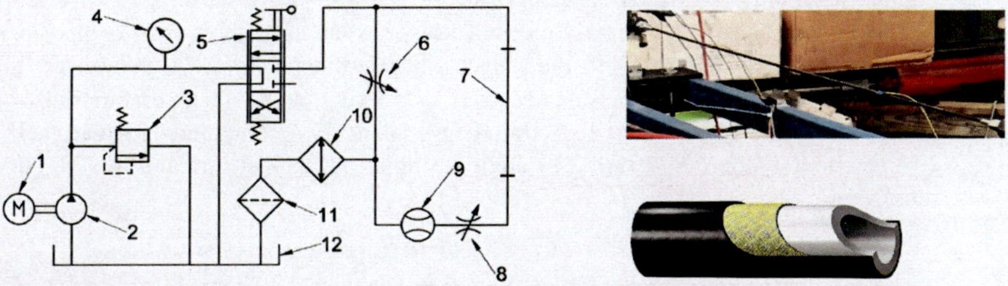

Fig. 2. Schematic diagram of an open hydraulic system: 1- motor; 2 – pump; 3 - safety valve; 4 - pressure gauge; 5 - hydraulic directional control valve; 6 - adjustable throttle valve; 7 - hose under test; 8 - adjustable throttle valve; 9 - flow meter; 10 – cooler; 11 – filter; 12 – tank.

Measuring lines

The schematic of the measurement path shown in Fig. 3. The task of the measuring part is to record hydrodynamic phenomena occurring during the test of the composite hose. The measuring tracks are divided into two segments according to their purpose. The first, provides us with the reading and recording of the hydrodynamic properties of the working fluid flowing through the test object, and the second - the response of the hose to the flow. Due to this solution, it is possible to observe the hydrodynamic phenomena of the fluid in the tested hose and see their effect on the test object.

The first segment consists of sensors for temperature (3, 13), static pressure (4, 12) and dynamic pressure (5, 11) of the fluid, they perform the function of receiving the basic physical parameters of the working fluid. They are located directly in front of and directly behind the test stand.

The second segment includes two temperature sensors (8, 10), which measure the external temperature of the test object at 250mm from the hose nut. A piezoelectric accelerometer (9) is located halfway along the test hose. An inductive displacement sensor (24) that measures the changes in the length of the hose under load (pressure inside the hose and pre-tension). The tested hose is preloaded through weights and the

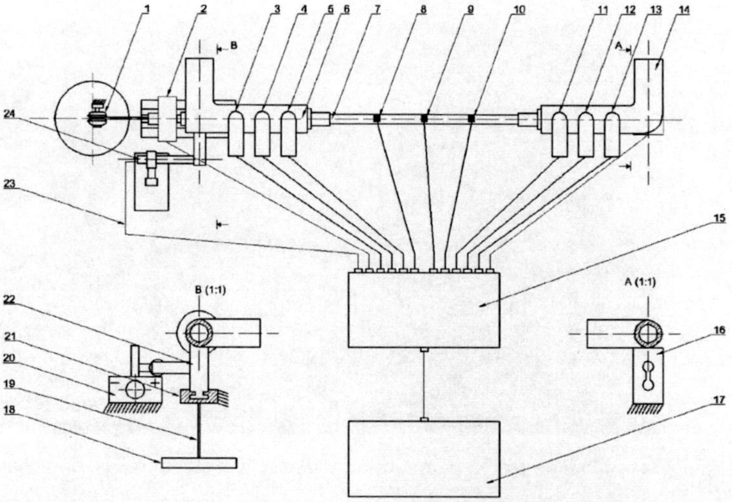

Fig. 3. Schematic of the measurement track: 1 – roller; 2 - force sensor; 3 - temperature sensor; 4 - static pressure sensor; 5 - dynamic pressure sensor; 6 - angle connector; 7 - composite micro hose under test; 8 - NTC thermistor; 9 - piezoelectric accelerometer; 10 - NTC thermistor; 11 - dynamic pressure sensor; 12 - static pressure sensor; 13 - working medium temperature sensor; 14 - angle fitting; 15 - signal conditioner; 16 - strain gauge beam; 17 – computer; 18 - load rack; 19 – wire; 20 - magnetic holder; 21 – rail; 22 - angle fitting stand; 23 - electrical hose; 24 - displacement sensor.

hose tension force is measured by the force sensor (2) and strain gauge beam (16). The force sensors measure the constant and variable components.

All measuring sensors were connected to two signal conditioners (15), which were connected to a computer (17). The task of the signal conditioners was to take input current, voltage and frequency signals and then give the proportional signals in output format. The data (after conversion from analog to digital) were sent to a computer, which recorded the results of the measurements and presented them in the form of graphs to make it easier to read and analyze the data.

3 Methodology and Findings

3.1 Methodology

The fundamental problem of hydrodynamic testing in a hydraulic microhose is the impossibility of locating measuring sensors inside hoses, that have such small nominal diameters that it is not feasible to locate the sensor inside in such a way that the sensor does not affect the characteristics of the tested object. One way to observe hydrodynamic phenomena in such small objects, is to locate measuring sensors directly in front of and directly behind the tested hose. This method, makes it possible to observe differences in the results, caused by physicomechanical phenomena, acting on the working fluid during its flow through the tested element.

The tests conducted on the vibration of the hose during hydrodynamic flow focused on three main measurement tracks used in the test stand. The first was the tension force, which was applied to the system to stabilize the test object and observe the change in stretch and deformation in the axis parallel to the direction of flow of the working fluid.

The second test track used an accelerometer to measure vibration, which allowed the determination of the vibration of the hose caused by the pulsatile flow of the working fluid. The amplitude-frequency spectrum made it possible to determine the components of the hose vibration spectrum and analyze the contribution of each component.

The third measurement track focused on measuring the axial displacement of the tested hose along its axis. This made it possible to record the axial displacement resulting from the pulsation of flow and pressure inside the tested hose. The use of an amplitude spectrum made it possible to determine the amplitude of the displacement pulsation, and the simultaneous use of an axial force sensor allowed the determination of the force causing this displacement.

The use of measuring devices, such as accelerometers or temperature sensors, directly on the surface of the tested hose allows to observe its motion and changes in internal energy. Additional force sensors in the axis of the hose, as well as in its base, allow monitoring the axial loads on the hose caused by the pulsating flow of the working fluid. All these measurement results provide valuable information about behavior of the hose during the flow of the working fluid.

3.2 Implementation of the Tests

Preliminary vibration tests of the hydraulic line were carried out on a composite micro-hydraulic line. These tests took place to obtain information on the dynamic properties of the line, in particular, its flexural vibration forced by pulsating flow. During the test, a constant tension force of 1 kg was applied to the hose along its axis. The average static pressure of the system was a parameter that have been changing between measurement series, while the temperature of the working fluid, the speed of the motor driving the pump and the flow rate were constant for each measurement series.

To carry out preliminary vibration tests of the composite hose, two measurement series were carried out, during which selected quantities were measured using the sensors of the measurement system with a measurement duration of one second. The sampling frequency was 19200 Hz. The tests were carried out for a hydraulic system flooded with RENOLIN VG 46 oil from FUCHS.

In the tests was used a piston pump driven by an electric motor with a speed of 960 rpm. The flow rate parameter was set to 1 dm^3/min, and the temperature of the working fluid oscillated in the range of 45–50 °C. Measurements were made for two values of static pressure, read from a sensor placed directly behind the test subject. The average pressure for the first series of measurements was 5 MPa, and for the second it was 10 MPa.

3.3 Test Results

During the preliminary hydrodynamic testing of the composite hose, its vibration was tested using a triaxial piezoelectric accelerometer, which was glued directly to the surface of the hose at the midpoint of its length. This sensor measured the axial accelerations of the hose in three axes (Fig. 4): along the axis of the hose (x-axis); perpendicular to the hose axis in the vertical direction (y-axis); perpendicular to the hose axis in the horizontal direction (z-axis).

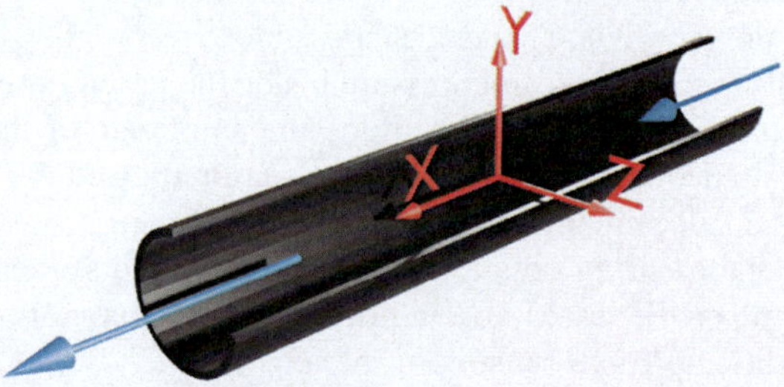

Fig. 4. Axes of triaxial piezoelectric accelerometer.

The test results are shown below using the amplitude-frequency spectrum of the test object's axial accelerations. In Figs. 5, 6, 7, 8, the spectrum for the second measurement test (measurement series) is upshifted by 0.2 m/s2.

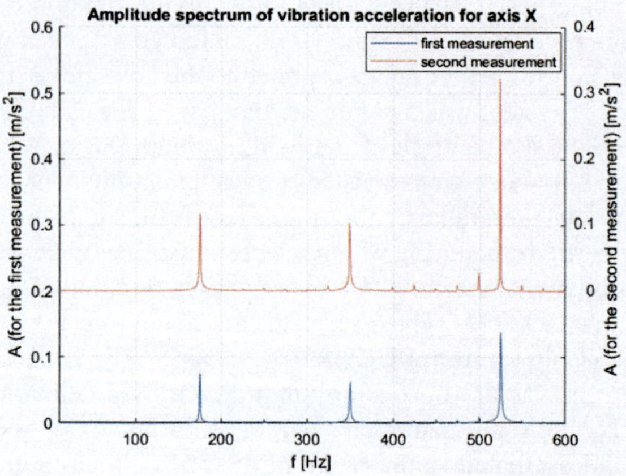

Fig. 5. Acceleration amplitude spectrum for two measurement series in the x-axis.

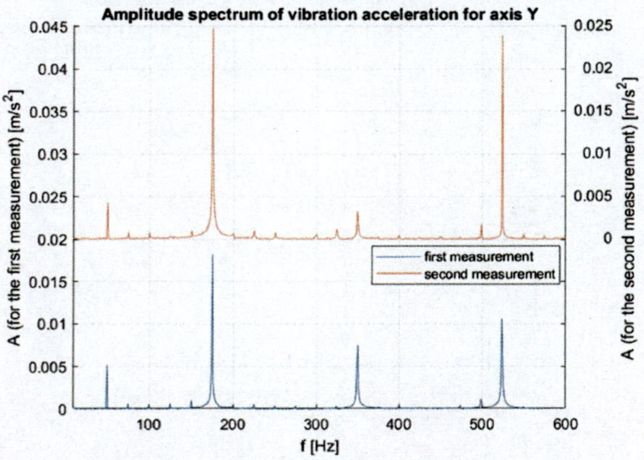

Fig. 6. Acceleration amplitude spectrum for two measurement series in the y-axis.

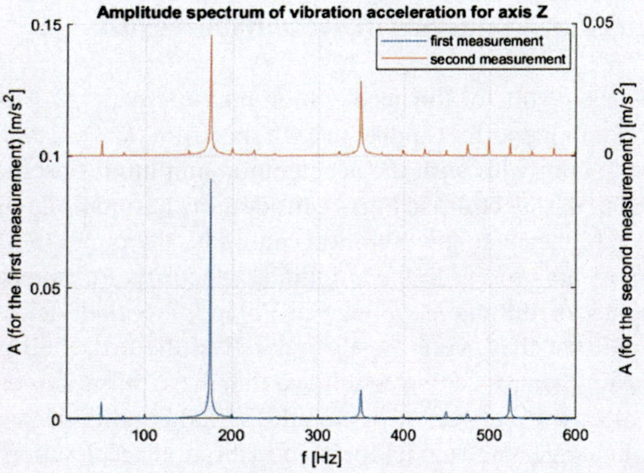

Fig. 7. Acceleration amplitude spectrum for two measurement series in the z-axis.

It was decided to use frequency characteristics instead of time waveforms in the presentation and analysis of measurement results. Frequency characteristics are a convenient form of presenting the dynamic characteristics of test objects. Such analysis of measurement data makes it possible to directly determine the vibration characteristics of an element.

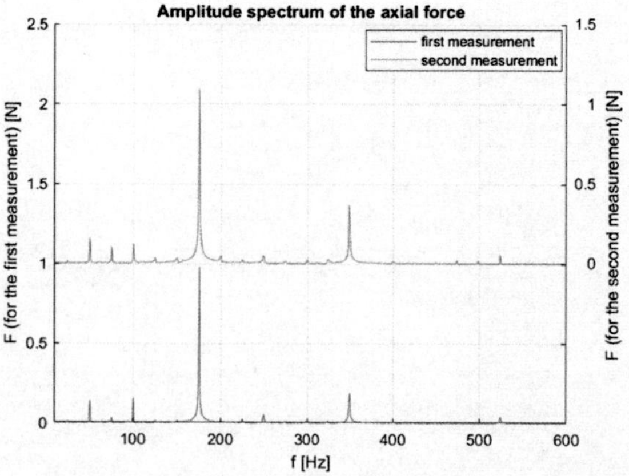

Fig. 8. Amplitude spectrum of the axial force for two measurement series.

4 Discussion. Analysis of Research Results

Analyzing the results of the acceleration measurements, it is possible to notice several harmonic components that appear in each spectrum. The pulsation frequency of the pump performance coincides with the acceleration amplitudes oscillating at 175 Hz (the first component of the spectrum in Fig. 4). In addition, harmonic components with frequencies many times higher than the vibration caused by the pump (350 Hz and 525 Hz) are the next components present in the amplitude spectrum. Pulsations of 50 Hz, equivalent to the frequency of the mains voltage in Poland, are also visible. Although their values are so small that they were mostly ignored in the test, their presence is noticeable in every measurement. It can be concluded that these pulsations are caused by interference from the electrical network to which the signal conditioner was connected. The detail analysis of changes in the axial force and vibration acceleration spectra of the composite microhose is presented in Table 1.

Table 1. Analysis of changes in the axial force and vibration acceleration spectra of the composite microhose

Physical size	Change in the amplitude of the first component f1_50/f1_100	Change in the amplitude of the second component f1_50/f1_100	Change in the amplitude of the third component f1_50/f1_100
X-axis acceleration	0.64	0.61	0.43
Y-axis acceleration	0.74	3.66	0.44
Z-axis acceleration	1.99	0.39	2.40
Axial force	0.90	0.52	0.69

The hydraulic line shows a predisposition to natural vibrations in each of its axes, which further contributes to an increase in the amplitude of the vibrations with the increasing system pressure. Spectral analysis revealed that the sources of vibration are mainly the pulsations of the piston pump, which reduces the culture of the hydraulic line. It can be noticed that for higher values of the average pressure (sample No. 2), the amplitudes of the presented spectral components increased significantly. An increase in pressure values from 50 bar to 100 bar (a 100% increase) resulted in an increasing the amplitude of the first harmonic component of the presented spectra for the x and y directions. For the x direction, the third component of the vibration spectrum is dominant.

5 Conclusions

The article presents a universal test stand for studying hydrodynamic processes in hydraulic lines. The stand was equipped with a multi-channel measurement track. As a test object, it was decided to use a composite micropipe, which was forced by pulsating flow generated by a positive displacement pump. From the posted results, for the direction of vibration X, as the mean pressure increases, the values of the amplitudes of the components of the spectrum - the first three harmonics presented - increase. A 100% increase in the value of the average pressure from 5 MPa to 10 MPa results in an increasing the values of the amplitudes of the first two harmonics by almost 50%. An even greater increase in the amplitude of vibration acceleration is observed for the third component. In addition, for all three directions, the increase in the amplitude of acceleration for the z axis is at the similar level - more than 100%. The relatively smallest increases in the amplitudes of hose vibrations with an increasing in the value of the average pressure are observed for the first component of the spectrum. On the other hand, the value of the axial force changes slightly with a 100% increase in the average pressure inside the hose for the first component of the spectrum.

The presented exemplary test results demonstrate the nonlinear characteristics of the composite microhose. It is important to notice that the tests were carried out at one fixed temperature of the working fluid and the tested microhose. The design of the stand allows to make the temperature value a variable parameter, stabilized in time. The results of the tests will allow the evaluation of the dynamic properties of the hydraulic line and will form the basis for further research and analysis. The obtained measurement results will be used to develop a mathematical model of the hose's vibrations. An important research challenge is the problem of reliably measuring the radial deformation of the microhose - the constant component resulting from the average pressure and the variable component resulting from the pressure pulsation.

From the results presented, it can be seen that the amplitude and frequency of flexural vibrations of the composite microhose are influenced, among other things, by the value of the mean pressure. Limitation of the vibration amplitude in practical cases is achieved by using additional supports for the hose, restraining it to the ground. In addition, the cable's ferrules are a critical point in a vibrating microhose. Prolonged vibration can lead to the formation and development of fatigue cracks in these areas and, in critical cases, to a loss of integrity and rupture of the hose.

References

1. Vacca, A., Franzoni, G.: Hydraulic Fluid Power: Fundamentals, Applications, and Circuit Design. WILEY, New Jersey, January 2020
2. Lubecki, M., Stosiak, M., Bocian, M., Urbanowicz, K.: Analysis of selected dynamic properties of the composite hydraulic microhose. Eng. Fail. Anal. **125**, 105431 (2021)
3. Moers, A., Reynaerts, D.: Flow control for high-pressure micro hydraulics. In: Menz, W., Dimov, S., Fillon, B. (eds.) 4M 2006 - Second International Conference on Multi-Material Micro Manufacture, pp. 393–396. Elsevier (2006)
4. Rodionov, L., Stryczek, J., Rekadze, P.: Challenges in design process of gear micropump from plastics. Arch. Civ. Mech. Eng. **21**(34), 1–14 (2021)
5. Skowrońska, J., Kosucki, A., Stawiński, Ł: Overview of materials used for the basic elements of hydraulic actuators and sealing systems and their surfaces modification methods. Materials **14**(6), 1422 (2021)
6. Deptuła, A.M., et al.: Risk assessment of innovation prototype for the example hydraulic cylinder. Sustainability **15**(1), 440 (2022)
7. Wei, D., An, C., Wu, C., Duan, M., Estefen, S.F.: Torsional structural behavior of composite rubber hose for offshore applications. Appl. Ocean Res. **128**, 103333 (2022)
8. Bhadane, G., Patil, S.: Mathematical modeling of multilayered composite material to obtain in plane elastic constants. Mater. Today Proc. **72**(3), 794–801 (2023)
9. Wanberg, J.: Composite materials fabrication Handbook #1. Anchor Books (2009)
10. Solazzi, L., Buffoli, A.: Fatigue design of hydraulic cylinder made of composite material. Compos. Struct. **277**, 114647 (2021)
11. Gao, Q., et al.: Investigation on structural behavior of ring-stiffened composite offshore rubber hose under internal pressure. Appl. Ocean Res. **79**, 7–19 (2018)
12. Parker Homepage. https://ph.parker.com/us/en/small-bore-mini-hydraulic-hose-high-pressure-2020n/2020n-012v30. Accessed 30 Mar 2023

Technology of Hydrogen Production from Aluminum Scrap by Using Ultrasonic Cavitation

Oleksandr Luhovskyi, Andrii Zilinskyi(✉), Alina Shulha, Ihor Gryshko, Ihor Nochnichenko, and Dmytro Kostiuk

National Technical University of Ukraine "Igor Sikorsky Kyiv Polytechnic Institute", Prosp. Peremohy 37, Kyiv 03056, Ukraine
{atoll-sonic,griwko,kostiukdv}@ukr.net, zilinski.andrew@gmail.com, a.liashok@kpi.ua, nochnichenko.ihor@lll.kpi.ua

Abstract. The world society is facing the threats of global energy and environmental crises. A significant decrease in natural resources, including various types of fuel, and the aggravation of environmental issues have led to the need to find alternative energy sources, one of which is hydrogen. Therefore, a large number of researchers are currently working on creating new or improving existing methods for the production of hydrogen gas. There are many developments in this area, but there is still no sufficiently efficient hydrogen production technology.

It is a well-known technology to produce hydrogen through the interaction of aluminum or its alloys with water, but this reaction quickly fades away due to the formation of an oxide film on the aluminum surface. Therefore, to maintain the high efficiency of the technology, it was proposed to carry out this reaction under ultrasonic cavitation.

The use of ultrasonic cavitation in the chemical process of water-aluminum interaction ensures, due to ultrasonic erosion, the permanent destruction of the oxide film on the aluminum surface, which allows for maximum efficiency and stability of the chemical reaction of hydrogen removal and aluminum hydroxide production. In the cavitation environment, aluminum particles are further dispersed into smaller ones, which leads to an additional increase in the contact surface of water and pure aluminum. As a result, it is possible to almost fully utilize the capabilities of pure aluminum in the chemical reaction of hydrogen production from water. The use of ultrasonic atomization in a thin layer device makes it possible to obtain monodispersed aluminum powder with the required dimensions, according to calculations.

The proposed technology allows not only to produce hydrogen in a way that is safe for humans and the environment, but also to eliminate one of the many causes of environmental pollution associated with the disposal of household and industrial waste.

Keywords: Ultrasound · Cavitation · Hydrolysis · Aluminum

J. Stryczek and U. Warzyńska (Eds.): NSHP 2023, LNME, pp. 291–301, 2024.
https://doi.org/10.1007/978-3-031-43002-2_27

1 Introduction

The world society is facing the threats of global energy and environmental crises. A significant decrease in natural resources, including various types of fuel, and the aggravation of environmental issues have led to the need to find alternative energy sources, one of which is hydrogen. Therefore, a large number of researchers are currently working on creating new or improving existing methods for the production of hydrogen gas. There are many developments in this area, but there is still no sufficiently efficient hydrogen production technology [1].

An analysis of information resources has shown that each aluminum beverage can weighing 15 g contains about 225 kJ, which is equivalent to 20 m of car mileage with a fuel consumption of 5 L per 100 km. It has been established that the volume of such containers for the territory of Ukraine is about 1 billion pieces, and for the territory of the European Union it is ten times more.

Chemical hydrogen production from aluminum scrap is promising for improving environmental safety in the world and in terms of utilizing aluminum scrap waste, which solves the issue of producing environmentally friendly hydrogen fuel. The Institute of Renewable Energy of the National Academy of Sciences of Ukraine is conducting research in this area [2, 3]. However, after analyzing their work, it should be noted that the proposed technical solutions are relevant, but the use of the proposed solutions for hydrogen production is associated with a large consumption of energy resources. A well-known scientific school dealing with the problems of cavitation processes in liquids and the environmental friendliness of waste processing is located at Ternopil National Technical University [4]. It should be noted that in their research, the school's specialists use the phenomenon of hydrodynamic cavitation [5, 6], which requires bulky and powerful equipment.

In the EU, this area of research has become more widespread. In addition to scientific institutions, such companies as: LLC "HYDROgen of Ukraine", UDP Renewables, Hydrogen Europe, Latvian Hydrogen Association, Svarog, Bandelina, Hielscher Ultrasonics GmbH, SONOTRONIC Nagel GmbH, INTERENERGORESURS LTD. All of them produce cavitation equipment for various technologies, including hydrogen production. However, the analysis of their designs and parameters indicate a number of shortcomings that do not allow the widespread use of this technology due to the lack of powerful ultrasonic equipment and the design features of the cavitators themselves. The identified shortcomings require the creation of a fundamentally new technology and high-performance ultrasonic equipment with high intensity of ultrasonic vibrations.

The technology of hydrogen production from waste, namely, from aluminum-containing raw materials, is a very promising area. The principle of operation of the mechanism is based on the chemical method, i.e. hydrogen is produced from water by decomposition of water by hydroreactive metals (aluminum, magnesium, gallium). This method has a high level of transformation efficiency in electricity generation. It can significantly increase the efficiency of Brown's or HHO gas generation.

2 Idea Exploration

To verify the chemical method of hydrogen production, an experiment was conducted according to the chemical formula:

$$2Al + 6H2O \rightarrow 2Al(OH)3 + 3H2 \quad (1)$$

For this experiment, 67.77 g of aluminum and a solution (KOH dissolved in distilled water at a ratio of 15 g/100 ml) were taken.

During the experiment, a number of data were obtained and processed, which made it possible to build a graph of the time versus temperature required to release 50 ml of gas (see Fig. 1).

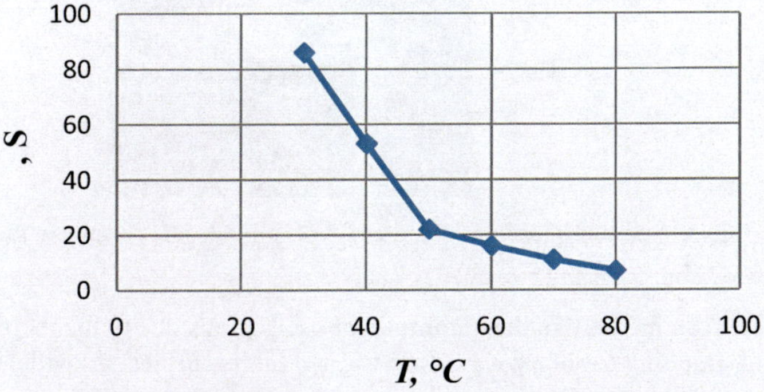

Fig. 1. Graph of temperature versus time required to release 50 ml of gas.

The experiment showed that at a temperature of 80 °C, gas was released from this mixture most rapidly.

The use of ultrasonic cavitation [7] in the chemical process of water-aluminum interaction ensures, due to ultrasonic erosion, the constant destruction [8] of the oxide film on the aluminum surface, which allows for maximum efficiency and stability of the chemical reaction of hydrogen removal and aluminum hydroxide production.

The effect of cavitation erosion occurs due to the impact action of spherical waves formed when the cavitation bubble collapses in the absence of nearby surfaces and energy-intensive cumulative jets directed towards nearby "solid" surfaces.

To significantly increase the surface area of water contact with aluminum, the technology under consideration uses a preliminary fine ultrasonic atomization of aluminum melt. As a result, it is possible to almost fully utilize the capabilities of pure aluminum in the chemical reaction of hydrogen production from water.

To produce the fine aerosol, the so-called thin-bed ultrasonic spraying technology is used. The ultrasonic atomization of aluminum melt is realized by applying a sinusoidal electrical signal to a piezoelectric transducer, which excites standing longitudinal vibrations in the dispersant resonant system. In this case, the spraying surface is placed in the void of the displacement wave. To achieve the oscillation amplitude required to start the spraying process, acoustic transformers of the oscillatory velocity are used. A liquid is fed to the spraying surface, which covers it with a layer of 0.3…0.8 mm thick.

It should be noted that when ultrasonic vibrations are excited in the dispersant resonant system, the force of surface friction of the liquid decreases and under the action of the cohesive force, the liquid is collected into a droplet. At a small amplitude of vibrations of the dispersant's vibrating surface, the liquid layer moves as one with the vibrating surface (Fig. 2).

Fig. 2. Spray surface vibrating with low amplitude and covered with a layer of liquid

With an increase in the amplitude of oscillations due to the inertial properties of the liquid, standing capillary waves are formed on its surface, as evidenced by the turbidity of the liquid layer (Fig. 3).

Fig. 3. A vibrating surface with a layer of liquid at the moment of capillary wave formation

The frequency of the capillary waves is half that of the dispersant (Fig. 4a). Capillary waves are uniformly distributed over the surface of the liquid (Fig. 4b). With the help of the created optical-acoustic setup, it was possible to experimentally confirm 94…96%

accuracy of the known analytical dependence for calculating the capillary wavelength:

$$\lambda_k = \sqrt[3]{\frac{8\pi\sigma}{\rho f^2}}, \quad (2)$$

where, σ - surface tension coefficient of the sprayed liquid;

ρ- density of the liquid; f- frequency of acoustic vibrations.

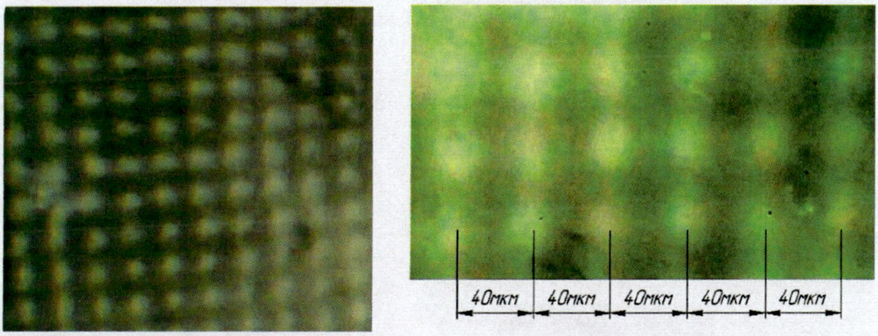

a) b)

Fig. 4. Location of capillary waves on a vibrating surface at $\sigma = 72,86 \cdot 10^{-3} H/m$, $\rho = 1000\,\mathrm{kg/m^3}$, $f = 60\,\mathrm{kHz}$ (a - 24 times magnification; b - 35 times magnification)

With a further increase in the amplitude of oscillations of the dispersant's working surface, a cavitation layer begins to form near the radiating surface, which further contributes to the destruction of the integrity of the molten metal layer (Fig. 5).

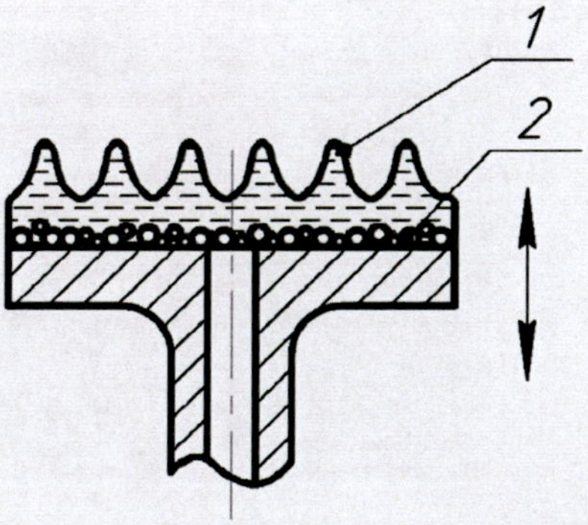

Fig. 5. Scheme of formation of capillary waves and cavitation layer at atomization in a thin layer (1 - capillary waves; 2 - cavitation layer)

When the surface tension forces are unable to maintain the integrity of the elongated capillary wave, its crest collapses and an aerosol droplet is torn off (Fig. 6).

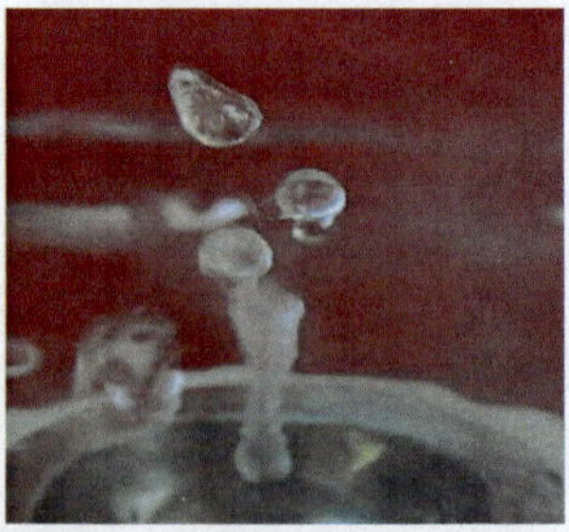

Fig. 6. Capillary wave destruction with droplet detachment

Aerosol droplets, detached in the dynamic process of capillary wave rise and having a certain mass, fly off in the direction of the capillary wave movement, i.e., normal to the spray surface (Fig. 7a). The kinetic energy possessed by the droplet allows it to fly a certain distance in a direction, i.e. to form the aerosol plume required for a given technological process (Fig. 7b, 7c).

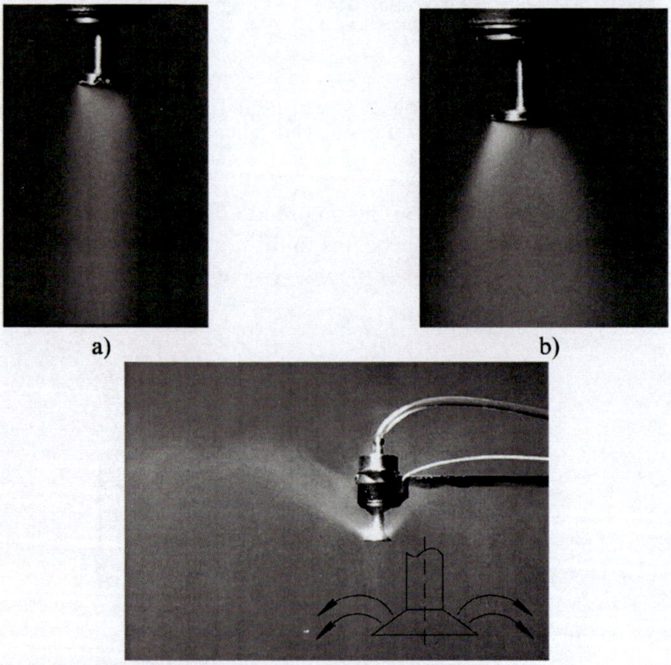

a) b)

c)

Fig. 7. Shapes of aerosol plumes when sprayed in a thin layer from different surfaces (a - flat surface; b - conical surface; c - reverse conical surface)

An experimental study of the dispersion of the resulting aerosol indicates the presence of some deviation from mono-dispersion, which negatively affects the quality of the aerosol and does not allow the use of this spraying method in technological processes with high requirements for aerosol monodispersity. Deviation from monodispersity is observed mainly along the perimeter of the liquid layer. In the case when there is little liquid and due to the cohesion force it collects into a large liquid spot drifting rapidly along the vibrating surface of the disperser due to the reduction of friction forces, the cavitation layer leads to the explosive ejection of differently sized liquid droplets around the spot perimeter, where it is closest to the liquid surface with the resulting capillary waves, which ensure the production of a monodisperse aerosol.

The performance of ultrasonic atomization in a thin layer of liquid with certain rheological parameters depends primarily on the aerosol droplet diameter and the spray surface area. For the case of ultrasonic atomization of molten metal, the authors proposed a special design of a dispersant with an increased spray surface area, which allows eliminating zones with zero amplitude of oscillations, in which the melt collects and cools, clogging the spray surface (Fig. 8) [10, 11]. Such zones are formed due to the excitation of radial-bending vibrations on the radiating surface.

Fig. 8. High performance ultrasonic atomizer

This high performance ultrasonic sprayer is designed for the complete plant for the production of hydrogen from aluminum scrap (Fig. 9a).

The equipment for the production of hydrogen from aluminum scrap (Fig. 9b) contains a closed body 1 with a bottom conical surface 2. A pipeline 3 for hydrogen outlet is connected to the upper part of the housing 1. On the lower bottom surface 2, there is an outlet 4 for discharging the residues of the reaction of water and aluminum.

Low-amplitude ultrasonic emitters 5 are mounted on the conical bottom surface. The housing 1 is partially filled with water via a pipeline 6. In the upper part of the housing, above the water level, there is a device for obtaining finely dispersed aluminum powder, made in the form of a vertically and sequentially installed hopper 7 for receiving coarsely crushed aluminum scrap 8, a mechanism 9 for feeding scrap into the melting chamber 10, which has an opening 11 for the molten aluminum to exit onto the vibrating conical surface 12 of the ultrasonic high-amplitude sprayer 13, which implements the method of ultrasonic spraying in a thin layer and ensures the formation of a finely dispersed aluminum aerosol, the droplets of which settle in the form of powder into the water that partially fills the housing 1. Ultrasonic vibrations with an intensity exceeding the threshold of cavitation in a mixture of water and fine powder are supplied to the ultrasonic emitters 5. The ultrasonic generators are not shown in the diagram.

The production of hydrogen according to the proposed method and, accordingly, the operation of the device is as follows.

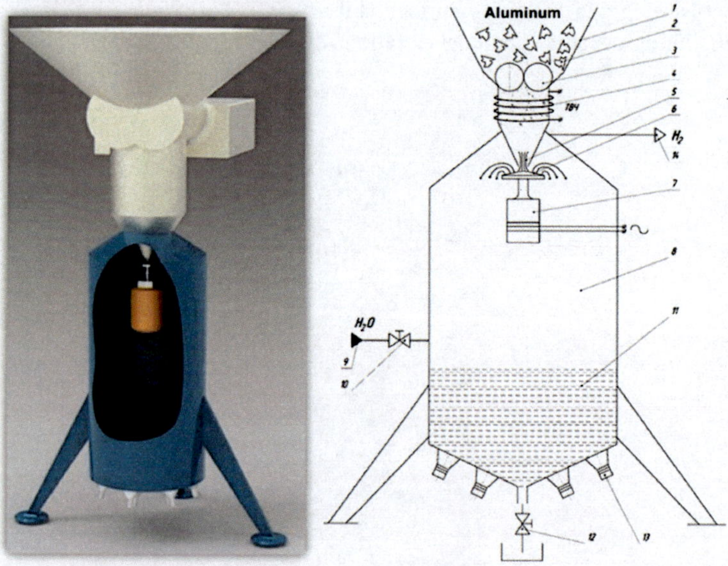

a) b)

Fig. 9. High Equipment for hydrogen production from aluminum scrap

Roughly shredded aluminum scrap 8 is fed into the receiving hopper 7. A rotating mechanism 9 feeds the scrap into the melting chamber 10. The melting process is carried out, for example, by high-frequency currents. The molten aluminum is fed through the opening 11 in the melting chamber by a stream to the vibrating conical surface 12 of the ultrasonic high-amplitude sprayer 13, which implements the method of ultrasonic spraying in a thin layer and ensures the formation of a finely dispersed aluminum aerosol, the droplets of which settle in the form of powder into water, partially filling the closed housing 1. The aluminum powder begins to react with water according to the above chemical equation, which leads to the formation of an oxide film on its surface, which first slows down and then stops this reaction. The hydrogen formed as a result of the chemical reaction is removed through nozzle 3. Since low-amplitude ultrasonic emitters are installed on the bottom surface of the closed housing 1, to which a high-frequency signal with an intensity exceeding the threshold of cavitation in the liquid under these conditions is supplied from the ultrasonic oscillation generator, ultrasonic cavitation occurs in the volume of water in the housing 1, which, due to the secondary effects accompanying the phenomenon of ultrasonic cavitation (cavitation erosion), actively destroys the oxide film and constantly cleans the surface of the aluminum powder, i.e., ensures a constant high efficiency of the chemical reaction of water with aluminum. Ultrasonic cavitation additionally heats the mixture continuously, which also increases the activity of the chemical reaction without additional external heating of the vessel 1. The mixture is also heated by the heat generated by the chemical reaction. The aluminum oxide formed as a result of the chemical reaction is deposited on the conical bottom surface of the housing 1. Since ultrasonic emitters are mounted on this surface, it vibrates continuously. Therefore, the oxide precipitate does not cake and can be easily discharged periodically through hole 4.

The authors tested the efficiency of the proposed method of hydrogen production. The essence of the experiment was the use of an ultrasonic cavitator during the chemical reaction of crushed aluminum in an alkaline medium.

A general view of the experimental setup with an ultrasonic sprayer and cavitator is shown in Fig. 11. The setup includes a closed housing, cavitator 1, atomizer 2, container 3, and control unit 4. As a result of the experiment, the temperature versus time dependence of the gas separation of 50 ml with and without an ultrasonic cavitator was obtained. Comparison of the two methods of hydrogen production Fig. 12. Showed that the use of ultrasonic cavitation increases the efficiency of the chemical process of hydrogen production due to the intensification of the chemical reaction and, accordingly, a decrease in the time of aluminum decomposition from 5…30%. The different decomposition time of aluminum is caused by the temperature of the working medium, since the cavitation efficiency significantly depends on the viscosity of the medium and the surface tension of the substance and its density (Fig. 10).

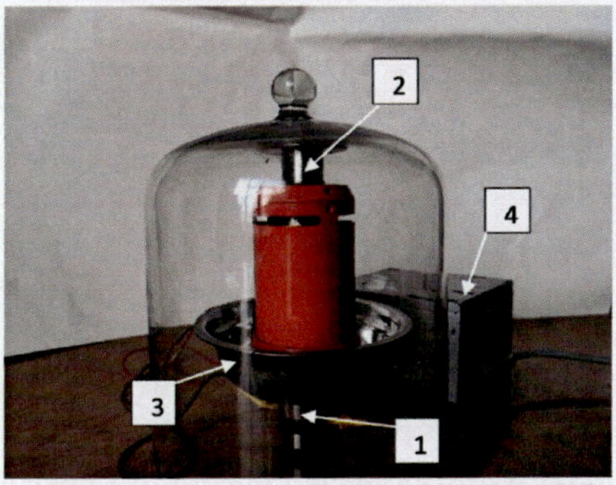

Fig. 10. General view of the experimental setup with ultrasonic sprayer and cavitator

Photo of the process of aluminum decomposition by ultrasonic cavitation Fig. 12.

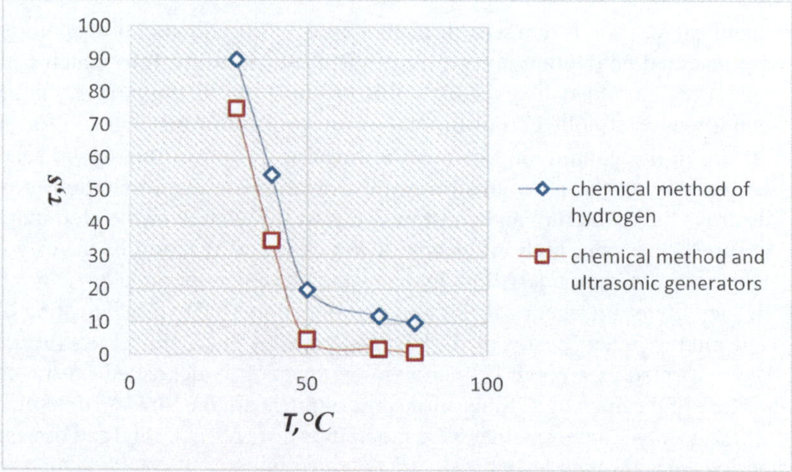

Fig. 11. Dependence of temperature on the time required for the release of 50 ml of gas with and without ultrasonication

Fig. 12. Photo of the process of ultrasonic cavitation with aluminum

3 Conclusion

The chemical reaction is very promising due to the fact that there is no need to supply electrical voltage to the electrodes and can be used to increase the efficiency of the hydrogen production process.

The proposed technological solution for hydrogen production makes it possible to carry out the reaction of aluminum with water with maximum efficiency without the addition of alkali, high pressure and temperature, since a process of separation of steam and hydrogen during cooling is required in the future. The main advantages of the new hydrogen production method are complete personnel safety, environmental friendliness and low costs.

Thus, the effectiveness of the use of ultrasonic cavitation in the chemical process of water-aluminum interaction and, accordingly, the reduction of the aluminum decomposition process time by 5…30%, with further intensification of the process by using ultrasonic spraying, was experimentally confirmed.

The proposed technology not only produces hydrogen in a way that is safe for humans and the environment, but also eliminates one of the many causes of environmental pollution associated with the disposal of household and industrial waste.

References

1. Zohuri, B.: Hydrogen-powered fuel cell and hybrid automobiles of the near future. In: Hydrogen Energy: Challenges and Solutions for a Cleaner Future, pp. 37–59. Springer, Cham (2019). https://doi.org/10.1007/978-3-319-93461-7_2
2. Fundamental aspects of renewable hydrogen energy and fuel cell technologies. (KIM, 2018)
3. Kudria, S., Riepkin, O., Yatsenko, L., Shynkarenko, L., Tkalenko, M.: The concept of the roadmap for the development of hydrogen energy in Ukraine for the period until 2035. Vidnovluvana Energ. 22–28 (2019)
4. Marynenko, N., Vitenko, T., Kramar, I.: European experience in waste management: selected aspects. In: Proceedings of the 8th International Science Conference on FORMATION Mechanical National Economic System Competitive POSITIONS Strength Global Registration LOCAL Dimensions, pp. 191–195 (2022)
5. Vitenko, T., Droždziel, P., Horodyskyi, N.: Hydrodynamics features in cavitation devices under tangent medium input. In: MATEC Web of Conferences, vol. 244 (2018)
6. Antoniak, P., et al.: Visualization research on the influence of an ultrasonic degassing system on the operation of a hydraulic gear pump. In: MATEC Web of Conferences, vol. 211 (2018)
7. Luhovskyi, O., Bernyk, I., Gryshko, I., Abdulina, D., Zilinskyi, A.: Mobile equipment for ultrasonic cavitation inactivation of microorganisms in the liquid environment. In: Stryczek, J., Warzyńska, U. (eds.) Advances in Hydraulic and Pneumatic Drives and Control 2020. NSHP 2020. LNME, vol. 24. Springer, Cham (2021). https://doi.org/10.1007/978-3-030-59509-8_24
8. Zilinskyi, A., Luhovskyi, O., Gryshko, I., Kovalev, V.: Study of the structural materials cavitation strength in ultrasonic technological equipment. In: Stryczek, J., Warzyńska, U. (eds.) Advances in Hydraulic and Pneumatic Drives and Control 2020, vol. 24. NSHP 2020. LNME. Springer, Cham (2021). https://doi.org/10.1007/978-3-030-59509-8_31
9. Luhovskyi, O., Shulha, A., Zilinskyi, A., Gryshko, I., Movchanuk, A.: Possibilities of increasing the productivity of the ultrasonic atomiser. J. Tech. Univ. Gabrovo **63**, 32–36 (2021)
10. Luhovskyi, O.F., et al.: Method for calculating a cylindrical ultrasonic cavitation filter chamber with a regeneration effect. Visnyk NTUU KPI Seriia - Radiotekhnika Radioaparatobuduvannia (2020)
11. Luhovskyi, O., et al.: Ultrasonic atomizer (2020)

Development and Testing of a Hydrotronic Control System for a Prototype Test Bench for Vibration Isolation Systems

Andrzej Kosiara(✉)

Wrocław University of Technology, Łukasiewicza 7/9, 50-371 Wrocław, Poland
andrzej.kosiara@pwr.edu.pl

Abstract. This paper presents the virtual prototyping of a hydrotronic hydraulic actuator position control system. The hydraulic actuator is part of a test bench for pneumatic vibration isolation systems. It should force a sinusoidally variable movement of the device platform. The amplitude and frequency of this movement are set and changed according to testing needs. The control system under development was based on the Danfoss Plus + 1 series proportional directional valves and controllers. The article focuses on discussing the synthesis of the control algorithm and presenting the simulation method used to predict the dynamic capabilities of the designed system.

Keywords: control system · hydrostatic drive · proportional directional valve

1 Introduction

Vibration excitation devices and systems are widely used in multiple fields of science and technology. They are used as actuators in various technological processes [1], to carry out experimental modal analysis [2, 3], in fatigue tests of engineering materials and real-scale structural elements [4–6], to verify the resistance of various devices to vibration [7], or to find the spring-damping characteristics of the elements and assemblies of suspension systems [8–10].

In practice, vibration excitation devices can be classified into the following groups: mechanical, hydraulic, pneumatic, and electrodynamic devices [1, 11]. Some of the state-of-the-art vibration excitation devices, e.g. the piezoelectric ones, cannot be assigned to any of the groups mentioned above; however, this is a very unique case.

A detailed classification and description of mechanical vibration excitation devices is presented in [1, 11]. The mechanical vibration excitation devices might be designed to deliver a dynamic excitation of some specified amplitude of displacement or force. The structure of these devices is simple. Unfortunately, the parameters of the oscillatory excitation delivered by the mechanical excitation devices cannot be freely adjusted.

Pneumatic vibration excitation devices are typically used in technological applications. They are the most noisy of all state-of-the-art excitation devices. The frequency

J. Stryczek and U. Warzyńska (Eds.): NSHP 2023, LNME, pp. 302–315, 2024.
https://doi.org/10.1007/978-3-031-43002-2_28

and amplitude of force excited with the pneumatic devices cannot be set independently of each other.

Electrodynamic vibration excitation devices can deliver a relatively wide range of excitation frequencies, while the remaining parameters of the vibrations excited with devices of this type can be easily adjusted. Unfortunately, the electrodynamic vibration excitation devices suffer from the following drawbacks; devices of this type can deliver a relatively small amplitude of displacement and a small amount of force, and they can hardly excite low-frequency vibrations.

Electrohydraulic vibration excitation devices exhibit superior performance in comparison to their counterparts of other types. They cannot deliver extremely high excitation frequencies and they are relatively expensive, which is their only drawback.

The performance of different vibration excitation devices has been compared in [1, 12, 13] in more detail.

Electrohydraulic vibration excitation devices are the most suitable for testing the elements and assemblies of suspension systems. They best simulate the operation conditions of real suspension systems in terms of excitation frequency and force. Consequently, they are the most frequently used in the application mentioned above. Hydraulic valves and control system are the elements that affect the most the dynamic properties and performance of every electrohydraulic vibration excitation device. For this reason, most of the literature on electrohydraulic vibration excitation devices focuses on either choosing/designing superior hydraulic valves or developing an effective control system. For example, an electrohydraulic vibration excitation device fitted with a high-performance proportional directional valve manufactured by the Parker company has been discussed in [14]. On the other hand, in [15], the dynamic performance of modern state-of-the-art electrohydraulic valves has been reviewed and discussed in terms of possible future improvements. In the most recent research, which has been discussed, for example, in [16, 17], an attempt has been made to improve the dynamic performance of electrohydraulic vibration excitation devices by replacing their conventional spool directional valves with new-generation rotary directional valves actuated by electric servomotors or stepper motors. On the other hand, in [18], a special electrohydraulic valve has been developed for electrohydraulic vibration excitation devices so that they can deliver a non-zero mean excitation force.

A comprehensive discussion on conventional and unconventional control systems for various electrohydraulic systems is presented in [19]. What is more, much scientific literature has been published on the control systems developed specifically for vibration excitation devices. It is worth noting that much of this specialized literature presents the so-called hybrid control systems. For example, in (9), a PID-based control system has been discussed whose settings are adjusted in real time by a fuzzy controller depending on the control error and the first-order time derivative of the control error.

It is usually very expensive to buy or develop a high-performance electrohydraulic directional valve. However, the dynamic performance of an electrohydraulic vibration excitation device can often be improved or optimized when a superior control system is developed for the excitation device, which is usually a much cheaper approach. This paper presents research and development of a new generation control system for an electrohydraulic vibration excitation device that consists of an affordable electrohydraulic proportional directional valve. The goal of this study is to maximize the dynamic performance of the vibration excitation device investigated in the article.

2 Subject and Scope of the Research

The subject of this paper is the construction and study of the dynamic properties of a virtual prototype of a hydrotronic test bench drive system. The bench is shown in Fig. 1. It is designed to test vibration isolation systems consisting of elastomeric air bags for vehicle suspension systems. The test procedure involves vibration excitation of the vibration isolation systems tested on the bench using the hydraulic actuator shown in Fig. 1. The piston rod of this actuator should perform a reciprocating (sinusoidally varying) motion relative to the cylinder with a set amplitude and frequency during testing. The parameters of this motion will be set using the Danfoss DP210 operator panel.

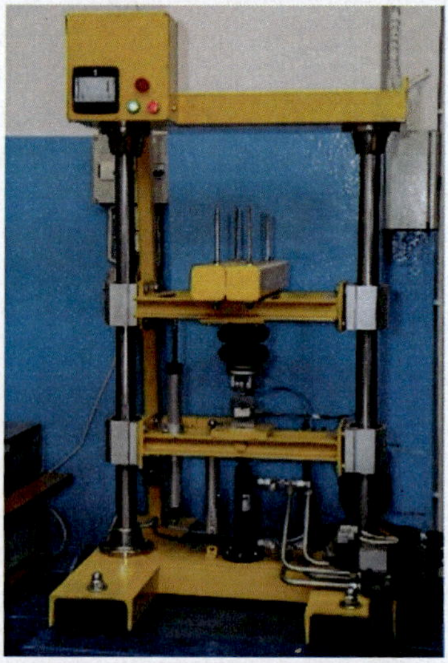

Fig. 1. The test bench whose drive system is the subject of the article.

The system that is the subject of this paper is the de facto supply system for the aforementioned hydraulic cylinder. The main goal of the simulation studies on this system was to tune the settings of the controller and to find the maximum amplitude and frequency the excitation that will be able to be implemented with the controller.

Among other things, it is expected to drive the systems under study to resonant vibration during bench testing. This will result in dynamic load changes in the drive system of the test bench. The author had doubts about the extent to which such load pulsations would interfere with the controlled movement of the hydraulic cylinder.

It was assumed that the hydraulic system of the bench would be a throttle system with a proportional directional valve USEB 6 from Ponar Wadowice. No other directional valve was considered because the budget of the project was very limited and the USEB 6 directional valve was available for free as it was left over from an earlier research project. As can be found in the literature [20], throttle systems have relatively good dynamics, which the author cared a lot about, as well as low efficiency. The low efficiency in this case, given the low power output of the drive system, seemed acceptable. The following major design assumptions were made:

- position feedback implemented using a potentiometric linear position sensor,
- control system built on the basis of the Danfoss SC024010 controller,
- communication between the controller and the operator panel using the CAN bus,
- pump with a capacity of $416{,}6 \cdot 10^{-6}$ m^3/s (25 l/min),
- 16 MPa supply pressure,
- a model of the drive system including control will be built using publicly available catalog data in Siemens' Simcenter Amesim software.

3 Proportional Directional Valve Model

In this study, it was decided to build the virtual prototype of the hydrotronic drive system using standard elements available in Simcenter Amesim software. It was assumed that heat transfer issues would not be considered during the study. With the assumed structure of the drive system, the hydraulic element that has a decisive impact on its dynamics is the proportional directional valve. In the Simcenter Amesim environment, a function block called HSV34_01 is available to model proportional directional valves, which we decided to use.

3.1 Model of Fluid Flow Through Directional Valve Channels

The flow rate of hydraulic oil through the directional valve paths in the functional block used is described by the equation [21]:

$$Q = C_q \cdot A \cdot \sqrt{\frac{2 \cdot \Delta p}{\rho}} \tag{1}$$

where Q is the volumetric flow rate, C_q is the flow coefficient, A the cross-section of the orifice, Δp the differential pressure and ρ the density. Flow coefficient C_q changes with the change in flow number λ according to the relationship [21]:

$$C_q = C_{qmax} \cdot tgh\left(\frac{2 \cdot \lambda}{\lambda_{crit}}\right) \tag{2}$$

where C_{qmax} is maximum flow coefficient and λ_{crit} is flow number at which transfer from laminar to turbulent characteristics occurs. The model assumes $\lambda_{crit}=1000$ [21].

Current flow number in the model is defined by the relationship [21]:

$$\lambda = \frac{D_h}{\nu} \cdot \sqrt{\frac{2 \cdot \Delta p}{\rho}} \tag{3}$$

where: D_h - hydraulic diameter and ν - kinematic viscosity.

In order to conduct tests using the HSV34_01 block, it was necessary to determine in advance how the cross-section A would change with the position of the spool of the directional valve. The possibility of describing this relationship through an array defined in an ASCII file was used here. This task turned out to be relatively simple because in the modeled USEB 6 directional valve this relationship was linear when the so-called dead zone of the valve was neglected.

All channels of the directional valve had also to be assigned with their respective values of the coefficient C_{qmax} before testing. This was done on the basis of the coordinates of individual points read from the Q = f (Δp) characteristics published in the catlog. The Simcenter Amesim program automatically converted this data into the appropriate coefficients C_{qmax}.

3.2 Valve Dynamics Model

Of the options available for the HSV34_01 block, a description of valve dynamics using a second-order system was chosen. It was the most elaborate mathematical description of the dynamics from those predefined for this block. The second-order system Eq. (4) has two parameters whose values had to be determined: the natural frequency f_n and the damping ratio ξ [22]. The symbol s in Eq. (4) denotes a Laplace variable, also called the Laplace operator [23].

$$\frac{x_{out}(s)}{x_{in}(s)} = \frac{(2\pi \cdot f_n)^2}{s^2 + 4\pi \cdot f_n \cdot \xi \cdot s + (2\pi \cdot f_n)^2} \tag{4}$$

To determine frequency f_n, the directional valve was disassembled and then the moving parts of the valve were weighed. They consisted primarily of the spool and cores of proportional electromagnets. The stiffness coefficient k of the spring of the directional valve was also determined by experiments, using a test stand whose schematic drawing is shown in Fig. 2.

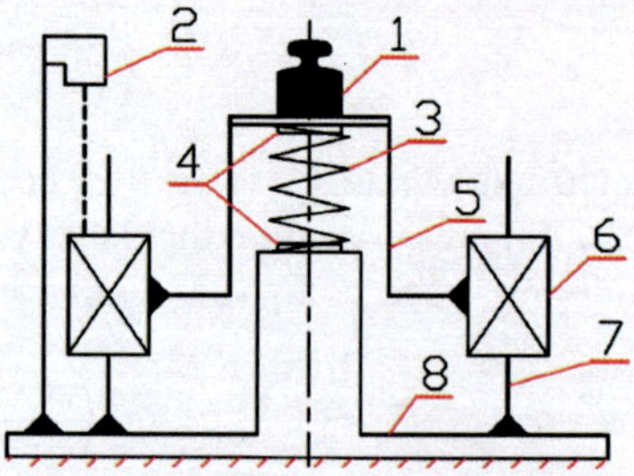

Fig. 2. A test stand for determining mechanical characteristics of coil springs: 1 – weight, 2 – optical linear position sensor (used to measure the spring deflection), 3 – tested spring, 4 - laterally spring guide, 5 – platform, 6, 7 – linear guides, 8 – base structure.

In these experiments, the deflection of the spring was measured under an a priori known load. The load was applied to the spring using arbitrary selected weights. The test results were summarized so that the load applied to the spring was plotted as a function of the spring deflection. After that, the best-fitting linear function was fitted to the experimental data and the slope of the best-fitting linear function was calculated and adopted as the stiffness coefficient k. Knowing stiffness k and mass of the directional valve's moving parts m, the natural frequency was calculated with Eq. (5).

$$f_n = \frac{1}{2\pi} \cdot \sqrt{\frac{k}{m}} \tag{5}$$

The damping ratio ξ for the valve was estimated based on the step response of the valve. Frequency characteristics were not available in the catalogue for this valve. No overshoot was observed on the available step responses. This meant that the system was overdamped. In such a case, according to the theory, the step response is described by Eq. (6)

$$x_{out} = \mathrm{K}\left\{1 - e^{-2\pi f_n \xi t}\left[cosh\left(2\pi f_n\sqrt{\xi^2 - 1} \cdot t\right) + \frac{\xi}{\sqrt{\xi^2 - 1}} \cdot sinh\left(2\pi f_n\sqrt{\xi^2 - 1} \cdot t\right)\right]\right\} \tag{6}$$

where K denotes the magnitude of step while t denotes time [22]. In order to estimate the damping ratio, the step responses were digitized and loaded into the Matlab environment. In the Matlab environment, the damping properties were estimated using the fit function. Namely, the damping was found that minimizes the deviation of the function (6) from the experimental data. Sample calculation results are shown in Fig. 3.

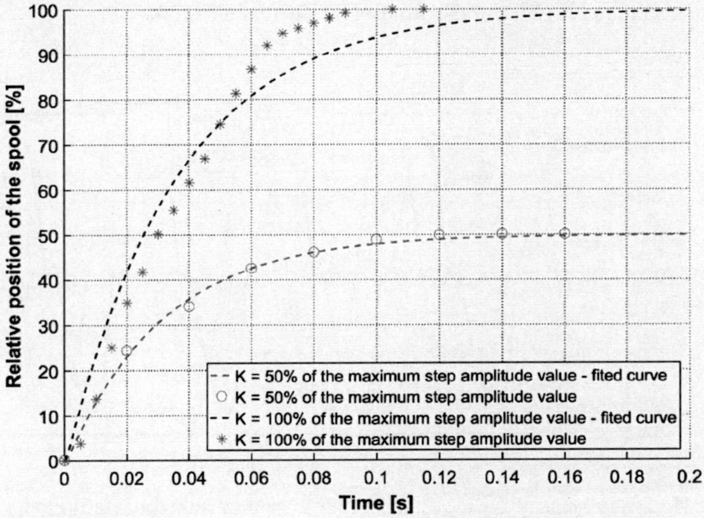

Fig. 3. Step responses of the valve. Measurement results and functions matched to the measurement results.

Depending on the magnitude of the step excitation, different results were obtained. For the step response representing 50% of the maximum step, $\zeta_{50\%} = 3.599$ damping was obtained. On the other hand, for step response representing 100% of the maximum step, $\zeta_{100\%} = 4.128$ damping was obtained. The maximum step value corresponds to the position of the spool forced by the max control voltage of 10 V. It was also observed that the larger the magnitude of the step excitation, the lower the quality of the model (6) fitted to the experimental data. It was hypothesized that this is due to the proportional properties of the valve's solenoids. According to the available literature [24], it is known that the force delivered by the electromagnet very often dependents on the amount of displacement. Typically, it decreases with an increase in the displacement of the spool. However, since the author did not have detailed information about the properties of the electromagnets investigated in this study, their properties were not included in the model. It was assumed that the internal feedback from the spool position satisfactorily linearizes the characteristics of the electromagnets and does not significantly affect the performance of the directional valve. A consequence of not including the properties of electromagnets in the model was that it also did not include feedback from the position of the directional valve spool. In the end, an intermediate amount of damping was assumed in the model, i.e. a value of 3.599.

4 Simulation Studies

4.1 Dead Zone Correction

The first results of simulation studies showed that regardless of the control algorithm used, the motion of the piston rod of the actuator relative to the cylinder deviated significantly from the requested motion trajectory. Sample issues of this type have been illustrated and highlighted with green ellipses in Fig. 4.

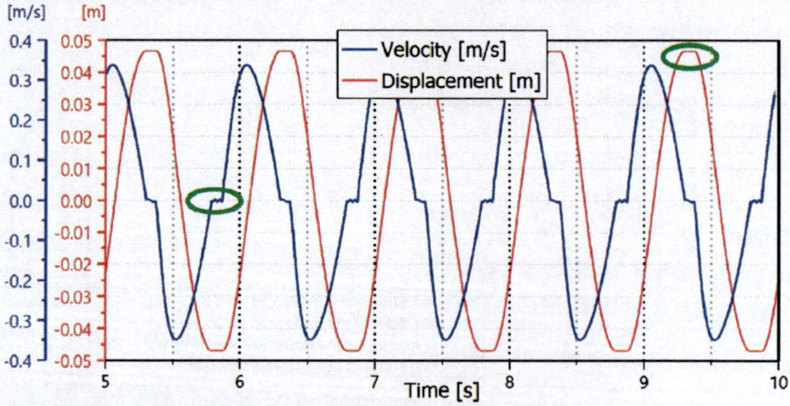

Fig. 4. Disturbance of the movement of the actuator piston rod relative to the cylinder caused by the dead zone of the proportional valve.

These discrepancies were mainly due to the so-called dead zone of the proportional valve (the valve had a so-called positive overlap). This disturbance was unacceptable.

To solve this problem, an additional USEB 6 valve was added to the system. In addition, the control system was modified so that individual USEB 6 valves operated alternately. Ultimately, one valve was responsible for the upward movement of the piston rod and the other for the downward movement of the piston rod. The valve, which was not being used at the moment, was kept by the control system at the border of the dead zone. Such a solution involves recalculating the control value coming out of the controller before it is converted into the appropriate control voltage for the directional valve. The structure of the hydrotronic drive system including the modified control system is shown in Fig. 5. The schematic in Fig. 5 differs from that of the system under study only in the way the setpoint is generated. This modification was made to improve numerical calculations.

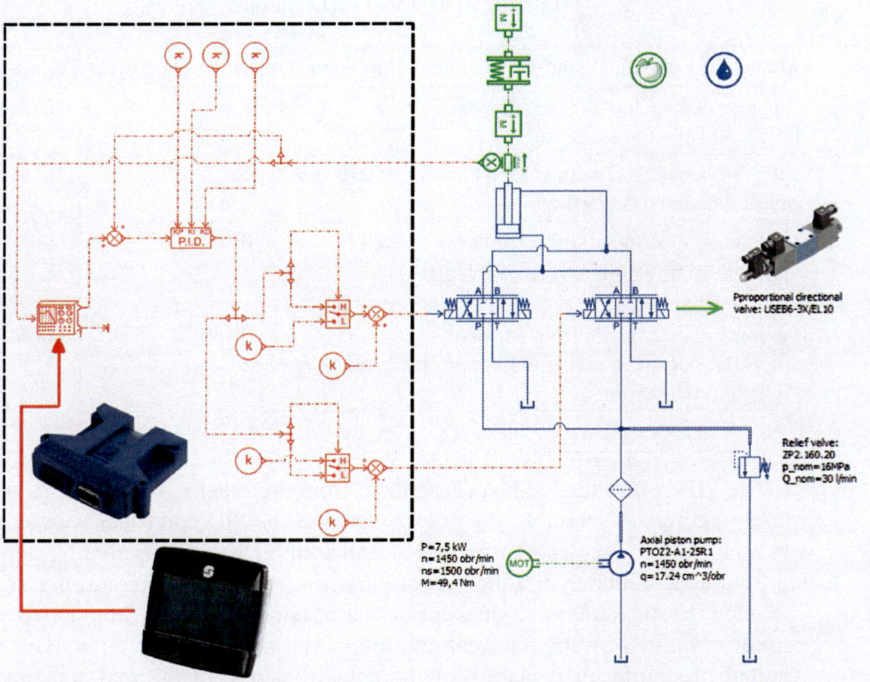

Fig. 5. Schematic of the hydrotronic drive system including the control system created in Siemens' Simcenter Amesim environment.

4.2 Determination of PID Controller Settings

Because of the large damping in the proportional valves of the control system under study, it was not possible to set such a gain in the PID controller that would destabilize operation of the system. The inability to induce unstable operation of the system was also equally due to the limited value of the maximum current that was safe for the coils of the electromagnet that actuated the spool of the directional valve. In the situation described, it was not possible to use the Ziegler-Nichols 1st method to determine reasonable PID controller settings. It was therefore decided to use the Ziegler-Nichols 2nd method. According to this method, the shape of the open-loop step response must be recorded for a control object to calculate the settings of the PID controller that will be used to control the object. The step response of the system investigated in this study was found based on the simulation model. The dynamic behavior of the drive system investigated herein depended on the direction of motion. For this reason, at the beginning of the PID tuning procedure, separate sets of PID settings were determined for each direction of motion, which are summarized in Table 1. According to Table 1, the results obtained for each direction of motion were similar in quantitative terms. Therefore, the averaged settings were implemented in the real control system presented in this paper.

Table 1. Designated PID controller settings.

Method of estimating controller settings	K_c gain	T_i doubling time	T_d advance time
Ziegler-Nichols 2nd method for astatic object – piston rod extension	5153–8588	0.065	0.016
Ziegler-Nichols 2nd method for astatic object – piston rod insertion	5270–8783	0.060	0.015
Parameters determined from Siemens controller documentation – 0% overshoot – piston rod extension	2791	0.163	0.008
Parameters determined from Siemens controller documentation – 0% overshoot – piston rod insertion	2855	0.150	0.007

The PID controller settings determined from the Ziegler Nichols method proved to be very aggressive. Namely, the output signal of the PID controller with such settings was so large for some certain frequency and amplitude of the oscillating motion of the real hydraulic actuator that the output current of the physical controller should have exceeded the maximum output rating of the physical controller. In practice, the output current is limited by the physical controller to assure safe operation. This limitation resulted in significant deviation of the actual position of the real actuator from the requested sinusoidal trajectory. Since this is not a new problem, many other alternative methods of selecting PID controller parameters can be found in the literature [25, 26]. In the end, it was decided to replace the Ziegler Nichols method with the method proposed by Siemens in the documentation of its PLCs. The concepts of the Siemens method and the Ziegler-Nichols method are similar in terms of the equations used to calculate the settings of the PID controller; however, different constants are involved in the equations for each method. PID controller settings determined based on the method developed by Siemens for the case when overshoot is to be avoided in the control system are given in Table 1. Numerical tests carried out with PID controller settings determined on the basis of the Siemens method allowed to produce the amplitude-frequency characteristics shown in Fig. 6.

These characteristics made it possible to evaluate the capabilities of the drive system along with its control system. In particular, it was necessary to determine for which excitation of the amplitude gain falls between −0.92 dB and 0.41 dB. Values within this range were considered acceptable. As can be seen in Fig. 6, this requirement was met with minor overshoots for the entire frequency range when the set amplitude did not exceed 10 mm. In the case of larger excitation amplitudes, it was possible to meet the requirements only at low frequencies, as described in detail in Table 2.

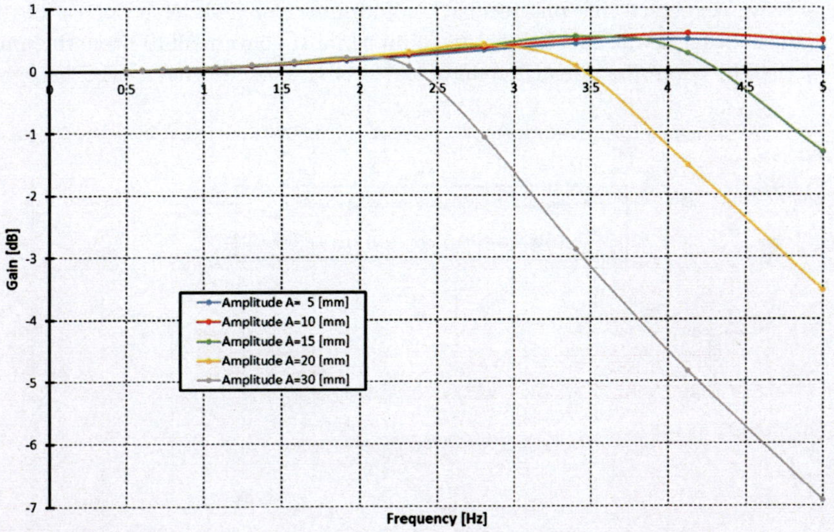

Fig. 6. Amplitude-frequency characteristics obtained at different forcing signal amplitudes and PID controller settings determined from Siemens' relationships.

Table 2. Ranges of feasible forcing parameters.

Forcing amplitude [mm]	Range of feasible forcing frequencies
5	form 0.5 Hz to 5.0 Hz
10	from 0.5 Hz to 5.0 Hz
15	from 0.5 Hz to about 4.5 Hz
20	from 0.5 Hz to about 3.5 Hz
30	from 0.5 Hz to about 2.5 Hz

The inability to realize the movement of the piston rod simultaneously with high amplitude and high frequency was due to the capabilities of the components of the hydraulic system. It was found that if these components were not replaced, there were no more significant opportunities to improve the dynamics of the analyzed system.

At high frequencies and large amplitudes, the PID controller generated control signals with values well above the maximum values. Of course, to a small extent, the quality of control could still have been improved, for example, by making the PID controller settings dependent on the set parameters of the excitation, i.e. amplitude and frequency. For example, Fig. 7 shows the amplitude characteristics for a system with a PID controller with fixed parameters and for a system with a PID controller in which the settings varied linearly with the excitation frequency. In the end, however, it was decided that such an optimization of the system did not make sense at this stage of the study because the model did not allow for the internal feedback in the proportional directional valves. It was expected that this feedback could introduce additional delays into the system. Therefore, it was decided that it would be most convenient to make the final tuning of the PID controller already on the actual facility. This turned out to be the right approach.

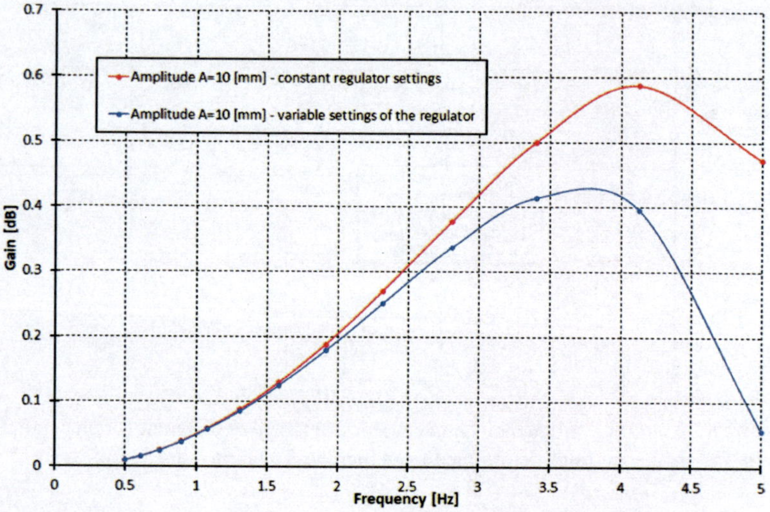

Fig. 7. Amplitude-frequency characteristics determined for the system with different variants of the PID controller.

5 Summary

After a preliminary determination of the potential capabilities of the designed drive system in the Amesim environment, the author proceeded with developing the in-house software. The software consisted of two independent programs. The first was uploaded to the DP210. Its purpose was to display the current settings of the drive system, i.e. the frequency and amplitude of the excitation, on the screen of the DP210. In addition, the program provided the ability to communicate between the operator and the drive system using four buttons directly below the screen. The operator-defined settings on the DP210 were transmitted using the CAN 2.0 bus to the SC024010 controller. There, in turn, the received settings were used to develop the appropriate signals to control the directional valves.

The program for the SC024010 controller was written in the Plus + 1 Guide environment. A distinctive feature of this environment is that it allows to create a program in a graphical language. The resulting program is then compiled and uploaded to the controller. The basic level of the created program is shown in Fig. 8. The more important modules of the program are marked in Fig. 8 with the numbers 1 through 5. The module marked 1 is responsible for receiving information from the CAN bus and decoding it. The module marked 2 is responsible for real-time reference input. The line marked 3 indicates the feedback signal generated by the position sensor of the hydraulic cylinder. The module marked 4 is the PID controller module. On the other hand, the number 5 denotes the module including dead zone corrections of the directional valves, separation of the signal from the PID controller into two signals controlling individual directional valves, and definition of the method to convert the digital signals into electrical ones.

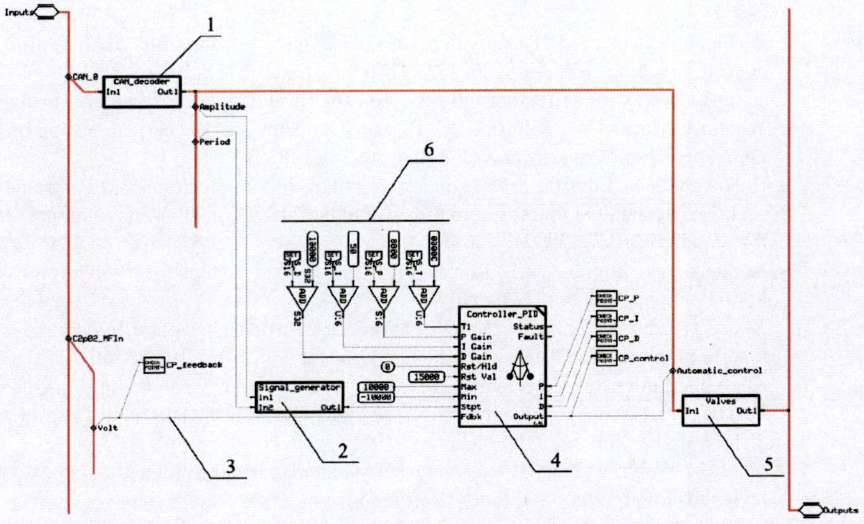

Fig. 8. Main program level of SC024010 controller in Plus + 1 Guide environment.

The author assumed the possibility of small discrepancies between the behavior of the virtual system and its real counterpart. This possibility was a consequence of several simplifications assumed by the simulation model. Therefore, a section has been built into the program that allows the PID controller to be tuned during operation. This section is marked with the number 6 in Fig. 8.

After writing the necessary programs, they were compiled and uploaded to the controller and monitor. A series of preliminary tests followed.

In general, the tests showed surprisingly good agreement between the calculation results and the actual behavior of the object. However, there were also some differences. The ranges determined by computer simulation (see Table 2) of feasible forcing frequencies have narrowed. The upper limits of these ranges turned out to be lower by about 0.5 Hz.

Further research is planned in the near future to refine the resulting mathematical model. As funding allows, consideration is also being given to replacing the proportional valves in place with others capable of expanding the testing capabilities of the vibration isolation systems test bench.

References

1. Michalczyk, J.: Vibrating machines - dynamic calculations, vibrations, noise. Wydawnictwa Naukowo-Techniczne, Warszawa, Poland (1995). (in Polish)
2. Uhl, T.: Computer-aided identification of models of mechanical structures. Wydawnictwa Naukowo-Techniczne, Warszawa, Poland (1997). (in Polish)
3. Zwolski, J., Krzyżanowski, J., Rawa, P., Skoczyński, W., Szymkowski, J.: Inertial exciter as a tool for dynamic assessment of railway Bridges. In: International Conference "Sustainable Bridges – Assessment for Future Traffic Demands and Longer Lives", Wrocław, Poland (2007)
4. Petřík, J., Lufinka A.: Hydrodynamic exciter control for durability analysis. In: Vibroengineering Procedia, vol. 3, pp. 29–32. JVE International LTD (2014)
5. Yung-Li Lee.: Fatigue Testing and Analysis, Theory and Practice. Elsevier, Amsterdam (2005)
6. Wojciech Macek, W., Zawiślak, S., Deptuła, A., Ulewicz, R.: Fatigue testing machines and apparatus. Qual. Prod. Improv. J. **1**(10), 80–108 (2019)
7. Road vehicles – Environmental conditions and testing for electrical and electronic equipment - part 3: Mechanical loads. International standard, ISO 16750–3. International Organization for Standardization (2007)
8. Konieczny, J., Sibielak, M., Rączka, W.: The control system for a vibration exciter. Solid State Phenom. **108**, 600–605 (2013)
9. Xiao, Q., Yang, L., Zhou, S., Guo, W., Zhou, X.: Control design of the hydraulic exciting servo system of the single wheel-set multi-function test-bed. Noise Vib. Contr. **41**(5), 65 (2021)
10. Koch, G., Pellegrini, E., Spirk, S., Lohmann, B.: Design and Modeling of a Quarter-Vehicle Test Rig for Active Suspension Control. Technical Reports on Automatic Control, vol. TRAC-5, Technische Universität München (2010)
11. Joshi, A., Wagh, V., Nalage, R., Kumar V.: Mechanisms of different vibratory exciters used – a review. J. Emerg. Technol. Innov. Res. **6**(4), 1–4 (2019)
12. Li, L.: Simulation and Control of Servo Hydraulic Actuators for Test Applications. Doctor's Thesis, Graz University of Technology (2015)
13. Augustyn, D.: Electro-hydraulic exciters - construction, operation, application. Hydraulika i Pneumatyka **3**, 57–60 (2021). (in Polish)
14. Cichoń, P., Stosiak, M.: Hydraulic exciter of mechanical vibrations in the proportional technique. Napęd i Sterowanie **3**, 142–148 (2011). (in Polish)
15. Liu, Y., Wang, T., Gong, G., Gao, R.: Present status and prospect of high-frequency electro-hydraulic vibration control technology. Chin. J. Mech. Eng. **32**(93), 1–16 (2019)
16. Wang, H., Chen, Z., Huang, J.: Improvement of vibration frequency and energy efficiency in the uniaxial electro-hydraulic shaking tables for sinusoidal vibration waveform. Energy **218**, 1–12 (2021)
17. Rena, Y., Ruan, J.: Theoretical and experimental investigations of vibration waveforms excited by an electro-hydraulic type exciter for fatigue with a two-dimensional rotary valve. Mechatronics **33**, 161–172 (2016)
18. Sahu, G.N., Singh, S., Singh, A., Law, M.: Static and dynamic characterization and control of a high-performance electro-hydraulic actuator. Actuators **9**(2), 46 (2020)

19. Jelali, M., Kroll, A.: Hydraulic Servo-systems - Modelling, Identification and Control, 2nd edn. Springer-Verlag, London (2004)
20. Stefański, T.: Identification of parameters of the mathematical model of a hydraulic drive with a proportional valve. Technika transportu szynowego 12/2015, pp. 1429–1435 (2015). (in Polish)
21. Simcenter Amesim, Help. Siemens Industry Software NW (2021)
22. Love, J.: Process Automation Handbook - A Guide to Theory and Practice. Springer-Verlag, London (2007)
23. Nof, S.Y.: (ed.) Springer Handbook of Automation. Springer, Berlin- Heidelberg (2009)
24. Yaguang, Z., Bo, J.: Analysis and modeling of a proportional directional valve with nonlinear solenoid. J. Braz. Soc. Mech. Sci. Eng. **38**, 507–514 (2016)
25. O'Dwyer, A.: Handbook of PI and PID Controller Tuning Rules, 3rd edn. Imperial College Press, London (2009)
26. Åström, K., Hägglund, T.: PID Controllers: Theory, Design and Tuning, 2nd edn. Instrument Society of America, Research Triangle Park, United States of America (1995)

Rule- Based Expert System as a Decision Support Tool in the Analysis of the Impact of Vibrations on a Microhydraulic Valve

Adam Deptuła[1(✉)], Marian A. Partyka[1], Kamil Urbanowicz[2], Krzysztof Towarnicki[3], Anna M. Deptuła[1], Rafał Łuszczyna[4], and Mariusz Łapka[5]

[1] Opole University of Technology, Prószkowska 76, 45-758 Opole, Poland
a.deptula@po.edu.pl

[2] West Pomeranian University of Technology, Piastów 17, 70-310 Szczecin, Poland

[3] Wrocław University of Science and Technology, 50-8 371 Wrocław, Poland

[4] Division of Machinery and Equipment, KOMAG Institute of Mining Technology, Pszczyńska 37, 44-101 Gliwice, Poland

[5] University of Agriculture in Kraków, Mickiewicza 21, 31-120 Kraków, Poland

Abstract. There is a relationship between design and operational parameters in the context of the influence of vibrations on the operation of micro-hydraulic components. The analysis of the author's and co-author's research work to date indicates the need to carry out research related to determining the influence of structural parameters, i.e. valve seat diameter, type and stiffness of the spring, different design of the side slope of the cone of the closing element, and operational parameters, i.e. flow rate, operating pressures, on the frequency of external mechanical vibrations acting on the microvalve. Previous analyses have been carried out using the method of multi- valued logic trees, and induction trees, which have tentatively determined the rank of importance of design parameters on pressure pulsations during the occurrence of external mechanical vibrations. For a comprehensive analysis, it was decided to use a rule-based expert system (BrES) as a decision support tool.

Keywords: Microhydraulic valve · Expert system · Logic of decision processes · Optimization

1 Introduction

The role played by integrated management systems and computer-aided manufacturing in Polish companies and production offices is significant. In the 1990s, an era of development for systems and tools used in decision-making support began (and continues to this day). The evolution in management support and design systems has led to the utilization of a field encompassing solutions based on artificial intelligence and machine learning [1, 2]. Such actions have enabled the automation of assessing proper management based on specific principles and designing with the application of knowledge transformation,

J. Stryczek and U. Warzyńska (Eds.): NSHP 2023, LNME, pp. 316–325, 2024.
https://doi.org/10.1007/978-3-031-43002-2_29

ultimately leading to the emergence of Intelligent Decision Systems and Expert Systems [3, 4]. The decision-making process must constantly leverage the latest available information technologies. In decision support systems for design methodologies and process management, the decision process follows a framework specification, which is often simplified and modified based on the importance and nature of the decision or design task. Frequently, unexpected problems cannot be detected and satisfactory results cannot be achieved through logical reasoning alone. The application of intelligent decision support methods in the broad field of mechanical engineering is of great importance for solving both structured and unstructured problems.

Among the widely used methods for decision support, it is worth mentioning group decision support systems [5, 6], ERP systems [7–9], and artificial neural networks [10, 11]. Other commonly employed methods for decision support and classification include Hidden Markov Models - Vector Quantization [12, 13], LVQ (Learning Vector Quantization) [14, 15], SOM (Self-Organizing Maps), GMM (Gaussian Mixture Models) [16, 17], and SVM (Support Vector Machines) [18, 19]. In particular, classification using inductive decision trees or HMMs (Hidden Markov Models) methods can be applied to fuzzy decision support systems.

A separate group of decision support systems are rule-based expert systems. They are programs designed to solve problems delegated to experts, characterized by a functional structure consisting of two main elements: a knowledge base and an inference engine, which is an executable program that draws conclusions based on the knowledge base and user-declared data. In these systems, the inference procedure plays a significant role by correctly deriving new data (inferences, results) from the knowledge base and the initial set of user-declared data (conditions, variable values) in the expert system.

In particular, expert systems can facilitate the resolution of various problems in mechanical engineering. The article presents an exemplary rule-based expert system applicable to the analysis of microvalves. The current work is another contribution by the authors concerning the analysis of vibrations in overflow microvalves, considering the methods of decision support systems. Previous works were presented in [14, 15]. The results obtained in these works have enabled the construction of an integrated decision support system.

2 The Analysis of Vibrations in Hydraulic Microvalves with the Consideration of Decision Support Systems and Previous Research Works.

Microhydraulic valves are special components used in microfluidic systems to control the flow of liquids or gases on a very small scale. They are characterized by their small size and precise flow regulation. Microhydraulic valves are particularly useful in fields such as microengineering, biotechnology, medicine, and laboratory applications [20, 21].

2.1 The Analyzed Microhydraulic Valve

During the conducted measurements, a microhydraulic overflow valve depicted in Fig. 1 was utilized. The total weight of this microvalve, including the associated cables, is

700 g. The design of the microvalve is based on a traditional overflow valve commonly employed in classical hydraulics. It has been miniaturized while retaining the classic structure with a conical closing element, known as the mushroom. Such a design finds application in microvalves with maximum flow capacity or in cases requiring precise flow control.

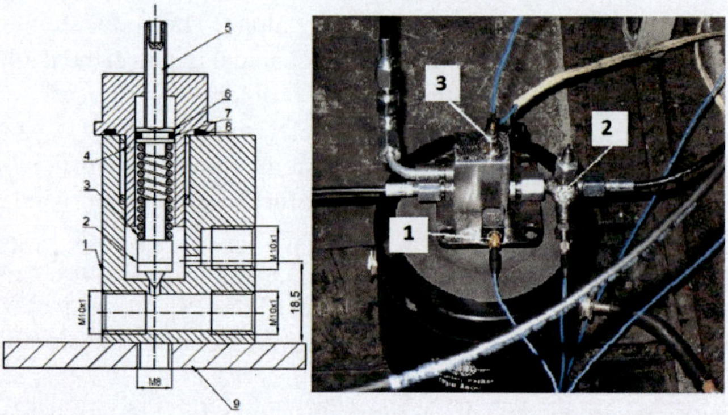

Fig.1. Hydraulic system valve-sensor: (a) Axial cross-section of the maximum microvalve: 1- body, 2- plug, 3- spring, 4- main screw, 5- set screw, 6- piston, 7- piston seal, 8- screw seal; (b) Overflow micro-valve with sensors installed: 1 – three-axis accelerometer, 2 – pressure sensor, 3– single-axis accelerometer

The conducted experimental studies determined the influence of external mechanical vibrations on changes in the amplitude-frequency spectrum of pressure pulsation in the system with the tested spill microvalve (Fig. 2).

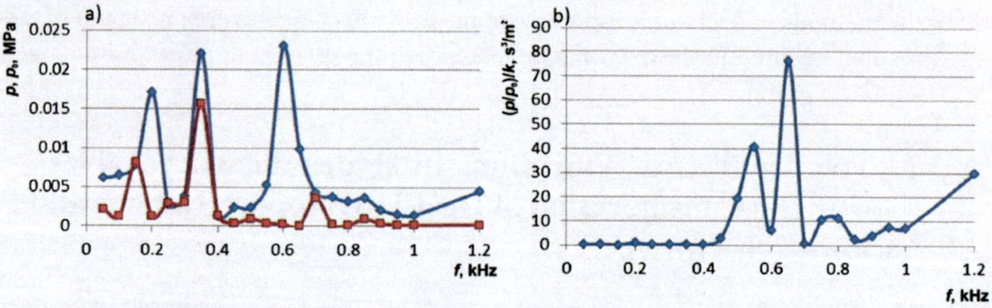

Fig. 2. a) A collective chart of pressure pulsation amplitude as a function of frequency. b) Relative change in pressure pulsation amplitude as a function of external valve operation frequency ($p_{śr}$ = 5 MPa, Q = 0.8 dm^3/min). Mechanical vibrations, Absence of external mechanical vibrations - exciter turned off. Where, *k*- stiffness constant. The expression: $(p/p_o)/k$ defines the intensity of mechaninic vibrations.

The influence of external mechanical vibrations on pressure pulsation in a micro hydraulic system is mainly observed in the frequency range of 550–600 Hz. This can lead to uneven operation of hydraulic receivers, decreased accuracy in their positioning, vibration excitation in micro hydraulic conduits, and the generation of noise. Therefore, a multi- parameter analysis of the valve's operational sensitivity is necessary to determine its optimal design.

2.2 Previous Works on Microhydraulic Valves with the Application of Selected Decision Support Methods.

Currently, there have been authored and co-authored works on the analysis of vibration effects on microhydraulic valves. In the initial study [23], an analysis of the vibration effects on the microhydraulic valve was conducted, taking into account neural network and decision tree induction methods. Measurements were carried out for flow rates through the valve equal to 1, 0.8 and 0.6 dm^3/min at the overflow valve opening pressure of 10 MPa. Figure 3 show results for a frequency of 220 Hz and flow rates of 0.8 dm^3/min and 0.6 dm^3/min.

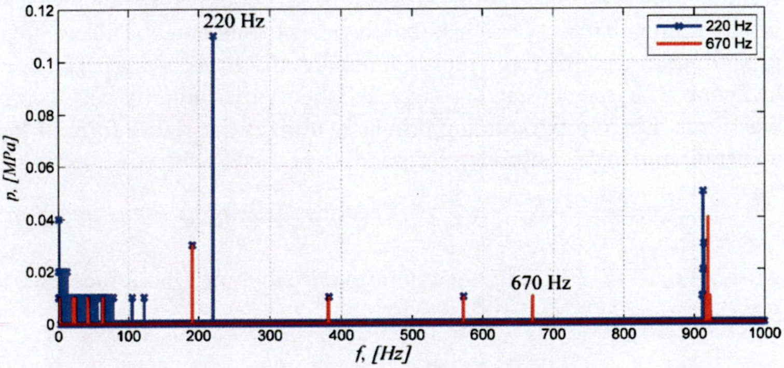

Fig. 3. Amplitude-and-frequency spectrum for a flow rate of 1 dm^3/min, with a mean pressure of 10 MPa and external driving force frequency of 220 and 670 Hz.

The preliminary analysis allowed for the division of the analysis area into intervals where the flow rate played a significant role in the entire amplitude-frequency spectrum (for its second and third ranges of variability). A total of 213 charts were generated (71 charts for each flow rate) [24].

3 Rule- Based Expert System

A rule-based expert system (BrES) is a computer-based system that utilizes a set of rules and knowledge to simulate the decision-making process of a human expert in a specific domain. It consists of a knowledge base, which stores the rules and facts, and an inference engine, which applies the rules to the given inputs and produces appropriate outputs or decisions. Typical (BrES) architecture is presented in Fig. 4.

The unique feature of expert systems is their ability to automatically generate various types of explanations during consultations with users, particularly explanations

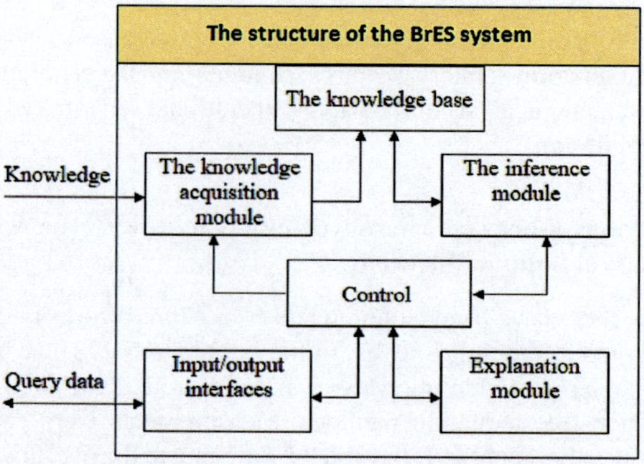

Fig. 4. Typical schema of an Expert System (ES)

regarding the problem-solving process [25]. In our expert system, we have applied logic programming. This approach is based on two fundamental ideas: treating logic as a programming language and representing algorithms as two separate components: logic and control. In the subsequent stage, the theory of similarity was incorporated into the classifiers. Logic programming primarily utilizes the clause form of logic. A clause is an expression in the following form:

$$\forall x_1, \ldots \forall x_m (A_1 \vee .. \vee A_m \vee \sim B_1 \vee .. \vee \sim B_n) \tag{1}$$

where: $A_1, \ldots, A_m$, $B_1, \ldots, B_n$ are atomic expressions, and it is represented in the following way:

$$A_1, .., A_m \leftarrow B_1, \ldots, B_n \tag{2}$$

If $m = 0$, such a clause is called a goal., If $n = 0$, the clause is an unconditional statement (assertion, fact). If $m = 0$ and $n = 0$, such a clause is called an empty clause.

Example of a logical clause:

job(valve, X) ← *optimization(X, decision generation) optimization(malgosia, decision generation)* ←

Our system will generate classification rules based on the dendrite structure.

3.1 Generating Classification Rules from Dendrites in Expert Systems

Generating rules from dendrites is the process of creating logical rules based on the dendrite structure in expert systems. A dendrite is a tree-like structure used for representing knowledge in rule-based systems [26, 27].

The space Z can be divided into sets of permissible bundles for each level of the structure:

$$Z = \left\{ Z^1, Z^2, Z^3, \ldots, Z^k, \ldots, Z^m \right\} \tag{3}$$

Meanwhile, the set Z_k:

$$Z^k = \left\{S_1^k, S_2^k, ..., S_j^k, ..., S_w^k\right\} \tag{4}$$

where any subset S_j^k describes the elements $z_j \in Z^k$, representing the outgoing edges from one vertex of the structure Q_j^{k-1}.

Figure 5 shows an exemplary diagram of rule-based dendritic searching for the parameter set m_1, c_1, and m_2, c_2.

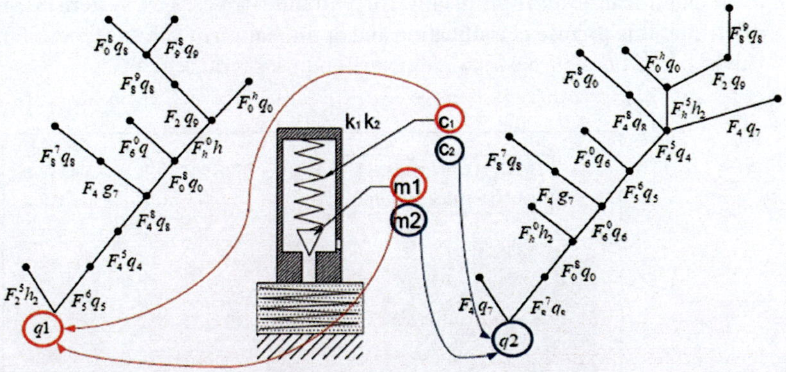

Fig. 5. An exemplary diagram of rule-based dendritic searching for the parameter set m1, c1, and m2, c2

In a general sense, the operation of a parametric structure can be described (analogously to a multi-stage automaton) by transition and output functions.

$$F(n) = f[F(n-1), z(n)]y(n) = \varphi[F(n)] \tag{5}$$

4 Rule- Based Expert System as a Decision Support Tool in the Analysis of the Impact of Vibrations on a Microhydraulic Valve

In the analysis of microhydraulic valves, the internal state of the dendrite can be determined on a set of:

$$F(n) = G[q(1), q(2), ..., q(k), ..., q(n-1), q(n)] \tag{6}$$

In our example, *q(1), q(2),..., q(n)* - describe the sets of design and operational parameters (spring constant, valve seat diameter, discharge pressure, etc.) of the microhydraulic valve, while *F(1), F(2),..., F(n)* - represent the output parameters (for example an amplitude and frequency spectrum for a flow rate). A total of approximately 900 measurement data were obtained for various combinations of changes in design and operating parameters. Table 1 presents an example combinatorial table of all measurements.

Table 1. Table captions should be placed above the tables.

p MPa	50			80			100		
c N/mm	3.44	7.49	0.76	3.44	7.49	0.76	3.44	7.49	0.76
f Hz	*< 0–900 >*								
Q dm^3/min	0.6			0.8			1		

The Fig. 6 presents the concept of operation of an advanced expert system dedicated to the classification and optimization of hydraulic valves. This system integrates dendritic rules, enabling precise classification and optimization of various types of microhydraulic valves based on their design parameters and user requirements.

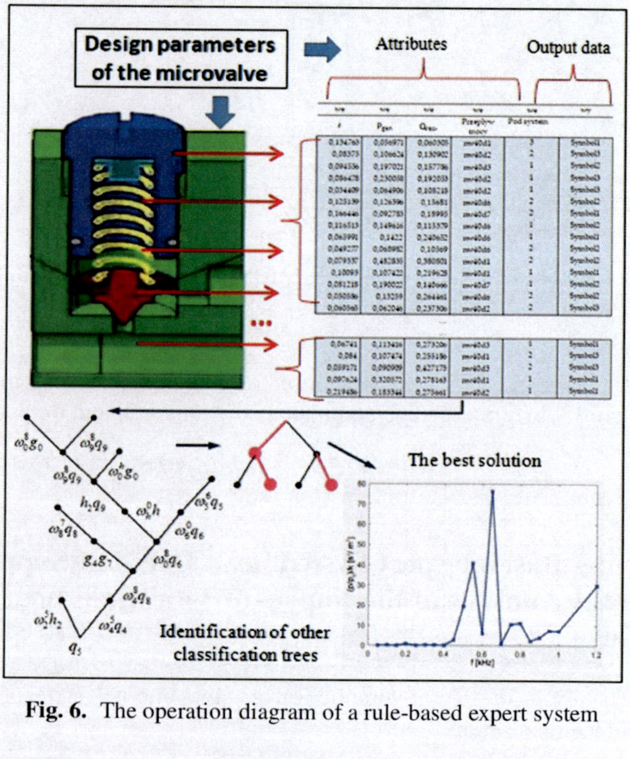

Fig. 6. The operation diagram of a rule-based expert system

An example of optimizing a microhydraulic valve using rule-based dendritic rules in the Python programming language:

Python ■ Copy code

```
def    optimize_valve(flow_rate,      pressure,      response_time,     re-
quired_precision,    required_stability,    energy_consumption,    vibra-
tions_detected, parameters):
    if flow_rate > desired_flow_rate:
        increase_valve_opening()
    if pressure > maximum_pressure:
        decrease_valve_opening()
    if response_time < desired_response_time:
        increase_valve_opening()
    if required_precision > desired_precision:
        increase_valve_opening()
    if required_stability:
        keep_valve_opening_constant()
    if energy_consumption > maximum_energy_consumption:
        minimize_valve_opening()
    if vibrations_detected:
        adjust_valve_opening_to_minimize_vibrations()
    if need_to_control_two_parameters:
        adjust_valve_opening_based_on_parameters(parameters)
    if specific_condition:
        customize_valve_opening()

def increase_valve_opening():
    # implementation of valve opening increase
    pass
def decrease_valve_opening():
    # implementation of valve opening reduction
    pass
def keep_valve_opening_constant():
    # implementation of maintenance of constant valve opening
    pass
def minimize_valve_opening():
    # implementation of valve opening minimization pass
def adjust_valve_opening_to_minimize_vibrations():
    # implementation of valve opening adjustment to minimize vibration
    pass
def adjust_valve_opening_based_on_parameters(parameters):
    # implementation of adjustment of valve opening based on other pa-
rameters
    pass
def customize_valve_opening():
    # implementation of valve opening adjustment according to a specif-
ic condition
    pass
```

In the above example, separate functions were introduced for each action related to valve optimization. These functions can be expanded with specific implementations to adapt the valve to optimal conditions. Rules are invoked based on conditions, and functions are responsible for performing the necessary actions on the valve.

5 Conclusions

The article presents the concept of an advanced expert system dedicated to the classification and optimization of hydraulic valves. The proposed solution is based on the use of dendritic rules, which enable precise determination of the design and operational parameters of these components. The system integrates various techniques, such as logic programming and the generation of classification rules from dendritic structures, allowing for accurate analysis and identification of optimal solutions.

In terms of further development, there are plans to prepare an article that includes practical examples of applying the discussed expert system to the classification and optimization of hydraulic valves. This article will present a specific set of cases where the system has been successfully applied, demonstrating its effectiveness and potential. Additionally, there are plans to expand the system by introducing an intuitive user interface that will make it even easier to use and interact with the system. In this way, the capabilities and utility of the system will be further enhanced, contributing to the continued development of the field of microhydraulic valves. The presented concept and future plans reflect the ongoing efforts to advance the capabilities of expert systems in the domain of hydraulic valves. The integration of various techniques and the development of user-friendly interfaces are aimed at improving the accessibility and practicality of these systems. The article and future work are expected to contribute valuable insights and practical applications, fostering advancements in the field.

References

1. Alam, A.: A digital game based learning approach for effective curriculum transaction for teaching-learning of artificial intelligence and machine learning. In: 2022 International Conference on Sustainable Computing and Data Communication Systems (ICSCDS), pp. 69–74 Erode, India (2022)
2. Ullah, Z., Al-Turjman, F., Mostarda, L., Gagliardi, R.: Applications of Artificial Intelligence and Machine learning in smart cities. Comput. Commun. **154**,313–323 (2020). ISSN 0140–3664
3. Chaves, M.L., Márquez, J.J., Pérez, H., Sánchez, L., Vizan, A.: Intelligent decision system based on fuzzy logic expert system to improve plastic injection molding process. In: Pérez García, H., Alfonso-Cendón, J., Sánchez González, L., Quintián, H., Corchado, E. (eds.) SOCO/CISIS/ICEUTE -2017. AISC, vol. 649, pp. 57–67. Springer, Cham (2018). https://doi.org/10.1007/978-3-319-67180-2_6
4. Dan, X.-P.: General frameworks for optimization of plastic injection molding process parameters. Simul. Model. Pract. Theory **41**, 15–27 (2014)
5. Poole, M., DeSanctis, G.: Use of group decision support systems as an appropriation process. In: Proceedings of the Twenty-Second Annual Hawaii International Conference on System Sciences, vol. IV, pp. 149–157, Emerging Technologies and Applications Track, Kailua-Kona, HI, USA (1989)
6. Vogel, D., Nunamaker, J.: Group decision support system impact: multi-methodological exploration. Inf. Manag. **18**(1), 15–28 (1990)
7. Höhne, J., Schreuder, M., Blankertz, B., Tangermann, M.: A novel 9-class auditory ERP paradigm driving a predictive text entry system. Front. Neurosci. Sec. Neuroprosthetics, **5**(99), 1–10 (2011)
8. Themistocleous, M., Irani, Z., O'Keefe, R.M., Paul, R.: ERP problems and application integration issues: an empirical survey. In: Proceedings of the 34th Annual Hawaii International Conference on System Sciences, Maui, pp. 10, HI, USA (2001)
9. Anguelov, K.: Applications of artificial intelligence for optimization of business processes in enterprise resource planning systems. In: 2021 12th National Conference with International Participation (ELECTRONICA), pp. 1–4. Sofia, Bulgaria (2021)

10. Mohandes, S.R., Zhang, X., Mahdiyar, A.: A comprehensive review on the application of artificial neural networks in building energy analysis. Neurocomputing **340**, 55–75 (2019)
11. Benedetti, M., Cesarotti, V., Introna, V., Serranti, J.: Energy consumption control automation using artificial neural networks and adaptive algorithms: proposal of a new methodology and case study. Appl. Energy **165**, 60–71 (2016)
12. Xuo, L., Jelinek, F.: Probabilistic classification of HMM states for large vocabulary. In: Proceedings of ICASSP, vol. 641, pp. 2044–2047. Phoenix, USA (1999)
13. Ultsch, A.: Self-organising neural networks for visualisation and classification. In: Opitz, O., Lausen, B., Klar R. (eds.) Information and Classification, vol. 643 , pp. 864–867. Springer, Berlin (1993)
14. Linde, Y., Buzo, A., Gray, R.M.: An algorithm for vector quantizer design. IEEE Trans. Commun. **COMM-28**, 84–95 (1980)
15. Huo-Ching, S., Yann-Chang, H., Chao-Ming, H.: Fault diagnosis of power transformers using computational intelligence: a review. Energy Procedia **14**, 1226–1231 (2012)
16. Wan, H., Wang, H., Scotney, B., Liu, J.: A novel Gaussian mixture model for classification. In: 2019 IEEE International Conference on Systems, Man and Cybernetics (SMC), Bari, Italy, pp. 3298–3303 (2019)
17. Wang, F., Liao, F., Li, Y., Wang, H.: A new prediction strategy for dynamic multi-objective optimization using Gaussian Mixture Model. Inf. Sci. **580**, 331–351 (2021)
18. Pisner, D.A., Schnyer, D.M.: Chapter 6 - Support vector machine. In: Mechelli, A., Vieira, S. (eds.) Machine Learning, pp. 101–121. Academic Press (2020)
19. Ghosh, S., Dasgupta, A., Swetapadma, A.: A study on support vector machine based linear and non-linear pattern classification. In: 2019 International Conference on Intelligent Sustainable Systems (ICISS), Palladam, India, pp. 24–28 (2019)
20. Scherrer, M., Scheidl, R.: An improved elastic and non-contact smart sealing concept for digital micro hydraulic valves. In: Proceedings of the BATH/ASME 2022 Symposium on Fluid Power and Motion Control. BATH/ASME 2022 Symposium on Fluid Power and Motion Control. Bath, UK. 14–16 September 2022. V001T01A036. ASME. https://doi.org/10.1115/FPMC2022-89847
21. Moers, A.J.M., De Volder, M.F.L., Reynaerts, D.: Integrated high pressure microhydraulic actuation and control for surgical instruments. Biomed. Microdevices **14**, 699–708 (2012). https://doi.org/10.1007/s10544-012-9650-y
22. https://www.fluidpowerworld.com/
23. Stosiak, M., Towarnicki, K., Partyka, M.A, Deptuła, A.: Analiza oddziaływania drgań na zawór mikrohydrauliczny wykorzystująca sieci neuronowe i drzewa indukcyjne. W T. Łagoda, M. Kurek, & A. Kurek, T. Łagoda, M. Kurek, & A. Kurek (Red.), Projektowanie i eksploatacja maszyn roboczych. Cz. 2 (ss. 17–38), 2020, 20 pkt. wg punktacji MNiSW w 2017
24. Stosiak, M., Towarnicki, K., Partyka, M.A., Deptuła, A.: Analysis of the impact of vibrations on a micro-hydraulic valve by sequence graph method. In: Lesiuk, G., Szata, M., Blazejewski, W., Jesus, A.M.d., Correia, J.A. (eds.) Structural Integrity and Fatigue Failure Analysis. VCMF 2020. Structural Integrity, vol. 25, pp. 357–367. Springer, Cham (2022). https://doi.org/10.1007/978-3-030-91847-7_34
25. Heragu, S.S., Kusiak, A.: Analysis of expert systems in manufacturing design. In: IEEE Transactions on Systems, Man, and Cybernetics, vol. 17, no. 6, pp. 898–912 (1987). https://doi.org/10.1109/TSMC.1987.6499302
26. Ji, J., Tang, C., Zhao, J., Tang, Z., Todo, Y.: A survey on dendritic neuron model: mechanisms, algorithms and practical applications. Neurocomputing **489**, 390–406 (2022). https://doi.org/10.1016/j.neucom.2021.08.153
27. Deptuła, A.: Zastosowanie graficznych struktur decyzyjnych w metodologii projektowania i zarządzania. Studia i Monografie / PO. T. 539. Studia i Monografie, PO, ISSN 1429-6063 (2020)

Author Index

J. Stryczek and U. Warzyńska (Eds.): NSHP 2023, LNME, pp. 327–328, 2024.
https://doi.org/10.1007/978-3-031-43002-2